QUANTUM COMPUTING EXPLAINED

THE WILEY BICENTENNIAL–KNOWLEDGE FOR GENERATIONS

Each generation has its unique needs and aspirations. When Charles Wiley first opened his small printing shop in lower Manhattan in 1807, it was a generation of boundless potential searching for an identity. And we were there, helping to define a new American literary tradition. Over half a century later, in the midst of the Second Industrial Revolution, it was a generation focused on building the future. Once again, we were there, supplying the critical scientific, technical, and engineering knowledge that helped frame the world. Throughout the 20th Century, and into the new millennium, nations began to reach out beyond their own borders and a new international community was born. Wiley was there, expanding its operations around the world to enable a global exchange of ideas, opinions, and know-how.

For 200 years, Wiley has been an integral part of each generation's journey, enabling the flow of information and understanding necessary to meet their needs and fulfill their aspirations. Today, bold new technologies are changing the way we live and learn. Wiley will be there, providing you the must-have knowledge you need to imagine new worlds, new possibilities, and new opportunities.

Generations come and go, but you can always count on Wiley to provide you the knowledge you need, when and where you need it!

WILLIAM J. PESCE
PRESIDENT AND CHIEF EXECUTIVE OFFICER

PETER BOOTH WILEY
CHAIRMAN OF THE BOARD

QUANTUM COMPUTING EXPLAINED

David McMahon

WILEY-INTERSCIENCE
A John Wiley & Sons, Inc., Publication

Published by John Wiley & Sons, Inc., Hoboken, New Jersey.
Published simultaneously in Canada.

Wiley Bicentennial Logo: Richard J. Pacifico

Library of Congress Cataloging-in-Publication Data:

McMahon, David (David M.)
Quantum computing expained / David McMahon.
p. cm.
Includes index.
ISBN 978-0-470-09699-4 (cloth)
1. Quantum computers. I. Title.
QA76.889.M42 2007
004.1–dc22

2007013725

10 9 8 7 6 5 4 3 2 1

CONTENTS

PREFACE

"In the twenty-first" century it is reasonable to expect that some of the most important developments in science and engineering will come about through interdisciplinary research. Already in the making is surely one of the most interesting and exciting development we are sure to see for a long time, *quantum computation*. A merger of computer science and physics, quantum computation came into being from two lines of thought. The first was the recognition that *information is physical*, which is an observation that simply states the obvious fact that information can't exist or be processed without a physical medium.

At the present time quantum computers are mostly theoretical constructs. However, it has been proved that in at least some cases quantum computation is much faster in principle than any done by classical computer. The most famous algorithm developed is Shor's factoring algorithm, which shows that a quantum computer, if one could be constructed, could quickly crack the codes currently used to secure the world's data. Quantum information processing systems can also do remarkable things not possible otherwise, such as teleporting the state of a particle from one place to another and providing unbreakable cryptography systems.

Our treatment is not rigorous nor is it complete for the following reason: this book is aimed primarily at two audiences, the first group being undergraduate physics, math, and computer science majors. In most cases these undergraduate students will find the standard presentations on quantum computation and information science a little hard to digest. This book aims to fill in the gap by providing undergraduate students with an easy to follow format that will help them grasp many of the fundamental concepts of quantum information science.

This book is also aimed at readers who are technically trained in other fields. This includes students and professionals who may be engineers, chemists, or biologists. These readers may not have the background in quantum physics or math that most people in the field of quantum computation have. This book aims to fill the gap here as well by offering a more "hand-holding" approach to the topic so that readers can learn the basics and a little bit on how to do calculations in quantum computation.

Finally, the book will be useful for graduate students in physics and computer science taking a quantum computation course who are looking for a calculationally oriented supplement to their main textbook and lecture notes.

The goal of this book is to open up and introduce quantum computation to these nonstandard audiences. As a result the level of the book is a bit lower than that found in the standard quantum computation books currently available. The presentation is informal, with the goal of introducing the concepts used in the field and then showing through explicit examples how to work with them. Some topics are left out entirely and many are not covered at the deep level that would be expected in a graduate level quantum computation textbook. An in-depth treatment of adiabatic quantum computation or cluster state computation is beyond this scope of this book. So this book could not be considered complete in any sense. However, it will give readers who are new to the field a substantial foundation that can be built upon to master quantum computation.

While an attempt was made to provide a broad overview of the field, the presentation is weighted more in the physics direction.

1

A BRIEF INTRODUCTION TO INFORMATION THEORY

In this chapter we will give some basic background that is useful in the study of quantum information theory. Our primary focus will be on learning how to quantify information. This will be done using a concept known as *entropy*, a quantity that can be said to be a measure of disorder in physics. Information is certainly the opposite of disorder, so we will see how entropy can be used to characterize the information content in a signal and how to determine how many bits we need to reliably transmit a signal. Later these ideas will be tied in with quantum information processing. In this chapter we will also briefly look at problems in computer science and see why we might find quantum computers useful. This chapter won't turn you into a computer engineer, we are simply going to give you the basic fundamentals.

CLASSICAL INFORMATION

Quantum computation is an entirely new way of information processing. For this reason traditional methods of computing and information processing you are familiar with are referred to as *classical information*. For those new to the subject, we begin with a simple and brief review of how information is stored and used in computers. The most basic piece of information is called a *bit*, and this basically represents a yes–no answer to a question. To represent this mathematically, we use the fact that we're dealing with a two-state system and choose to represent information using base 2 or *binary* numbers. A binary number can be 0 or 1, and a bit can assume one or the other of these values. Physically we can implement a bit with an electrical circuit that is either at ground or zero volts (binary 0), or at say +5 volts (binary 1). The physical implementation of a computing system is not our concern in this book; we are only worried about the mathematics and logic of the system. As a first step in getting acquainted with the binary world we might want to learn how to count using base 2.

Before we do that, we need to know that the number of bits required to represent something can be determined in the following way: Suppose that some quantity can assume one of m different states. Then

$$2^n \geq m \tag{1.1}$$

for some n. The smallest n for which this holds tells us the number of bits we need to represent or encode that quantity.

To see how this works, suppose that we want to represent the numbers 0, 1, 2, 3 in binary. We have four items, and $2^2 = 4$. Therefore we need at least two bits to represent these numbers. The representation is shown in Table 1.1.

To represent the numbers 0 through 7, we have $2^3 = 8$, so we need three bits. The binary representation of the numbers 0 through 7 is shown in Table 1.2.

INFORMATION CONTENT IN A SIGNAL

Now that we know how to encode information, we can start thinking about how to quantify it. That is, given a message m, how much information is actually contained in that message?

A clue about how this quantification might be done can be found by looking at (1.1). Considering the case where we take the equal sign, let's take the base two logarithm of both sides. That is, we start with

$$m = 2^n$$

TABLE 1.1 Binary representation of the numbers 0–3

Decimal	Binary
0	00
1	01
2	10
3	11

TABLE 1.2 Binary representation of the numbers 0–7

Decimal	Binary
0	000
1	001
2	010
3	011
4	100
5	101
6	110
7	111

Taking the base 2 log of both sides, we find that

$$\log_2 m = n \tag{1.2}$$

Equation (1.2) was proposed by Ralph Hartley in 1927. It was the first attempt at quantifying the amount of information in a message. What (1.2) tells us is that n bits can store m different messages. To make this more concrete, notice that

$$\log_2 8 = 3$$

That tells us that 3 bits can store 8 different messages. In Table 1.2 the eight messages we encoded were the numbers 0 through 7. However, the code could represent anything that had eight different possibilities.

You're probably familiar with different measurements of information storage capacity from your computer. The most basic word or unit of information is called a *byte*. A byte is a string of eight bits linked together. Now

$$\log_2 256 = 8$$

Therefore a byte can store 256 different messages. Measuring information in terms of logarithms also allows us to exploit the fact that logarithms are additive.

ENTROPY AND SHANNON'S INFORMATION THEORY

The Hartley method gives us a basic characterization of information content in a signal. But another scientist named Claude Shannon showed that we can take things a step further and get a more accurate estimation of the information content in a signal by thinking more carefully. The key step taken by Shannon was that he asked how *likely* is it that we are going to see a given piece of information? This is an important insight because it allows us to characterize how much information we actually *gain* from a signal.

If a message has a very high probability of occurrence, then we don't gain all that much new information when we come across it. On the other hand, if a message has a low probability of occurrence, when we are made of aware of it, we gain a significant amount of information. We can make this concrete with an example. A major earthquake occurred in the St. Louis area way back in 1812. Generally speaking, earthquakes in that area are relatively rare—after all, when you think of earthquakes, you think of California, not Missouri.

So most days people in Missouri aren't waiting around for an earthquake. Under typical conditions the probability of an earthquake occurring in Missouri is low, and the probability of an earthquake *not* occurring is high. If our message is that tomorrow there will *not* be an earthquake in Missouri, our message is a high probability message, and it conveys very little new information—for the last two hundred years day after day there hasn't been an earthquake. On the other hand, if the message is that tomorrow there will be an earthquake, this is dramatic news for Missouri residents. They gain *a lot* of information in this case.

Shannon quantified this by taking the base 2 logarithm of the probability of a given message occurring. That is, if we denote the information content of a message by I, and the probability of its occurrence by p, then

$$I = -\log_2 p \tag{1.3}$$

The negative sign ensures that the information content of a message is positive, and that the less probable a message, the higher is the information content. Let's suppose that the probability of an earthquake not happening tomorrow in St. Louis is 0.995. The information content of this fact is

$$I = -\log_2 0.995 = 0.0072$$

Now the probability that an earthquake does happen tomorrow is 0.005. The information content of this piece of information is

$$I' = -\log_2 0.005 = 7.6439$$

So let's summarize the use of logarithms to characterize the information content in a signal by saying:

- A message that is unlikely to occur has a low probability and therefore has a large information content.
- A message that is very likely to occur has a high probability and therefore has a small information content.

Next let's develop a more formal definition. Let X be a random variable characterized by a probability distribution $\vec{p}$, and suppose that it can assume one of the values $x_1, x_2, \ldots, x_n$ with probabilities $p_1, p_2, \ldots, p_n$. Probabilities satisfy $0 \leqslant p_i \leqslant 1$ and $\sum_i p_i = 1$.

The Shannon entropy of X is defined as

$$H(X) = -\sum_i p_i \log_2 p_i \tag{1.4}$$

If the probability of a given x_j is zero, we use $0 \log 0 = 0$. Notice that if we are saying that the logarithm of the probability of x gives the information content, we can also view the Shannon entropy function as a measure of the amount of uncertainty or randomnessin a signal.

We can look at this more concretely in terms of transmitted message signals as follows: Suppose that we have a signal that always transmits a "2," so that the signal is the string 22222222222.... What is the entropy in this case? The probability of obtaining a 2 is 1, so the entropy or disorder is

$$H = -\log_2 1 = 0$$

The Shannon entropy works as we expect—a signal that has all the same characters with no changes has no disorder and hence no entropy.

Now let's make a signal that's a bit more random. Suppose that the probability of obtaining a "1" is 0.5 and the probability of obtaining a "2" is 0.5, so the signal looks something like 11212221212122212121112... with approximately half the characters 1's and half 2's. What is the entropy in this case? It's

$$H = -\frac{1}{2}\log_2 \frac{1}{2} - \frac{1}{2}\log_2 \frac{1}{2} = \frac{1}{2} + \frac{1}{2} = 1$$

Suppose further that there are three equally likely possibilities. In that case we would have

$$H = -\frac{1}{3}\log_2 \frac{1}{3} - \frac{1}{3}\log_2 \frac{1}{3} = 0.528 + 0.528 + 0.528 = 1.585$$

In each case that we have examined here, the uncertainty in regard to what character we will see next in the message has increased each time—so the entropy also increases each time. In this view we can see that Shannon entropy measures the amount of uncertainty or randomness in the signal. That is:

- If we are certain what the message is, the Shannon entropy is zero.
- The more uncertain we are as to what comes next, the higher the Shannon entropy.

We can summarize Shannon entropy as

Decrease uncertainty ⇒ Increase information

Increase uncertainty ⇒ Increase entropy

Now suppose that we require l_i bits to represent each x_i in X. Then the *average bit rate* required to encode X is

$$R_X = \sum_{i=1}^{n} l_i p_i \tag{1.5}$$

The Shannon entropy is the lower bound of the average bit rate

$$H(X) \leqslant R_X \tag{1.6}$$

The worst-case scenario in which we have the least information is a distribution where the probability of each item is the same—meaning a uniform distribution. Again, suppose that it has n elements. The probability of finding each x_i if the distribution is uniform is $1/n$. So sequence X with n elements occurring with uniform probabilities $1/n$ has entropy $-\sum \frac{1}{n} \log_2 \frac{1}{n} = \sum \frac{1}{n} \log n = \log n$. This tells us that the Shannon entropy has the bounds

$$0 \leqslant H(X) \leqslant \log_2 n \tag{1.7}$$

The *relative entropy* of two variables X and Y characterized by probability distributions p and q is

$$H(X\|Y) = \sum p \log_2 \frac{p}{q} = -H(X) - \sum p \log_2 q \tag{1.8}$$

Suppose that we take a fixed value y_i from Y. From this we can get a conditional probability distribution $p(X|y_i)$ which are the probabilities of X given that we have y_i with certainty. Then

$$H(X|Y) = -\sum_j p(x_j|y_i) \log_2(p(x_j|y_i)) \tag{1.9}$$

This is known as the *conditional entropy*. The conditional entropy satisfies

$$H(X|Y) \leqslant H(X) \tag{1.10}$$

To obtain equality in (1.10), the variables X and Y must be independent. So

$$H(X,Y) = H(Y) + H(X|Y) \tag{1.11}$$

We are now in a position to define *mutual information* of the variables X and Y. In words, this is the difference between the entropy of X and the entropy of X

given knowledge of what value Y has assumed, that is,

$$I(X|Y) = H(X) - H(X|Y) \tag{1.12}$$

This can also be written as

$$I(X|Y) = H(X) + H(Y) - H(X, Y) \tag{1.13}$$

PROBABILITY BASICS

Before turning to quantum mechanics in the next chapter, it's a good idea to quickly mention the basics of probability. Probability is heavily used in quantum theory to predict the possible results of measurement.

We can start by saying that the probability p_i of an event x_i falls in the range

$$0 \leqslant p_i \leqslant 1 \tag{1.14}$$

The two extremes of this range are characterized as follows: The probability of an event that is *impossible* is 0. The probability of an event that is *certain to happen* is 1. All other probabilities fall within this range.

The probability of an event is simply the relative frequency of its occurrence. Suppose that there are n total events, the *jth* event occurs n_j times, and we have $\sum_{j=1}^{\infty} n_j = n$. Then the probability that the *jth* event occurs is

$$p_j = \frac{n_j}{n} \tag{1.15}$$

The sum of all the probabilities is 1, since

$$\sum_{j=1}^{\infty} p_j = \sum_{j=1}^{\infty} \frac{n_j}{n} = \frac{1}{n} \sum_{j=1}^{\infty} n_j = \frac{n}{n} = 1 \tag{1.16}$$

The average value of a distribution is referred to as the *expectation value* in quantum mechanics. This is given by

$$\langle j \rangle = \sum_{j=1}^{\infty} \frac{j n_j}{n} = \sum_{j=1}^{\infty} j p_j \tag{1.17}$$

The *variance* of a distribution is

$$\langle (\Delta j)^2 \rangle = \langle j^2 \rangle - \langle j \rangle^2 \tag{1.18}$$

Example 1.1

A group of students takes an exam. The number of students associated with each score is

Score	Students
95	1
85	3
77	7
71	10
56	3

What is the most probable test score? What is the expectation value or average score?

Solution

First we write down the total number of students

$$n = \sum n_j = 1 + 3 + 7 + 10 + 3 = 24$$

The probability of scoring 95 is

$$p_1 = \frac{n_1}{n} = \frac{1}{24} = 0.04$$

and the other probabilities are calculated similarly. The most probable score is 71 with probability

$$p_4 = \frac{n_4}{n} = \frac{10}{24} = 0.42$$

The expectation value is found using (1.17):

$$\langle j \rangle = \sum j\, p_j = 95(0.04) + 85(0.13) + 77(0.29) + 71(0.42) + 56(0.13) = 74.3$$

In the next chapter we will see how to quantum mechanics uses probability.

EXERCISES

1.1. *How many bits are necessary to represent the alphabet using a binary code if we only allow uppercase characters? How about if we allow both uppercase and lowercase characters?*

1.2. *Describe how you can create an OR gate using NOT gates and AND gates.*

1.3. *A kilobyte is 1024 bytes. How many messages can it store?*

1.4. *What is the entropy associated with the tossing of a fair coin?*

1.5. *Suppose that X consists of the characters A, B, C, D that occur in a signal with respective probabilities 0.1, 0.4, 0.25, and 0.25. What is the Shannon entropy?*

1.6. *A room full of people has incomes distributed in the following way:*

$$n(25.5) = 3$$
$$n(30) = 5$$
$$n(42) = 7$$
$$n(50) = 3$$
$$n(63) = 1$$
$$n(75) = 2$$
$$n(90) = 1$$

What is the most probable income? What is the average income? What is the variance of this distribution?

2

QUBITS AND QUANTUM STATES

In this chapter we will expand on our discussion of the qubit and learn some basic facts and notation that are necessary when learning how to work with quantum states.

THE QUBIT

In the last chapter we saw that the basic unit of information processing in a modern-day computer is the bit, which can assume one of two states that we label 0 and 1. In an analogous manner, we can define a basic unit of information processing that can be used in quantum computation. This basic unit of information in quantum computing is called the *qubit*, which is short for *quantum bit*. While a qubit is going to look in some way superficially similar to a bit, we will see as we go along that it is fundamentally different and that its fundamental difference allows us to do information processing in new and interesting ways.

Like a bit, a qubit can also be in one of two states. In the case of a qubit, for reasons that for the moment will seem utterly obscure, we label these two states

Quantum Computing Explained, by David McMahon

by $|0\rangle$ and $|1\rangle$. In quantum theory an object enclosed using the notation $|\ \rangle$ can be called a *state*, a *vector*, or a *ket*.

So how is a qubit any different than an ordinary bit? While a bit in an ordinary computer can be in the state 0 *or* in the state 1, a qubit is somewhat more general. A qubit can exist in the state $|0\rangle$ or the state $|1\rangle$, but it can also exist in what we call a *superposition* state. This is a state that is a linear combination of the states $|0\rangle$ and $|1\rangle$. If we label this state $|\psi\rangle$, a superposition state is written as

$$|\psi\rangle = \alpha|0\rangle + \beta|1\rangle \tag{2.1}$$

Here α, β are complex numbers. That is, the numbers are of the form $z = x + iy$, where $i = \sqrt{-1}$.

While a qubit can exist in a superposition of the states $|0\rangle$ and $|1\rangle$, whenever we make a measurement we aren't going to find it like that. In fact, when a qubit is measured, it is only going to be found to be in the state $|0\rangle$ or the state $|1\rangle$. The laws of quantum mechanics tell us that the modulus squared of α, β in (2.1) gives us the probability of finding the qubit in state $|0\rangle$ or $|1\rangle$, respectively. In other words:

$|\alpha|^2$: Tells us the probability of finding $|\psi\rangle$ in state $|0\rangle$
$|\beta|^2$: Tells us the probability of finding $|\psi\rangle$ in state $|1\rangle$

The fact that probabilities must sum to one puts some constraints on what the multiplicative coefficients in (2.1) can be. Since the squares of these coefficients are related to the probability of obtaining a given measurement result, α and β are constrained by the requirement that

$$|\alpha|^2 + |\beta|^2 = 1 \tag{2.2}$$

Generally speaking, if an event has N possible outcomes and we label the probability of finding result i by p_i, the condition that the probabilities sum to one is written as

$$\sum_{i=1}^{N} p_i = p_1 + p_2 + \cdots + p_N = 1 \tag{2.3}$$

When this condition is satisfied for the squares of the coefficients of a qubit, we say that the qubit is *normalized*.

We can calculate the modulus of these numbers in the following way:

$$|\alpha|^2 = (\alpha)(\alpha^*)$$
$$|\beta|^2 = (\beta)(\beta^*)$$

where α^* is the complex conjugate of α and β^* is the complex conjugate of β. We recall that to form the complex conjugate of $z = x + iy$, we let $i \to -i$. Therefore the modulus of a complex number z is

$$|z|^2 = (x+iy)(x-iy) = x^2 + ixy - ixy + y^2$$
$$\Rightarrow |z| = \sqrt{x^2+y^2}$$

Right now, it might seem as if we have traded in the reliability of a classical computer for a probabilistic guessing game. Later we will see that this isn't the case and that these strange propeies actually make a qubit an asset, rather than a liability, in information processing. For now let's look at a couple of examples to reinforce the basic ideas.

Example 2.1

For each of the following qubits, if a measurement is made, what is the probability that we find the qubit in state $|0\rangle$? What is the probability that we find the qubit in the state $|1\rangle$?

(a) $|\psi\rangle = \frac{1}{\sqrt{3}}|0\rangle + \sqrt{\frac{2}{3}}|1\rangle$

(b) $|\phi\rangle = \frac{i}{2}|0\rangle + \frac{\sqrt{3}}{2}|1\rangle$

(c) $|\chi\rangle = \frac{(1+i)}{\sqrt{3}}|0\rangle - \frac{i}{\sqrt{3}}|1\rangle$

Solution

To find the probability that each qubit is found in the state $|0\rangle$ or the state $|1\rangle$, we compute the modulus squared of the appropriate coefficient.

(a) In this case the probability of finding $|\psi\rangle$ in state $|0\rangle$ is

$$\left|\frac{1}{\sqrt{3}}\right|^2 = \frac{1}{3}$$

The probability that we find $|\psi\rangle$ in state $|1\rangle$ is

$$\left|\frac{\sqrt{2}}{\sqrt{3}}\right|^2 = \frac{2}{3}$$

When doing calculations in quantum mechanics or quantum computation, it is always a good idea to verify that the probabilities you find add to 1. We can label the probability that the system is in state $|0\rangle$ p_0 and the probability that the system is in state $|1\rangle$ p_1. In the context of this example $p_0 = 1/3$ and $p_1 = 2/3$, hence

$$\sum p_i = p_0 + p_1 = \frac{1}{3} + \frac{2}{3} = 1$$

(b) The next state has coefficients that are complex numbers. Remember we need to use the complex conjugate when calculating the modulus squared. We find that the probability of the system being in state $|0\rangle$ is

$$|\phi\rangle = \frac{i}{2}|0\rangle + \frac{\sqrt{3}}{2}|1\rangle, \Rightarrow p_0 = \left|\frac{i}{2}\right|^2 = \left(\frac{i}{2}\right)^* \left(\frac{i}{2}\right) = \left(\frac{-i}{2}\right)\left(\frac{i}{2}\right) = \frac{1}{4}$$

The probability that the system is in state $|1\rangle$ is

$$\left|\frac{\sqrt{3}}{2}\right|^2 = \frac{3}{4}$$

Again, we check that the probabilities sum to one:

$$\sum p_i = p_0 + p_1 = \frac{1}{4} + \frac{3}{4} = 1$$

(c) Finally, for the last state, the probability the system is in state $|0\rangle$ is

$$\left|\frac{1+i}{\sqrt{3}}\right|^2 = \left(\frac{1-i}{\sqrt{3}}\right)\left(\frac{1+i}{\sqrt{3}}\right) = \frac{1-i+i+1}{3} = \frac{2}{3}$$

The probability that the system is in state $|1\rangle$ is

$$\left|\frac{-i}{\sqrt{3}}\right|^2 = \left(\frac{-i}{\sqrt{3}}\right)^* \left(\frac{-i}{\sqrt{3}}\right) = \left(\frac{i}{\sqrt{3}}\right)\left(\frac{-i}{\sqrt{3}}\right) = \frac{1}{3}$$

Again, these probabilities sum to 1:

$$p_0 + p_1 = \frac{2}{3} + \frac{1}{3} = 1$$

VECTOR SPACES

While the basic unit of information in quantum computation is the qubit, the arena in which quantum computation takes place is a mathematical abstraction called a *vector space*. If you've taken elementary physics or engineering classes, then you know what a vector is. It turns out that quantum states behave mathematically in an analogous way to physical vectors—hence the term vector space. This type of space is one that shares with physical vectors the most basic properties that vectors have—for example, a length. In this section we are going to look at state vectors more generally and talk a little bit about the spaces they inhabit. We aren't going to be rigorous here, the purpose of this section is just to introduce some basic ideas and terminology. To avoid further confusion, let's just get down to the basic definitions. A vector space V is a nonempty set with elements u, v called *vectors* for which the following two operations are defined:

1. Vector addition: An operation that assigns the sum $w = u + v$, which is also an element of V; in other words, w is another vector belonging to the same space

2. Scalar multiplication: Defines multiplication of a vector by a number α such that the vector $\alpha u \in V$

In addition the following axioms hold for the vector space V:

Axiom 1: Associativity of addition. Given vectors u, v, and w,

$$(u + v) + w = u + (v + w)$$

Axiom 2: There is a vector belonging to V called the *zero vector* that satisfies

$$u + 0 = 0 + u = u$$

for any vector $u \in V$.

Axiom 3: For every $u \in V$ there exists an additive inverse of u such that

$$u + (-u) = (-u) + u = 0$$

Axiom 4: Addition of vectors is commutative:

$$u + v = v + u$$

There are other axioms that apply to vectors spaces, but these should suffice for most of our purposes.

One particular vector space that is important in quantum computation is the vector space $\mathbb{C}^n$, which is the vector space of "n-tuples" of complex numbers. When we say "n-tuple" this is just a fancy way of referring to an ordered collection of numbers. Following the notation we have introduced for qubits, we label the elements of $\mathbb{C}^n$ by $|a\rangle$, $|b\rangle$, $|c\rangle$. Then we write down an element of this vector space as an n-dimensional *column vector* or simply a list of numbers $a_1, a_2, \ldots, a_n$ arranged in the following way:

$$|a\rangle = \begin{pmatrix} a_1 \\ a_2 \\ \vdots \\ a_n \end{pmatrix} \tag{2.4}$$

This type of notation can be used with qubits. Look back at the general notation used for qubit given in (2.1). We write this in a column vector format by putting the coefficient of $|0\rangle$ in the first row and the coefficient of $|1\rangle$ in the second row:

$$|\psi\rangle = \begin{pmatrix} \alpha \\ \beta \end{pmatrix}$$

For a more concrete example, let's take a look back at the qubits we examined in Example 2.1. The first qubit was

$$|\psi\rangle = \frac{1}{\sqrt{3}}|0\rangle + \sqrt{\frac{2}{3}}|1\rangle$$

Hence the column vector representation of this qubit is

$$|\psi\rangle = \begin{pmatrix} \frac{1}{\sqrt{3}} \\ \sqrt{\frac{2}{3}} \end{pmatrix} = \frac{1}{\sqrt{3}}\begin{pmatrix} 1 \\ \sqrt{2} \end{pmatrix}$$

In general, the numbers a_i in (2.4), which we call the *components* of the vector, are complex—something we've already mentioned.

Multiplication of a vector by a scalar proceeds as

$$\alpha|a\rangle = \alpha \begin{pmatrix} a_1 \\ a_2 \\ \vdots \\ a_n \end{pmatrix} = \begin{pmatrix} \alpha a_1 \\ \alpha a_2 \\ \vdots \\ \alpha a_n \end{pmatrix} \tag{2.5}$$

It's easy to see that this produces another column vector with n complex numbers, so the result is another element in $\mathbb{C}^n$. So $\mathbb{C}^n$ is closed under scalar multiplication.

Vector addition is carried out component by component, producing a new, third vector:

$$|a\rangle + |b\rangle = \begin{pmatrix} a_1 \\ a_2 \\ \vdots \\ a_n \end{pmatrix} + \begin{pmatrix} b_1 \\ b_2 \\ \vdots \\ b_n \end{pmatrix} = \begin{pmatrix} a_1 + b_1 \\ a_2 + b_2 \\ \vdots \\ a_n + b_n \end{pmatrix} \tag{2.6}$$

This should all be pretty straightforward, but let's illustrate it with an example.

Example 2.2

We want to define the vectors

$$|u\rangle = \begin{pmatrix} -1 \\ 7i \\ 2 \end{pmatrix}, \quad |v\rangle = \begin{pmatrix} 0 \\ 2 \\ 4 \end{pmatrix}$$

and then compute $7|u\rangle + 2|v\rangle$.

Solution

First we compute $7|u\rangle$ by multiplying each of the components of the vector by 7:

$$7|u\rangle = 7\begin{pmatrix} -1 \\ 7i \\ 2 \end{pmatrix} = \begin{pmatrix} -7 \\ 49i \\ 14 \end{pmatrix}$$

Next we compute the scalar multiplication for the other vector:

$$2|v\rangle = 2\begin{pmatrix} 0 \\ 2 \\ 4 \end{pmatrix} = \begin{pmatrix} 0 \\ 4 \\ 8 \end{pmatrix}$$

Now we follow the vector addition rule, adding the vectors component by component:

$$7|u\rangle + 2|v\rangle = \begin{pmatrix} -7 \\ 49i \\ 14 \end{pmatrix} + \begin{pmatrix} 0 \\ 4 \\ 8 \end{pmatrix} = \begin{pmatrix} -7+0 \\ 49i+4 \\ 14+8 \end{pmatrix} = \begin{pmatrix} -7 \\ 4+49i \\ 22 \end{pmatrix}$$

LINEAR COMBINATIONS OF VECTORS

In the previous example we calculated an important quantity, a *linear combination* of vectors. Generally speaking, let α_i be a set of complex coefficients and $|v_i\rangle$ be a set of vectors. A linear combination of these vectors is given by

$$\alpha_1|v_1\rangle + \alpha_2|v_2\rangle + \cdots + \alpha_n|v_n\rangle = \sum_{i=1}^{n} \alpha_i|v_i\rangle \tag{2.7}$$

We've seen a linear combination before, when we considered a superposition state of a qubit.

Now, if a given set of vectors $|v_1\rangle, |v_2\rangle, \ldots, |v_n\rangle$ can be used to represent any vector $|u\rangle$ that belongs to the vector space V, we say that the set $\{|v_i\rangle\}$ *spans* the given vector space. For example, consider the three-dimensional vector space $\mathbb{C}^3$. We can write any three-dimensional column vector in the following way:

$$|u\rangle = \begin{pmatrix} \alpha \\ \beta \\ \gamma \end{pmatrix} = \begin{pmatrix} \alpha \\ 0 \\ 0 \end{pmatrix} + \begin{pmatrix} 0 \\ \beta \\ 0 \end{pmatrix} + \begin{pmatrix} 0 \\ 0 \\ \gamma \end{pmatrix} = \alpha\begin{pmatrix} 1 \\ 0 \\ 0 \end{pmatrix} + \beta\begin{pmatrix} 0 \\ 1 \\ 0 \end{pmatrix} + \gamma\begin{pmatrix} 0 \\ 0 \\ 1 \end{pmatrix}$$

Therefore we see that the set of vectors defined as

$$|v_1\rangle = \begin{pmatrix} 1 \\ 0 \\ 0 \end{pmatrix}, \quad |v_2\rangle = \begin{pmatrix} 0 \\ 1 \\ 0 \end{pmatrix}, \quad |v_3\rangle = \begin{pmatrix} 0 \\ 0 \\ 1 \end{pmatrix}$$

Spans the space $\mathbb{C}^3$. We've already seen a spanning set for qubits when we considered the basic states a qubit can be found in, $\{|0\rangle, |1\rangle\}$. Recall from (2.1) that an arbitrary qubit can be written as

$$|\psi\rangle = \alpha|0\rangle + \beta|1\rangle$$

Writing this in column vector notation, and then using the properties of vector addition and scalar multiplication, we have

$$|\psi\rangle = \begin{pmatrix} \alpha \\ \beta \end{pmatrix} = \begin{pmatrix} \alpha \\ 0 \end{pmatrix} + \begin{pmatrix} 0 \\ \beta \end{pmatrix} = \alpha \begin{pmatrix} 1 \\ 0 \end{pmatrix} + \beta \begin{pmatrix} 0 \\ 1 \end{pmatrix}$$

This allows us to identify the vectors $|0\rangle$ and $|1\rangle$ as

$$|0\rangle = \begin{pmatrix} 1 \\ 0 \end{pmatrix}, \quad |1\rangle = \begin{pmatrix} 0 \\ 1 \end{pmatrix}$$

An important notion involving the linear combination of a set of vectors is that of *linear independence*. If

$$\alpha_1|v_1\rangle + \alpha_2|v_2\rangle + \cdots + \alpha_n|v_n\rangle = 0$$

and at least one of the $\alpha_i \neq 0$, we say the set $\{|v_i\rangle\}$ is linearly dependent. Another way of saying this is that if one vector of the set can be written as a linear combination of the other vectors in the set, then the set is *linearly dependent*.

Example 2.3

Show that the set

$$|a\rangle = \begin{pmatrix} 1 \\ 2 \end{pmatrix}, \quad |b\rangle = \begin{pmatrix} -1 \\ 1 \end{pmatrix}, \quad |c\rangle = \begin{pmatrix} 5 \\ 4 \end{pmatrix}.$$

is linearly dependent; that is, one of the vectors can be written as a linear combination of the other two.

Solution

We can show that $|c\rangle$ can be expressed as a linear combination of the other two vectors. Let's start out by writing $|c\rangle$ as some arbitrary combination of the other two vectors:

$$\begin{pmatrix} 5 \\ 4 \end{pmatrix} = \alpha \begin{pmatrix} 1 \\ 2 \end{pmatrix} + \beta \begin{pmatrix} -1 \\ 1 \end{pmatrix}$$

Now we actually have two equations. These are

$$\alpha - \beta = 5, \quad \Rightarrow \beta = \alpha - 5$$
$$2\alpha + \beta = 4$$

Substitution of $\beta = \alpha - 5$ into the second equation gives

$$2\alpha + (\alpha - 5) = 3\alpha - 5 = 4, \Rightarrow \alpha = 3$$

So immediately we see that $\beta = -2$. Using these results, we see that the following relationship holds, demonstrating the linear dependence of this set of vectors:

$$3|a\rangle - 2|b\rangle + |c\rangle = 0$$

UNIQUENESS OF A SPANNING SET

A spanning set of vectors for a given space V is not unique. Once again, consider the complex vector space $\mathbb{C}^2$, consisting of column vectors with two elements. Earlier we saw that we can write any arbitrary column vector with two elements in the following way:

$$\begin{pmatrix}\alpha\\\beta\end{pmatrix} = \begin{pmatrix}\alpha\\0\end{pmatrix} + \begin{pmatrix}0\\\beta\end{pmatrix} = \alpha\begin{pmatrix}1\\0\end{pmatrix} + \beta\begin{pmatrix}0\\1\end{pmatrix}$$

And therefore the set

$$|0\rangle = \begin{pmatrix}1\\0\end{pmatrix}, \quad |1\rangle = \begin{pmatrix}0\\1\end{pmatrix}$$

spans $\mathbb{C}^2$, the vector space in which qubits live. Now consider the set:

$$|u_1\rangle = \begin{pmatrix}1\\1\end{pmatrix}, \quad |u_2\rangle = \begin{pmatrix}1\\-1\end{pmatrix}$$

This set also spans the space $\mathbb{C}^2$, and this means that we can also use these vectors to represent qubits. We can write any element of this space in terms of these vectors:

$$\begin{aligned}|\psi\rangle &= \begin{pmatrix}\alpha\\\beta\end{pmatrix}\\ &= \begin{pmatrix}\alpha\\0\end{pmatrix} + \begin{pmatrix}0\\\beta\end{pmatrix} = \frac{1}{2}\begin{pmatrix}\alpha+\alpha\\\alpha-\alpha\end{pmatrix} + \frac{1}{2}\begin{pmatrix}\beta-\beta\\\beta+\beta\end{pmatrix}\\ &= \frac{1}{2}\begin{pmatrix}\alpha\\\alpha\end{pmatrix} + \frac{1}{2}\begin{pmatrix}\alpha\\-\alpha\end{pmatrix} + \frac{1}{2}\begin{pmatrix}\beta\\\beta\end{pmatrix} - \frac{1}{2}\begin{pmatrix}\beta\\-\beta\end{pmatrix}\end{aligned}$$

Now we use the scalar multiplication rule to pull the constants α, β outside of the vectors:

$$\frac{1}{2}\begin{pmatrix} \alpha \\ \alpha \end{pmatrix} + \frac{1}{2}\begin{pmatrix} \alpha \\ -\alpha \end{pmatrix} + \frac{1}{2}\begin{pmatrix} \beta \\ \beta \end{pmatrix} - \frac{1}{2}\begin{pmatrix} \beta \\ -\beta \end{pmatrix}$$

$$= \frac{\alpha}{2}\begin{pmatrix} 1 \\ 1 \end{pmatrix} + \frac{\alpha}{2}\begin{pmatrix} 1 \\ -1 \end{pmatrix} + \frac{\beta}{2}\begin{pmatrix} 1 \\ 1 \end{pmatrix} - \frac{\beta}{2}\begin{pmatrix} 1 \\ -1 \end{pmatrix}$$

Rearranging a bit, we find that

$$\left(\frac{\alpha+\beta}{2}\right)\begin{pmatrix} 1 \\ 1 \end{pmatrix} + \left(\frac{\alpha-\beta}{2}\right)\begin{pmatrix} 1 \\ -1 \end{pmatrix} = \left(\frac{\alpha+\beta}{2}\right)|u_1\rangle + \left(\frac{\alpha-\beta}{2}\right)|u_2\rangle$$

BASIS AND DIMENSION

When a set of vectors is linearly independent and they span the space, the set is known as a *basis*. We can express any vector in the space V in terms of a linear expansion on a basis set. Moreover the *dimension* of a vector space V is equal to the number of elements in the basis set. We have already seen a basis set for $\mathbb{C}^2$, with the qubit states $|0\rangle$ and $|1\rangle$:

$$|0\rangle = \begin{pmatrix} 1 \\ 0 \end{pmatrix}, \quad |1\rangle = \begin{pmatrix} 0 \\ 1 \end{pmatrix}$$

As we have shown, any vector in $\mathbb{C}^2$ can be written as a linear combination of these two vectors; hence they span the space. It is also pretty easy to show that these two vectors are linearly independent. Therefore they constitute a basis of $\mathbb{C}^2$.

A vector space can have many basis sets. We have already seen another basis set that can be used for $\mathbb{C}^2$ (and hence for qubits):

$$|u_1\rangle = \begin{pmatrix} 1 \\ 1 \end{pmatrix}, \quad |u_2\rangle = \begin{pmatrix} 1 \\ -1 \end{pmatrix}$$

In term of basis vectors, quantum states in n dimensions are straightforward generalizations of qubits. A quantum state $|\psi\rangle$ can be written as a linear combination of a basis set $|v_i\rangle$ with complex coefficients of expansion c_i as

$$|\psi\rangle = \sum_{i=1}^{n} c_i|v_i\rangle = c_1|v_1\rangle + c_2|v_2\rangle + \cdots + c_n|v_n\rangle \tag{2.8}$$

The modulus squared of a given coefficient c_i gives the probability that measurement finds the system in the state $|v_i\rangle$.

INNER PRODUCTS

To compute the length of a vector, even if it's a length in an abstract sense, we need a way to find the *inner product*. This is a generalization of the dot product used with ordinary vectors in Euclidean space as you may already be familiar with. While the dot product takes two vectors and maps them into a real number, in our case the inner product will take two vectors from $\mathbb{C}^n$ and map them to a *complex* number. We write the inner product between two vectors $|u\rangle$, $|v\rangle$ with the notation $\langle u|v\rangle$. If the inner product between two vectors is zero,

$$\langle u|v\rangle = 0$$

We say that $|u\rangle$, $|v\rangle$ are *orthogonal* to one another. The inner product is a complex number. The conjugate of this complex number satisfies

$$\langle u|v\rangle^* = \langle v|u\rangle \tag{2.9}$$

We can use the inner product to define a norm (or length)—by computing the inner product of a vector with itself:

$$\|u\| = \sqrt{\langle u|u\rangle} \tag{2.10}$$

Notice that the norm is a real number, and hence can define a length. For any vector $|u\rangle$ we have

$$\langle u|u\rangle \geq 0 \tag{2.11}$$

with equality if and only if $|u\rangle = 0$. When considering the inner product of a vector with a superposition or linear combination of vectors, the following linear and *anti-linear* relations hold:

$$\begin{aligned} \langle u|\alpha v + \beta w\rangle &= \alpha\langle u|v\rangle + \beta\langle u|w\rangle \\ \langle \alpha u + \beta v|w\rangle &= \alpha^*\langle u|w\rangle + \beta^*\langle v|w\rangle \end{aligned} \tag{2.12}$$

To compute the inner product between two vectors, we must calculate the Hermitian conjugate of a vector

$$(|u\rangle)^\dagger = \langle u|$$

In quantum physics $\langle u|$ is sometimes called the *dual vector* or *bra* corresponding to $|u\rangle$.

If a ket is a column vector, the dual vector or bra is a row vector whose elements are the complex conjugates of the elements of the column vector. In other words, when working with column vectors, the Hermitian conjugate is computed in two steps:

1. Write the components of the vector as a row of numbers.
2. Take the complex conjugate of each element and arrange them in a row vector.

Generally, we have

$$\begin{pmatrix} a_1 \\ a_2 \\ \vdots \\ a_n \end{pmatrix}^{\dagger} = \begin{pmatrix} a_1^* & a_2^* & \dots & a_n^* \end{pmatrix} \tag{2.13}$$

With qubits, this is fairly easy. Let's go back to the state

$$|\phi\rangle = \frac{i}{2}|0\rangle + \frac{\sqrt{3}}{2}|1\rangle$$

The column vector representation of this state is

$$|\phi\rangle = \begin{pmatrix} \dfrac{i}{2} \\ \dfrac{\sqrt{3}}{2} \end{pmatrix}$$

The bra or dual vector is found by computing the complex conjugate of each element and then arranging the result as a row vector. In this case

$$\langle\phi| = \begin{pmatrix} -\dfrac{i}{2} & \dfrac{\sqrt{3}}{2} \end{pmatrix}$$

Using the dual vector to find the inner product makes the calculation easy. The inner product $\langle a|b\rangle$ is calculated in the following way:

$$\langle a|b\rangle = \begin{pmatrix} a_1^* & a_2^* & \dots & a_n^* \end{pmatrix} \begin{pmatrix} b_1 \\ b_2 \\ \vdots \\ b_n \end{pmatrix} = a_1^* b_1 + a_2^* b_2 + \dots + a_n^* b_n = \sum_{i=1}^{n} a_i^* b_i \tag{2.14}$$

Example 2.4

Two vectors in $\mathbb{C}^3$ are given by

$$|a\rangle = \begin{pmatrix} -2 \\ 4i \\ 1 \end{pmatrix}, \quad |b\rangle = \begin{pmatrix} 1 \\ 0 \\ i \end{pmatrix}$$

Find

(a) $\langle a|$, $\langle b|$

(b) $\langle a|b\rangle$, $\langle b|a\rangle$

(c) $|c\rangle = |a\rangle + 2|b\rangle$, $\langle c|a\rangle$

Solution

(a) We consider $|a\rangle$ first. Begin by rewriting the elements in a row vector format. This is known as the *transpose* operation:

$$(|a\rangle)^T = (-2 \quad 4i \quad 1)$$

The Hermitian conjugate is then found by computing the complex conjugate of each element:

$$\langle a| = (|a\rangle)^\dagger = \left(|a\rangle^T\right)^* = (-2 \quad -4i \quad 1)$$

A similar procedure for $|b\rangle$ yields

$$\langle b| = (1 \quad 0 \quad -i)$$

(b) The inner product, from (2.14), is given by

$$\langle a|b\rangle = (-2 \quad 4i \quad 1)\begin{pmatrix}1\\0\\i\end{pmatrix} = -2^*1 + 4i^*0 + 1^*i = -2 + i$$

Remember, the inner product on a complex vector space obeys $\langle a|b\rangle = \langle b|a\rangle^*$; therefore we should find $\langle b|a\rangle = -2 - i$. We verify this with an explicit calculation

$$\langle b|a\rangle = (1 \quad 0 \quad -i)\begin{pmatrix}-2\\4i\\1\end{pmatrix} = 1^* - 2 + 0^*4i + (-i)^*1 = -2 - i$$

(c) We apply the rules of vector addition and scalar multiplication to obtain

$$|c\rangle = |a\rangle + 2|b\rangle = \begin{pmatrix}-2\\4i\\1\end{pmatrix} + 2\begin{pmatrix}1\\0\\i\end{pmatrix} = \begin{pmatrix}-2\\4i\\1\end{pmatrix} + \begin{pmatrix}2\\0\\2i\end{pmatrix} = \begin{pmatrix}-2+2\\4i+0\\1+2i\end{pmatrix} = \begin{pmatrix}0\\4i\\1+2i\end{pmatrix}$$

Therefore the inner product is

$$\langle c|a\rangle = (0 \quad -4i \quad 1-2i)\begin{pmatrix}-2\\4i\\1\end{pmatrix} = 0^*(-2) + (-4i)(4i) + (1-2i)^*1$$

$$= 16 + 1 - 2i = 17 - 2i$$

As an exercise, verify that the form of $\langle c|$ is correct by computing the Hermitian conjugate of $|c\rangle$.

Example 2.5

Compute the norm of the following vectors:

$$|u\rangle = \begin{pmatrix} 2 \\ 4i \end{pmatrix}, \quad |v\rangle = \begin{pmatrix} -1 \\ 3i \\ i \end{pmatrix}$$

Solution

We start by computing the Hermitian conjugate of each vector. Remember, first take the transpose of the vector and write the list of numbers as a row; then compute the complex conjugate of each element:

$$(|u\rangle)^T = (2 \quad 4i), \quad \Rightarrow \langle u| = (|u\rangle)^\dagger = (|u\rangle^T)^* = (2 \quad -4i)$$
$$\langle v| = (-1 \quad 3i \quad i)^* = (-1 \quad -3i \quad -i)$$

We find that

$$\langle u|u\rangle = (2 \quad -4i)\begin{pmatrix} 2 \\ 4i \end{pmatrix} = 2 * 2 + (-4i)^* 4i = 4 + 16 = 20$$

$$\langle v|v\rangle = (-1 \quad -3i \quad -i)\begin{pmatrix} -1 \\ 3i \\ i \end{pmatrix} = -1^*(-1) + (-3i) * 3i + (-i) * i = 1 + 9 + 1 = 11$$

The norm is found by taking the square root of these quantities:

$$\|u\| = \sqrt{\langle u|u\rangle} = \sqrt{20}$$
$$\|v\| = \sqrt{\langle v|v\rangle} = \sqrt{11}$$

ORTHONORMALITY

When the norm of a vector is unity, we say that vector is *normalized*. That is, if

$$\langle a|a\rangle = 1$$

then we say that $|a\rangle$ is normalized. If a vector is not normalized, we can generate a normalized vector by computing the norm (which is just a number) and dividing the vector by it. For the vectors in the previous example, $|u\rangle$, $|v\rangle$ are not normalized, since we found that $\|u\| = \sqrt{20}$ and $\|v\| = \sqrt{11}$. But the vectors

$$|\tilde{u}\rangle = \frac{|u\rangle}{\|u\|} = \frac{1}{\sqrt{20}}|u\rangle$$

$$|\tilde{v}\rangle = \frac{|v\rangle}{\|v\|} = \frac{1}{\sqrt{11}}|v\rangle$$

are normalized. This is easy to see. Check the first case:

$$\langle\tilde{u}|\tilde{u}\rangle = \left(\frac{1}{\sqrt{20}}\langle u|\right)\left(\frac{1}{\sqrt{20}}|u\rangle\right) = \frac{1}{20}\langle u|u\rangle = \frac{20}{20} = 1$$

If each element of a set of vectors is normalized and the elements are orthogonal with respect to each other, we say the set is *orthonormal*. For example, consider the set $\{|0\rangle, |1\rangle\}$. Remember, we have the following definition:

$$|0\rangle = \begin{pmatrix}1\\0\end{pmatrix}, \quad |1\rangle = \begin{pmatrix}0\\1\end{pmatrix}$$

Therefore, using (2.14), we have

$$\langle 0|0\rangle = (1 \quad 0)\begin{pmatrix}1\\0\end{pmatrix} = 1*1+0*0 = 1$$

$$\langle 0|1\rangle = (1 \quad 0)\begin{pmatrix}0\\1\end{pmatrix} = 1*0+0*1 = 0$$

$$\langle 1|0\rangle = (0 \quad 1)\begin{pmatrix}1\\0\end{pmatrix} = 0*1+1*0 = 0$$

$$\langle 1|1\rangle = (0 \quad 1)\begin{pmatrix}0\\1\end{pmatrix} = 0*0+1*1 = 1$$

By showing that $\langle 0|0\rangle = \langle 1|1\rangle = 1$, we showed the vectors are normalized, while showing that $\langle 0|1\rangle = \langle 1|0\rangle = 0$, we showed they were orthogonal. Hence the set is orthonormal. Earlier we saw that any qubit could also be written as a linear combination of the vectors:

$$|u_1\rangle = \begin{pmatrix}1\\1\end{pmatrix}, \quad |u_2\rangle = \begin{pmatrix}1\\-1\end{pmatrix}$$

Is this set orthonormal? Well the vectors are orthogonal, since

$$\langle u_1|u_2\rangle = (1 \quad 1)\begin{pmatrix}1\\-1\end{pmatrix} = 1-1 = 0$$

But the vectors are not normalized. It's easy to see that $\|u_1\| = \|u_2\| = \sqrt{2}$. We can create an orthonormal set from these vectors by normalizing them. We denote these vectors by $|+\rangle$ and $|-\rangle$. Therefore another orthonormal basis set for $\mathbb{C}^2$ (and hence for qubits) is

$$|+\rangle = \frac{1}{\sqrt{2}}\begin{pmatrix}1\\1\end{pmatrix}, \quad |-\rangle = \frac{1}{\sqrt{2}}\begin{pmatrix}1\\-1\end{pmatrix}$$

GRAM-SCHMIDT ORTHOGONALIZATION

An orthonormal basis can be produced from an *arbitrary* basis by application of the *Gram-Schmidt orthogonalization* process. Let $\{|v_1\rangle, |v_2\rangle, \ldots\}|v_n\rangle$, be a basis for an inner product space V. The Gram-Schmidt process constructs an orthogonal basis $|w_i\rangle$ as follows:

$$|w_1\rangle = |v_1\rangle$$

$$|w_2\rangle = |v_2\rangle - \frac{\langle w_1|v_2\rangle}{\langle w_1|w_1\rangle}|w_1\rangle$$

$$\vdots$$

$$|w_n\rangle = |v_n\rangle - \frac{\langle w_1|v_n\rangle}{\langle w_1|w_1\rangle}|w_1\rangle - \frac{\langle w_2|v_n\rangle}{\langle w_2|w_2\rangle}|w_2\rangle - \cdots - \frac{\langle w_{n-1}|v_n\rangle}{\langle w_{n-1}|w_{n-1}\rangle}|w_{n-1}\rangle$$

To form an orthonormal set using the Gram-Schmidt procedure, divide each vector by its norm. For example, the normalized vector we can use to construct $|w_2\rangle$ is

$$|w_2\rangle = \frac{|v_2\rangle - \langle w_1|v_2\rangle|w_1\rangle}{\||v_2\rangle - \langle w_1|v_2\rangle|w_1\rangle\|}$$

Many readers might find this a bit abstract, so let's illustrate with a concrete example.

Example 2.6

Use the Gram-Schmidt process to construct an orthonormal basis set from

$$|v_1\rangle = \begin{pmatrix}1\\2\\-1\end{pmatrix}, \quad |v_2\rangle = \begin{pmatrix}0\\1\\-1\end{pmatrix}, \quad |v_3\rangle = \begin{pmatrix}3\\-7\\1\end{pmatrix}$$

Solution

We use a tilde character to denote the unnormalized vectors. The first basis vector is

$$|\tilde{w}_1\rangle = |v_1\rangle$$

Now let's normalize this vector

$$\langle v_1|v_1\rangle = \begin{pmatrix} 1 & 2 & -1 \end{pmatrix} \begin{pmatrix} 1 \\ 2 \\ -1 \end{pmatrix} = 1*1+2*2+(-1)*(-1) = 1+4+1 = 6$$

$$\Rightarrow |w_1\rangle = \frac{|\tilde{w}_1\rangle}{\sqrt{\langle v_1|v_1\rangle}} = \frac{1}{\sqrt{6}} \begin{pmatrix} 1 \\ 2 \\ -1 \end{pmatrix}$$

Looking back at the algorithm for the Gram-Schmidt process, we construct the second vector by using the formula

$$|\tilde{w}_2\rangle = |v_2\rangle - \frac{\langle \tilde{w}_1|v_2\rangle}{\langle \tilde{w}_1|\tilde{w}_1\rangle}|\tilde{w}_1\rangle$$

First we compute

$$\langle \tilde{w}_1|v_2\rangle = \begin{pmatrix} 1 & 2 & -1 \end{pmatrix} \begin{pmatrix} 0 \\ 1 \\ -1 \end{pmatrix} = [1*0+2*1+(-1)*(-1)] = 3$$

We already normalized $|w_1\rangle$, and so

$$|\tilde{w}_2\rangle = |v_2\rangle - \frac{\langle \tilde{w}_1|v_2\rangle}{\langle \tilde{w}_1|\tilde{w}_1\rangle}|\tilde{w}_1\rangle = \begin{pmatrix} 0 \\ 1 \\ -1 \end{pmatrix} - \frac{3}{6}\begin{pmatrix} 1 \\ 2 \\ -1 \end{pmatrix} = \begin{pmatrix} -\frac{1}{2} \\ 0 \\ -\frac{1}{2} \end{pmatrix}$$

Now we normalize this vector

$$\langle \tilde{w}_2|\tilde{w}_2\rangle = \begin{pmatrix} -\frac{1}{2} & 0 & -\frac{1}{2} \end{pmatrix} \begin{pmatrix} -\frac{1}{2} \\ 0 \\ -\frac{1}{2} \end{pmatrix} = \frac{1}{4} + 0 + \frac{1}{4} = \frac{1}{2}$$

So a second normalized vector is

$$|w_2\rangle = \frac{1}{\sqrt{\langle \tilde{w}_2|\tilde{w}_2\rangle}}|\tilde{w}_2\rangle = \sqrt{2}\begin{pmatrix} -\frac{1}{2} \\ 0 \\ -\frac{1}{2} \end{pmatrix} = \begin{pmatrix} -\frac{1}{\sqrt{2}} \\ 0 \\ -\frac{1}{\sqrt{2}} \end{pmatrix}$$

We check to see that this orthogonal to $|w_1\rangle$:

$$\langle w_1|w_2\rangle = \frac{1}{\sqrt{6}}(1 \quad 2 \quad -1)\begin{pmatrix} -\frac{1}{\sqrt{2}} \\ 0 \\ -\frac{1}{\sqrt{2}} \end{pmatrix} = \frac{1}{\sqrt{6}}\left[-\frac{1}{\sqrt{2}} + \frac{1}{\sqrt{2}}\right] = 0$$

Finally, the third vector is found from

$$|\tilde{w}_3\rangle = |v_3\rangle - \frac{\langle \tilde{w}_1|v_n\rangle}{\langle \tilde{w}_1|\tilde{w}_1\rangle}|\tilde{w}_1\rangle - \frac{\langle \tilde{w}_2|v_3\rangle}{\langle \tilde{w}_2|\tilde{w}_2\rangle}|\tilde{w}_2\rangle$$

Now we have

$$\langle \tilde{w}_2|v_3\rangle = \begin{pmatrix} -\frac{1}{2} & 0 & -\frac{1}{2} \end{pmatrix}\begin{pmatrix} 3 \\ -7 \\ 1 \end{pmatrix} = -\frac{3}{2} - \frac{1}{2} = -\frac{4}{2} = -2$$

Therefore

$$|\tilde{w}_3\rangle = |v_3\rangle - \frac{\langle \tilde{w}_1|v_3\rangle}{\langle \tilde{w}_1|\tilde{w}_1\rangle}|\tilde{w}_1\rangle - \frac{\langle \tilde{w}_2|v_3\rangle}{\langle \tilde{w}_2|\tilde{w}_2\rangle}|\tilde{w}_2\rangle = \begin{pmatrix} 3 \\ -7 \\ 1 \end{pmatrix} + \frac{12}{6}\begin{pmatrix} 1 \\ 2 \\ -1 \end{pmatrix} + \frac{2}{\left(\frac{1}{2}\right)}\begin{pmatrix} -\frac{1}{2} \\ 0 \\ -\frac{1}{2} \end{pmatrix}$$

$$= \begin{pmatrix} 3 \\ -7 \\ 1 \end{pmatrix} + \begin{pmatrix} 2 \\ 4 \\ -2 \end{pmatrix} + \begin{pmatrix} -2 \\ 0 \\ -2 \end{pmatrix} = \begin{pmatrix} 3 \\ -3 \\ -3 \end{pmatrix}$$

Normalizing, we find that

$$\langle \tilde{w}_3|\tilde{w}_3\rangle = (3 \quad -3 \quad -3)\begin{pmatrix} 3 \\ -3 \\ -3 \end{pmatrix} = 9 + 9 + 9 = 27$$

Therefore

$$|w_3\rangle = \frac{1}{\sqrt{\langle \tilde{w}_3|\tilde{w}_3\rangle}}|\tilde{w}_3\rangle = \frac{1}{\sqrt{27}}\begin{pmatrix} 3 \\ -3 \\ -3 \end{pmatrix} = \frac{1}{3\sqrt{3}}\begin{pmatrix} 3 \\ -3 \\ -3 \end{pmatrix} = \frac{1}{\sqrt{3}}\begin{pmatrix} 1 \\ -1 \\ -1 \end{pmatrix}$$

BRA-KET FORMALISM

In quantum mechanics, once you have a state specified in terms of a basis it is not necessary to work directly with the components of each vector. An alternative way of doing calculations is to represent states in an orthonormal basis and do calculations using "bras" and "kets." We are already familiar with this formalism, remember a ket is just another label for the vector notation we have been using. Examples of kets are

$$|\psi\rangle, \quad |\phi\rangle, \quad |0\rangle$$

We saw above that "bras" or dual vectors are the Hermitian conjugates of the corresponding kets. Abstractly, the bras corresponding to the kets above are

$$\langle\psi|, \quad \langle\phi|, \quad \langle 0|$$

Let's demonstrate how to work in this formalism using an example.

Example 2.7

Suppose that $\{|u_1\rangle, |u_2\rangle, |u_3\rangle\}$ is an orthonormal basis for a three-dimensional Hilbert space. A system is in the state

$$|\psi\rangle = \frac{1}{\sqrt{5}}|u_1\rangle - i\sqrt{\frac{7}{15}}|u_2\rangle + \frac{1}{\sqrt{3}}|u_3\rangle$$

(a) Is this state normalized?

(b) If a measurement is made, find the probability of finding the system in each of the states $\{|u_1\rangle, |u_2\rangle, |u_3\rangle\}$

Solution

(a) We know that $\{|u_1\rangle, |u_2\rangle, |u_3\rangle\}$ is an orthonormal basis. To do our calculations, we rely on the following:

$$\langle u_1|u_1\rangle = \langle u_2|u_2\rangle = \langle u_3|u_3\rangle = 1$$

$$\langle u_1|u_2\rangle = \langle u_1|u_3\rangle = \langle u_2|u_3\rangle = 0$$

To show the state is normalized, we must show that

$$\langle\psi|\psi\rangle = 1$$

First we calculate the Hermitian conjugate of the ket $|\psi\rangle$:

$$|\psi\rangle^\dagger = \langle\psi| = \left(\frac{1}{\sqrt{5}}|u_1\rangle\right)^\dagger - \left(i\sqrt{\frac{7}{15}}|u_2\rangle\right)^\dagger + \left(\frac{1}{\sqrt{3}}|u_3\rangle\right)^\dagger$$

$$= \frac{1}{\sqrt{5}}\langle u_1| - \left(-i\sqrt{\frac{7}{15}}\right)\langle u_2| + \frac{1}{\sqrt{3}}\langle u_3|$$

Next we compute the inner product

$$\langle\psi|\psi\rangle = \left(\frac{1}{\sqrt{5}}\langle u_1| + i\sqrt{\frac{7}{15}}\langle u_2| + \frac{1}{\sqrt{3}}\langle u_3|\right)\left(\frac{1}{\sqrt{5}}|u_1\rangle - i\sqrt{\frac{7}{15}}|u_2\rangle + \frac{1}{\sqrt{3}}|u_3\rangle\right)$$

$$= \frac{1}{5}\langle u_1|u_1\rangle + \left(\frac{1}{\sqrt{5}}\right)\left(-i\sqrt{\frac{7}{15}}\right)\langle u_1|u_2\rangle + \left(\frac{1}{\sqrt{5}}\right)\left(\frac{1}{\sqrt{3}}\right)\langle u_1|u_3\rangle$$

$$+ \left(i\sqrt{\frac{7}{15}}\right)\left(\frac{1}{\sqrt{5}}\right)\langle u_2|u_1\rangle + \left(i\sqrt{\frac{7}{15}}\right)\left(-i\sqrt{\frac{7}{15}}\right)\langle u_2|u_2\rangle$$

$$+ \left(i\sqrt{\frac{7}{15}}\right)\left(\frac{1}{\sqrt{3}}\right)\langle u_2|u_3\rangle + \left(\frac{1}{\sqrt{3}}\right)\left(\frac{1}{\sqrt{5}}\right)\langle u_3|u_1\rangle$$

$$+ \left(\frac{1}{\sqrt{3}}\right)\left(-i\sqrt{\frac{7}{15}}\right)\langle u_3|u_2\rangle + \left(\frac{1}{3}\right)\langle u_3|u_3\rangle$$

Because of the orthonormality relations the only terms that survive are

$$\langle u_1|u_1\rangle = \langle u_2|u_2\rangle = \langle u_3|u_3\rangle = 1$$

and so

$$\langle\psi|\psi\rangle = \frac{1}{5}\langle u_1|u_1\rangle + \left(i\sqrt{\frac{7}{15}}\right)\left(-i\sqrt{\frac{7}{15}}\right)\langle u_2|u_2\rangle + \frac{1}{3}\langle u_3|u_3\rangle$$

$$= \frac{1}{5} + \frac{7}{15} + \frac{1}{3} = \frac{3}{15} + \frac{7}{15} + \frac{5}{15} = \frac{15}{15} = 1$$

Therefore the state is normalized. Another easy way to verify this is to check that the probabilities sum to one.

(b) The probability that the system is found to be in state $|u_1\rangle$ upon measurement is

$$|\langle u_1|\psi\rangle|^2$$

So we obtain

$$\langle u_1|\psi\rangle = \langle u_1|\left(\frac{1}{\sqrt{5}}|u_1\rangle + i\sqrt{\frac{7}{15}}|u_2\rangle + \frac{1}{\sqrt{3}}|u_3\rangle\right)$$

$$= \frac{1}{\sqrt{5}}\langle u_1|u_1\rangle + i\sqrt{\frac{7}{15}}\langle u_1|u_2\rangle + \frac{1}{\sqrt{3}}\langle u_1|u_3\rangle$$

$$= \frac{1}{\sqrt{5}} * 1 + i\sqrt{\frac{7}{15}} * 0 + \frac{1}{\sqrt{3}} * 0 = \frac{1}{\sqrt{5}}$$

The probability is found by calculating the modulus squared of this term

$$p_1 = |\langle u_1|\psi\rangle|^2 = \left|\frac{1}{\sqrt{5}}\right|^2 = \frac{1}{5}$$

For $|u_2\rangle$ we obtain

$$\langle u_2|\psi\rangle = \langle u_2|\left(\frac{1}{\sqrt{5}}|u_1\rangle + i\sqrt{\frac{7}{15}}|u_2\rangle + \frac{1}{\sqrt{3}}|u_3\rangle\right)$$

$$= \frac{1}{\sqrt{5}}\langle u_2|u_1\rangle + i\sqrt{\frac{7}{15}}\langle u_2|u_2\rangle + \frac{1}{\sqrt{3}}\langle u_2|u_3\rangle$$

$$= \frac{1}{\sqrt{5}} * 0 + i\sqrt{\frac{7}{15}} * 1 + \frac{1}{\sqrt{3}} * 0 = i\sqrt{\frac{7}{15}}$$

The probability is

$$p_2 = |\langle u_2|\psi\rangle|^2 = \left|i\sqrt{\frac{7}{15}}\right|^2 = \left(-i\sqrt{\frac{7}{15}}\right)\left(i\sqrt{\frac{7}{15}}\right) = \frac{7}{15}$$

Finally, for $|u_3\rangle$ we find that

$$\langle u_3|\psi\rangle = \langle u_2|\left(\frac{1}{\sqrt{5}}|u_1\rangle + i\sqrt{\frac{7}{15}}|u_2\rangle + \frac{1}{\sqrt{3}}|u_3\rangle\right)$$

$$= \frac{1}{\sqrt{5}}\langle u_3|u_1\rangle + i\sqrt{\frac{7}{15}}\langle u_3|u_2\rangle + \frac{1}{\sqrt{3}}\langle u_3|u_3\rangle$$

$$= \frac{1}{\sqrt{5}} * 0 + i\sqrt{\frac{7}{15}} * 0 + \frac{1}{\sqrt{3}} * 1 = \frac{1}{\sqrt{3}}$$

Therefore the probability the system is found to be in state $|u_3\rangle$ is

$$p_3 = |\langle u_3|\psi\rangle|^2 = \left|\frac{1}{\sqrt{3}}\right|^2 = \frac{1}{3}$$

THE CAUCHY-SCHWARTZ AND TRIANGLE INEQUALITIES

Two important identities are the Cauchy-Schwarz inequality

$$|\langle\psi|\phi\rangle|^2 \le \langle\psi|\psi\rangle\langle\phi|\phi\rangle$$

and the triangle inequality

$$\sqrt{\langle\psi+\phi|\psi+\phi\rangle} \le \sqrt{\langle\psi|\psi\rangle} + \sqrt{\langle\phi|\phi\rangle}$$

Example 2.8

Suppose

$$|\psi\rangle = 3|0\rangle - 2i|1\rangle, \quad |\phi\rangle = |0\rangle + 5|1\rangle$$

(a) Show that these states obey the Cauchy-Schwarz and triangle inequalities.

(b) Normalize the states.

Solution

(a) First we compute $\langle\psi|\psi\rangle$, $\langle\phi|\phi\rangle$:

$$\begin{aligned}
\langle\psi| &= (3)^*\langle 0| + (-2i)^*\langle 1| = 3\langle 0| + 2i\langle 1| \\
\Rightarrow \langle\psi|\psi\rangle &= (3\langle 0| + 2i\langle 1|)(3|0\rangle - 2i|1\rangle) \\
&= 9\langle 0|0\rangle + (2i)(-2i)\langle 1|1\rangle \\
&= 9 + 4 = 13
\end{aligned}$$

For $|\phi\rangle$ we obtain

$$\begin{aligned}
\langle\phi| &= \langle 0| + 5\langle 1| \\
\Rightarrow \langle\phi|\phi\rangle &= (\langle 0| + 5\langle 1|)(|0\rangle + 5|1\rangle) \\
&= \langle 0|0\rangle + 25\langle 1|1\rangle \\
&= 1 + 25 = 26
\end{aligned}$$

The inner product $\langle\psi|\phi\rangle$ is given by

$$\begin{aligned}
\langle\psi|\phi\rangle &= (3\langle 0| + 2i\langle 1|)(|0\rangle + 5|1\rangle) \\
&= 3\langle 0|0\rangle + (2i)(5)\langle 1|1\rangle \\
&= 3 + 10i \\
\Rightarrow |\langle\psi|\phi\rangle|^2 &= \langle\psi|\phi\rangle\langle\psi|\phi\rangle^* = (3 + 10i)(3 - 10i) = 9 + 100 = 109
\end{aligned}$$

Now we have

$$\langle\psi|\psi\rangle\langle\phi|\phi\rangle = (13)(26) = 338 > 109$$

Therefore the Cauchy-Schwarz inequality is satisfied. Next we check the triangle inequality:

$$\sqrt{\langle\psi|\psi\rangle} + \sqrt{\langle\phi|\phi\rangle} = \sqrt{13} + \sqrt{26} \cong 8.7$$

For the left side we have

$$|\psi + \phi\rangle = 3|0\rangle - 2i|1\rangle + |0\rangle + 5|1\rangle = (3 + 1)|0\rangle + (5 - 2i)|1\rangle = 4|0\rangle + (5 - 2i)|1\rangle$$

The bra corresponding to this ket is

$$\langle\psi+\phi| = 4\langle 0| + (5-2i)^*\langle 1| = 4\langle 0| + (5+2i)\langle 1|$$

Then the norm of the state is found to be

$$\begin{aligned}\langle\psi+\phi|\psi+\phi\rangle &= [4\langle 0| + (5+2i)\langle 1|][4|0\rangle + (5-2i)|1\rangle] \\ &= 16\langle 0|0\rangle + (5+2i)(5-2i)\langle 1|1\rangle \\ &= 16+10+4 = 30\end{aligned}$$

With this result in hand we find that

$$\sqrt{\langle\psi+\phi|\psi+\phi\rangle} = \sqrt{30} \simeq 5.5 < 8.7$$

Therefore the triangle inequality is satisfied.

(b) The inner products of these states are

$$\langle\psi|\psi\rangle = 13, \quad \Rightarrow \||\psi\rangle\| = \sqrt{\langle\psi|\psi\rangle} = \sqrt{13}$$

$$\langle\phi|\phi\rangle = 26, \quad \Rightarrow \||\phi\rangle\| = \sqrt{\langle\phi|\phi\rangle} = \sqrt{26}$$

So we obtain the normalized states

$$|\tilde{\psi}\rangle = \frac{|\psi\rangle}{\sqrt{\langle\psi|\psi\rangle}} = \frac{1}{\sqrt{13}}(3|0\rangle - 2i|1\rangle) = \frac{3}{\sqrt{13}}|0\rangle - \frac{2i}{\sqrt{13}}|1\rangle$$

$$|\tilde{\phi}\rangle = \frac{|\phi\rangle}{\sqrt{\langle\phi|\phi\rangle}} = \frac{1}{\sqrt{26}}(|0\rangle + 5|1\rangle) = \frac{1}{\sqrt{26}}|0\rangle + \frac{5}{\sqrt{26}}|1\rangle$$

Example 2.9

The qubit trine is defined by the states

$$\begin{aligned}|\psi_0\rangle &= |0\rangle \\ |\psi_1\rangle &= -\frac{1}{2}|0\rangle - \frac{\sqrt{3}}{2}|1\rangle \\ |\psi_2\rangle &= -\frac{1}{2}|0\rangle + \frac{\sqrt{3}}{2}|1\rangle\end{aligned}$$

Find states $|\overline{\psi}_0\rangle$, $|\overline{\psi}_1\rangle$, $|\overline{\psi}_2\rangle$ that are normalized, superposition states of $\{|0\rangle, |1\rangle\}$, and orthogonal to $|\psi_0\rangle, |\psi_1\rangle, |\psi_2\rangle$, respectively.

Solution

We write the states as superpositions with unknown coefficients

$$|\overline{\psi}_0\rangle = \alpha_0|0\rangle + \beta_0|1\rangle$$

$$|\overline{\psi}_1\rangle = \alpha_1|0\rangle + \beta_1|1\rangle$$

$$|\overline{\psi}_2\rangle = \alpha_2|0\rangle + \beta_2|1\rangle$$

For the first of these states, the requirement that it be orthogonal to $|\psi_0\rangle$ allows us to write

$$\langle\overline{\psi}_0|\psi_0\rangle = (\alpha_0^*\langle 0| + \beta_0^*\langle 1|)(|0\rangle) = \alpha_0^*\langle 0|0\rangle = \alpha_0^*$$

The requirement that $\langle\overline{\psi}_0|\psi_0\rangle = 0$ tells us that $\alpha_0 = 0$. So we have reduced the state $|\overline{\psi}_0\rangle$ to

$$|\overline{\psi}_0\rangle = \beta_0|1\rangle$$

The states are normalized, and we obtain

$$1 = \langle\overline{\psi}_0|\overline{\psi}_0\rangle = \beta_0^*\langle 1|(\beta_0|1\rangle) = |\beta_0|^2\langle 1|1\rangle = |\beta_0|^2$$

$$\Rightarrow \beta_0 = 1$$

The orthogonality requirement for $|\overline{\psi}_1\rangle$ gives us

$$\langle\overline{\psi}_1|\psi_1\rangle = (\alpha_1^*\langle 0| + \beta_1^*\langle 1|)\left(-\frac{1}{2}|0\rangle - \frac{\sqrt{3}}{2}|1\rangle\right)$$

$$= -\frac{\alpha_1^*}{2}\langle 0|0\rangle - \beta_1^*\frac{\sqrt{3}}{2}\langle 1|1\rangle$$

$$= -\frac{\alpha_1^*}{2} - \beta_1^*\frac{\sqrt{3}}{2}$$

This term must be zero, so

$$\alpha_1^* = -\sqrt{3}\beta_1^*$$

The requirement that the state be normalized gives

$$1 = \langle\overline{\psi}_1|\overline{\psi}_1\rangle = (\alpha_1^*\langle 0| + \beta_1^*\langle 1|)(\alpha_1|0\rangle + \beta_1|1\rangle)$$

$$= |\alpha_1|^2\langle 0|0\rangle + |\beta_1|^2\langle 1|1\rangle$$

$$= |\alpha_1|^2 + |\beta_1|^2$$

$$\alpha_1^* = -\sqrt{3}\beta_1^*, \Rightarrow \alpha_1 = -\sqrt{3}\beta_1$$

$$1 = 3|\beta_1|^2 + |\beta_1|^2 = 4|\beta_1|^2$$

We take

$$\beta_1 = -\frac{1}{2}$$

So the state is

$$|\overline{\psi}_1\rangle = \frac{\sqrt{3}}{2}|0\rangle - \frac{1}{2}|1\rangle$$

For the final state the orthogonality requirement gives us

$$\begin{aligned}\langle\overline{\psi}_2|\psi_2\rangle &= (\alpha_2^*\langle 0| + \beta_2^*\langle 1|)\left(-\frac{1}{2}|0\rangle + \frac{\sqrt{3}}{2}|1\rangle\right)\\ &= -\frac{\alpha_2^*}{2}\langle 0|0\rangle + \beta_2^*\frac{\sqrt{3}}{2}\langle 1|1\rangle\\ &= -\frac{\alpha_2^*}{2} + \beta_2^*\frac{\sqrt{3}}{2}\end{aligned}$$

Again, if these states are orthogonal, this term must be equal to zero

$$\begin{aligned}-\frac{\alpha_2^*}{2} + \frac{\sqrt{3}}{2}\beta_2^* &= 0\\ \Rightarrow \alpha_2^* &= \sqrt{3}\beta_2^*\end{aligned}$$

From normalization we find that

$$\begin{aligned}1 = \langle\overline{\psi}_2|\overline{\psi}_2\rangle &= (\alpha_2^*\langle 0| + \beta_2^*\langle 1|)(\alpha_2|0\rangle + \beta_2|1\rangle)\\ &= |\alpha_2|^2\langle 0|0\rangle + |\beta_2|^2\langle 1|1\rangle\\ &= |\alpha_2|^2 + |\beta_2|^2\\ \alpha_2^* = \sqrt{3}\beta_2^*, &\quad \Rightarrow \alpha_2 = \sqrt{3}\beta_2\\ 1 = 3|\beta_2|^2 &+ |\beta_2|^2 = 4|\beta_2|^2\end{aligned}$$

Because $\beta_2 = \frac{1}{2}$, the state is

$$|\overline{\psi}_2\rangle = \frac{\sqrt{3}}{2}|0\rangle + \frac{1}{2}|1\rangle$$

SUMMARY

In this chapter we introduced the notion of a quantum state, with particular attention paid to the two-state system known as a qubit. An arbitrary qubit can be written as a superposition of the basis states $|0\rangle$ and $|1\rangle$ as $|\psi\rangle = \alpha|0\rangle + \beta|1\rangle$, where $|\alpha|^2$ gives the probability of finding $|\psi\rangle$ in the state $|0\rangle$ and $|\beta|^2$ gives the probability of finding the system in the state $|1\rangle$. We then learned what a spanning set was and when a set of vectors is linearly independent and constitutes a basis. In the next chapter we will see how to work with qubits and other quantum states by considering the notion of operators and measurement.

EXERCISES

2.1. *A quantum system is in the state*

$$\frac{(1-i)}{\sqrt{3}}|0\rangle + \frac{1}{\sqrt{3}}|1\rangle$$

If a measurement is made, what is the probability the system is in state $|0\rangle$ or in state $|1\rangle$?

2.2. *Two quantum states are given by*

$$|a\rangle = \begin{pmatrix} -4i \\ 2 \end{pmatrix}, \quad |b\rangle = \begin{pmatrix} 1 \\ -1+i \end{pmatrix}$$

(A) Find $|a+b\rangle$.
(B) Calculate $3|a\rangle - 2|b\rangle$.
(C) Normalize $|a\rangle$, $|b\rangle$.

2.3. *Another basis for $\mathbb{C}^2$ is*

$$|+\rangle = \frac{|0\rangle + |1\rangle}{\sqrt{2}}, \quad |-\rangle = \frac{|0\rangle - |1\rangle}{\sqrt{2}}$$

Invert this relation to express $\{|0\rangle, |1\rangle\}$ in terms of $\{|+\rangle, |-\rangle\}$.

2.4. *A quantum system is in the state*

$$|\psi\rangle = \frac{3i|0\rangle + 4|1\rangle}{5}$$

(A) Is the state normalized?
(B) Express the state in the $|+\rangle$, $|-\rangle$ basis.

2.5. *Use the Gram-Schmidt process to find an orthonormal basis for a subspace of the four-dimensional space $\mathbb{R}^4$ spanned by*

$$|u_1\rangle = \begin{pmatrix} 1 \\ 1 \\ 1 \\ 1 \end{pmatrix}, \quad |u_2\rangle = \begin{pmatrix} 1 \\ 2 \\ 4 \\ 5 \end{pmatrix}, \quad |u_3\rangle = \begin{pmatrix} 1 \\ -3 \\ -4 \\ -2 \end{pmatrix}$$

2.6. *Photon horizontal and vertical polarization states are written as* $|h\rangle$ *and* $|v\rangle$*, respectively. Suppose*

$$|\psi_1\rangle = \frac{1}{2}|h\rangle + \frac{\sqrt{3}}{2}|v\rangle$$

$$|\psi_2\rangle = \frac{1}{2}|h\rangle - \frac{\sqrt{3}}{2}|v\rangle$$

$$|\psi_3\rangle = |h\rangle$$

Find

$$|\langle\psi_1|\psi_2\rangle|^2, \quad |\langle\psi_1|\psi_3\rangle|^2, \quad |\langle\psi_3|\psi_2\rangle|^2$$

3

MATRICES AND OPERATORS

An *operator* is a mathematical rule that can be applied to a function to transform it into another function. Operators are often indicated by placing a caret above the symbol used to denote the operator. For example, we can define the derivative operator as

$$\hat{D} = \frac{d}{dx} \tag{3.1}$$

We can apply the derivative operator to a function, say, $f(x) = x \cos x$:

$$\hat{D}f = \hat{D}(x \cos x) = \frac{d}{dx}(x \cos x) = \cos x - x \sin x$$

This idea can be extended to vector spaces. In this case an operator $\hat{A}$ is a mathematical rule that transforms a ket $|\psi\rangle$ into another ket that we will call $|\phi\rangle$:

$$\hat{A}|\psi\rangle = |\phi\rangle \tag{3.2}$$

Operators can also act on bras. The result is another bra:

$$\langle\mu|\hat{A} = \langle\nu| \tag{3.3}$$

Quantum Computing Explained, by David McMahon

In many cases of interest, an operator transforms vectors into other vectors that belong to the *same* space.

We say that an operator is *linear* if the following relationship holds given complex numbers α and β and state vectors $|\psi_1\rangle$ and $|\psi_2\rangle$:

$$\hat{A}(\alpha|\psi_1\rangle + \beta|\psi_2\rangle) = \alpha(\hat{A}|\psi_1\rangle) + \beta(\hat{A}|\psi_2\rangle) \tag{3.4}$$

More generally, a linear operator $\hat{A}$ acts on a state vector as follows:

$$\hat{A}\left(\sum_i \alpha_i|u_i\rangle\right) = \sum_i \alpha_i\left(\hat{A}|u_i\rangle\right) \tag{3.5}$$

The simplest operator is the identity operator $\hat{I}$. As its name implies, this operator leaves a state unchanged:

$$\hat{I}|\psi\rangle = |\psi\rangle \tag{3.6}$$

The zero operator transforms a vector into the zero vector:

$$\hat{0}|\psi\rangle = 0 \tag{3.7}$$

We will often denote operators by italicized capital letters. So A and B can be used to represent two operators as well as $\hat{A}$ and $\hat{B}$.

OBSERVABLES

In quantum theory, dynamical variables like position, momentum, angular momentum, and energy are called *observables*. This is because observables are things we measure in order to characterize the quantum state of a particle. It turns out that an important postulate of quantum theory is that there is an operator that corresponds to each physical observable.

Keep in mind that sometimes people get tired of writing carets all over the place and omit them when writing down operators. The fact that a certain object is an operator should be clear from the context.

THE PAULI OPERATORS

Operators can act on qubits, which as we have seen are just kets or state vectors in a two-dimensional vector space. A set of operators that turns out to be of fundamental importance in quantum computation is known as *Pauli operators*. Including the identity operator, there are four Pauli operators, but sometimes the identity operator is omitted. Unfortunately, there are a few different notations used for these operators and you will just have to learn them all. The Pauli operators are sometimes denoted by σ_0, σ_1, σ_2, and σ_3, by σ_0, σ_x, σ_y, and σ_z, or by I, X, Y, and Z.

Starting with the first of the Pauli operators, we have the identity operator. In this context, as noted in the previous paragraph, the identity operator is sometimes denoted by σ_0. Looking at (3.6), we see that this operator acts on the computational basis states as

$$\sigma_0|0\rangle = |0\rangle, \quad \sigma_0|1\rangle = |1\rangle \tag{3.8}$$

The next Pauli operator is denoted by $\sigma_1 = \sigma_x = X$. It acts as follows:

$$\sigma_1|0\rangle = |1\rangle, \quad \sigma_1|1\rangle = |0\rangle \tag{3.9}$$

For this reason X is sometimes known as the *NOT operator*.

Continuing, we have $\sigma_2 = \sigma_y = Y$, which acts on the computational basis in the following way:

$$\sigma_2|0\rangle = -i|1\rangle, \quad \sigma_2|1\rangle = i|0\rangle \tag{3.10}$$

Finally, $\sigma_3 = \sigma_z = Z$ acts as

$$\sigma_3|0\rangle = |0\rangle, \quad \sigma_3|1\rangle = -|1\rangle \tag{3.11}$$

OUTER PRODUCTS

The product of a ket $|\psi\rangle$ with a bra $\langle\phi|$, which is written as $|\psi\rangle\langle\phi|$, is sometimes known as the *outer product*. This quantity is an operator. To see this, we apply it to an arbitrary ket $|\chi\rangle$:

$$(|\psi\rangle\langle\phi|)|\chi\rangle = |\psi\rangle\langle\phi|\chi\rangle = (\langle\phi|\phi|\chi\rangle)|\psi\rangle \tag{3.12}$$

Remember, the inner product $\langle\phi|\chi\rangle$ is just a complex number. Hence the result of this operation has been to transform the vector $|\chi\rangle$ into one proportional to $|\psi\rangle$. The number $\langle\phi|\chi\rangle$ is the constant of proportionality.

Example 3.1

Describe the action of the operator $\hat{A} = |0\rangle\langle 0| - |1\rangle\langle 1|$ on qubits.

Solution

We let this operator act on an arbitrary qubit, which we write as $|\psi\rangle = \alpha|0\rangle + \beta|1\rangle$. Using linearity (3.4) together with the fact that $\langle 0|0\rangle = \langle 1|1\rangle = 1$ and $\langle 0|1\rangle = \langle 1|0\rangle = 0$, we find that

$$\begin{aligned}\hat{A}|\psi\rangle &= (|0\rangle\langle 0| - |1\rangle\langle 1|)(\alpha|0\rangle + \beta|1\rangle) \\ &= \alpha(|0\rangle\langle 0| - |1\rangle\langle 1|)|0\rangle + \beta(|0\rangle\langle 0| - |1\rangle\langle 1|)|1\rangle\end{aligned}$$

$$= \alpha\,(|0\rangle\langle 0|0\rangle - |1\rangle\langle i|0\rangle) + \beta\,(|0\rangle\langle 0|\rangle - |1\rangle\langle 1|1\rangle)$$
$$= \alpha|0\rangle - \beta|1\rangle$$

This operator has mapped $|0\rangle \mapsto |0\rangle$ and $\||1\rangle \mapsto -|1\rangle$. From (3.11) we see that $\hat{A}$ is the outer product representation of the Z operator.

You Try It

Consider the action of $\hat{A} = |0\rangle\langle 0| + |1\rangle\langle 1|$ on an arbitrary qubit and deduce that this is the outer product representation of the identity operator.

THE CLOSURE RELATION

The previous exercise you tried was an illustration of the *completeness* or *closure* relation. This states that given a basis set $\{|u_i\rangle\}$ in n dimensions, the identity operator can be written as

$$\sum_{i=1}^{n} |u_i\rangle\langle u_i| = \hat{I} \tag{3.13}$$

The closure relation is often useful in manipulating expressions. For example, if we denote the inner product as $\langle u_i|\psi\rangle = c_i$, an arbitrary state $|\psi\rangle$ can be expanded in terms of the basis $\{|u_i\rangle\}$ in the following way:

$$|\psi\rangle = \hat{I}|\psi\rangle = \left(\sum_{i=1}^{n} |u_i\rangle\langle u_i|\right)|\psi\rangle = \sum_{i=1}^{n} |u_i\rangle\langle u_i|\psi\rangle = \sum_{i=1}^{n} c_i|u_i\rangle \tag{3.14}$$

Therefore in the world of qubits which is a two-dimensional Hilbert space, we can write the identity operator using of the computational basis as

$$\hat{I} = |0\rangle\langle 0| + |1\rangle\langle 1| \tag{3.15}$$

REPRESENTATIONS OF OPERATORS USING MATRICES

We saw in the last chapter that a ket could be represented as a column vector. It turns out that operators can be represented by matrices, meaning that the action of an operator on a vector is reduced to simple matrix multiplication. This makes doing computations relatively simple. In an n-dimensional vector space, operators are represented by $n \times n$ matrices. If we know the action of an operator $\hat{A}$ on a basis set $\{|u_i\rangle\}$, then we can find the matrix elements of the operator using the closure relation:

$$\hat{A} = \hat{I}\hat{A}\hat{I} = \left(\sum_i |u_i\rangle\langle u_i|\right)\hat{A}\left(\sum_j |u_j\rangle\langle u_j|\right) = \sum_{i,j}\langle u_i|\hat{A}|u_j\rangle|u_i\rangle\langle u_j| \quad (3.16)$$

The quantity $\langle u_i|\hat{A}|u_j\rangle = A_{ij}$ is a number that represents the matrix element of the operator $\hat{A}$ located at row i and column j in the operators matrix representation with respect to the basis $\{|u_i\rangle\}$:

$$\hat{A} = \begin{pmatrix} \langle u_1|\hat{A}|u_1\rangle & \langle u_1|\hat{A}|u_2\rangle & \cdots & \langle u_1|\hat{A}|u_n\rangle \\ \langle u_2|\hat{A}|u_1\rangle & \langle u_2|\hat{A}|u_2\rangle & & \vdots \\ & & \ddots & \\ \langle u_n|\hat{A}|u_1\rangle & \cdots & & \langle u_n|\hat{A}|u_n\rangle \end{pmatrix} \quad (3.17)$$

It is important to keep in mind that when the matrix representation of an operator is given, it is written *with respect to a certain basis.* If a different basis is used, then the operator will be represented by a different matrix. For example, suppose that we also have a basis set $\{|v_i\rangle\}$ at our disposal. The operator $\hat{A}$ can also be written as a matrix using this basis:

$$\hat{A} = \begin{pmatrix} \langle v_1|\hat{A}|v_1\rangle & \langle v_1|\hat{A}|v_2\rangle & \cdots & \langle v_1|\hat{A}|v_n\rangle \\ \langle v_2|\hat{A}|v_1\rangle & \langle v_2|\hat{A}|v_2\rangle & & \vdots \\ & & \ddots & \\ \langle v_n|\hat{A}|v_1\rangle & \cdots & & \langle v_n|\hat{A}|v_n\rangle \end{pmatrix}$$

The numbers in each matrix are not, in general, equal; that is, $\langle u_i|\hat{A}|u_j\rangle \neq \langle v_i|\hat{A}|v_j\rangle$. Later we will see how to transform between different matrix representations using a *unitary transformation.*

OUTER PRODUCTS AND MATRIX REPRESENTATIONS

Earlier we stated that the outer product of a ket with a bra is an operator. Let's examine the outer product using matrices. Using straightforward matrix multiplication we consider two qubits given by

$$|\psi\rangle = \begin{pmatrix} a \\ b \end{pmatrix}, \quad |\phi\rangle = \begin{pmatrix} c \\ d \end{pmatrix}$$

The outer product $|\psi\rangle\langle\phi|$ is calculated in the following way:

$$|\psi\rangle\langle\phi| = \begin{pmatrix} a \\ b \end{pmatrix} \begin{pmatrix} c^* & d^* \end{pmatrix} = \begin{pmatrix} ac^* & ad^* \\ bc^* & bd^* \end{pmatrix}$$

Meanwhile

$$|\phi\rangle\langle\psi| = \begin{pmatrix} c \\ d \end{pmatrix} \begin{pmatrix} a^* & b^* \end{pmatrix} = \begin{pmatrix} ca^* & cb^* \\ da^* & db^* \end{pmatrix}$$

You Try It

Let $|\psi\rangle = a|1\rangle + b|2\rangle + c|3\rangle, |\phi\rangle = e|1\rangle + f|2\rangle + g|3\rangle$. Show that the outer product $|\psi\rangle\langle\phi|$ is given by

$$|\psi\rangle\langle\phi| = \begin{pmatrix} ae^* & af^* & ag^* \\ be^* & bf^* & bg^* \\ ce^* & cf^* & cg^* \end{pmatrix}$$

MATRIX REPRESENTATION OF OPERATORS IN TWO-DIMENSIONAL SPACES

For the most part we are interested in the two-dimensional Hilbert space $\mathbb{C}^2$ where qubits live. For operators that act on qubits, we represent the operators by 2×2 matrices

$$A = \begin{pmatrix} a & b \\ c & d \end{pmatrix} \tag{3.18}$$

To write the matrix representation of an operator with respect to the computational basis, we use the following convention to arrange the matrix elements:

$$A = \begin{pmatrix} \langle 0|A|0\rangle & \langle 0|A|1\rangle \\ \langle 1|A|0\rangle & \langle 1|A|1\rangle \end{pmatrix} \tag{3.19}$$

Example 3.2

Write down the matrix representation of the Z operator, using (3.11).

Solution

First we restate (3.11) here for convenience:

$$Z|0\rangle = |0\rangle, \quad Z|1\rangle = -|1\rangle$$

Using (3.19), we have

$$Z = \begin{pmatrix} \langle 0|Z|0\rangle & \langle 0|Z|1\rangle \\ \langle 1|Z|0\rangle & \langle 1|Z|1\rangle \end{pmatrix} = \begin{pmatrix} \langle 0|(Z|0\rangle) & \langle 0|(Z|1\rangle) \\ \langle 1|(Z|0\rangle) & \langle 1|(Z|1\rangle) \end{pmatrix}$$

$$= \begin{pmatrix} \langle 0|(|0\rangle) & \langle 0|(-|1\rangle) \\ \langle 1|(|0\rangle) & \langle 1|(-|1\rangle) \end{pmatrix} = \begin{pmatrix} \langle 0|0\rangle & -\langle 0|1\rangle \\ \langle 1|0\rangle & -\langle 1|1\rangle \end{pmatrix}$$

$$= \begin{pmatrix} 1 & 0 \\ 0 & -1 \end{pmatrix}$$

You Try It

The identity matrix has 1s along the diagonal and 0s everywhere else. Using (3.8), verify that the matrix representation of σ_0 is the 2×2 identity matrix.

Definition: The Pauli Matrices

With respect to the computational basis the matrix representation of the Pauli operators X, Y, and Z are given by

$$X = \begin{pmatrix} 0 & 1 \\ 1 & 0 \end{pmatrix}, \quad Y = \begin{pmatrix} 0 & -i \\ i & 0 \end{pmatrix}, \quad Z = \begin{pmatrix} 1 & 0 \\ 0 & -1 \end{pmatrix} \tag{3.20}$$

Example 3.3

The Pauli operators are given by

$$X = \begin{pmatrix} 0 & 1 \\ 1 & 0 \end{pmatrix}, \quad Y = \begin{pmatrix} 0 & -i \\ i & 0 \end{pmatrix}, \quad Z = \begin{pmatrix} 1 & 0 \\ 0 & -1 \end{pmatrix}$$

in the $\{|0\rangle, |1\rangle\}$ basis. Find the action of these operators on the basis states by considering the column vector representation of the $\{|0\rangle, |1\rangle\}$ basis.

Solution

We recall that the basis states are given by

$$|0\rangle = \begin{pmatrix} 1 \\ 0 \end{pmatrix}, \quad |1\rangle = \begin{pmatrix} 0 \\ 1 \end{pmatrix}$$

For X, we find that

$$X|0\rangle = \begin{pmatrix} 0 & 1 \\ 1 & 0 \end{pmatrix}\begin{pmatrix} 1 \\ 0 \end{pmatrix} = \begin{pmatrix} 0*1 + 1*0 \\ 1*1 + 0*0 \end{pmatrix} = \begin{pmatrix} 0 \\ 1 \end{pmatrix} = |1\rangle$$

$$X|1\rangle = \begin{pmatrix} 0 & 1 \\ 1 & 0 \end{pmatrix}\begin{pmatrix} 0 \\ 1 \end{pmatrix} = \begin{pmatrix} 0*0 + 1*1 \\ 1*0 + 0*1 \end{pmatrix} = \begin{pmatrix} 1 \\ 0 \end{pmatrix} = |0\rangle$$

As noted above, since X transforms one basis state into the other, it is sometimes called the NOT operator or *bit flip* operator.

For Y, we obtain

$$Y|0\rangle = \begin{pmatrix} 0 & -i \\ i & 0 \end{pmatrix} \begin{pmatrix} 1 \\ 0 \end{pmatrix} = \begin{pmatrix} 0*1 - i*0 \\ i*1 + 0*0 \end{pmatrix} = \begin{pmatrix} 0 \\ i \end{pmatrix} = i|1\rangle$$

$$Y|1\rangle = \begin{pmatrix} 0 & -i \\ i & 0 \end{pmatrix} \begin{pmatrix} 0 \\ 1 \end{pmatrix} = \begin{pmatrix} 0*0 - i*1 \\ i*0 + 0*1 \end{pmatrix} = \begin{pmatrix} -i \\ 0 \end{pmatrix} = -i|0\rangle$$

And for Z, we have

$$Z|0\rangle = \begin{pmatrix} 1 & 0 \\ 0 & -1 \end{pmatrix} \begin{pmatrix} 1 \\ 0 \end{pmatrix} = \begin{pmatrix} 1*1 + 0*0 \\ 0*1 - 1*0 \end{pmatrix} = \begin{pmatrix} 1 \\ 0 \end{pmatrix} = |0\rangle$$

$$Z|1\rangle = \begin{pmatrix} 1 & 0 \\ 0 & -1 \end{pmatrix} \begin{pmatrix} 0 \\ 1 \end{pmatrix} = \begin{pmatrix} 1*0 + 0*1 \\ 0*0 - 1*1 \end{pmatrix} = \begin{pmatrix} 0 \\ -1 \end{pmatrix} = -|1\rangle$$

HERMITIAN, UNITARY, AND NORMAL OPERATORS

Two special types of operators play a fundamental role in quantum theory and hence in quantum computation. These are *Hermitian* and *unitary* operators. The *Hermitian adjoint* of an operator $\hat{A}$ is denoted by $\hat{A}^\dagger$ and is defined as follows:

$$\langle a|\hat{A}^\dagger|b\rangle = \langle b|\hat{A}|a\rangle^* \tag{3.21}$$

To compute the Hermitian adjoint of any expression, we take the complex conjugate of all constants in the expression, replace all kets by bras and bras by kets, and replace operators by their adjoints. When faced with a product of operators, the order of the operators must be reverse when computing the Hermitian adjoint. These ideas are clarified in the following equations:

$$(\alpha\hat{A})^\dagger = \alpha^* A^\dagger \tag{3.22}$$

$$(|\psi\rangle)^\dagger = \langle\psi| \tag{3.23}$$

$$(\langle\psi|)^\dagger = |\psi\rangle \tag{3.24}$$

$$(\hat{A}\hat{B})^\dagger = \hat{B}^\dagger\hat{A}^\dagger \tag{3.25}$$

$$(\hat{A}|\psi\rangle)^\dagger = \langle\psi|\hat{A}^\dagger \tag{3.26}$$

$$(\hat{A}\hat{B}|\psi\rangle) = \langle\psi|\hat{B}^\dagger A^\dagger \tag{3.27}$$

If an operator is written in outer product notation, we can use (3.23), (3.24), and (3.25) to compute it's adjoint. If $\hat{A}$

$$\hat{A}^\dagger = |\phi\rangle\langle\psi| \tag{3.28}$$

A final rule to note is that the adjoint of a sum is equal to the sum of the adjoints:

$$(\hat{A}+\hat{B}+\hat{C})^\dagger = \hat{A}^\dagger + \hat{B}^\dagger + \hat{C}^\dagger \tag{3.29}$$

Example 3.4

Find the adjoint of the operator $\hat{A} = 2|0\rangle\langle 1| - i|1\rangle\langle 0|$.

Solution

First we note that (3.29) tells us that

$$\hat{A}^\dagger = (2|0\rangle\langle 1|)^\dagger - (i|1\rangle\langle 0|)^\dagger$$

We can compute the adjoint of each term by taking the complex conjugate of the constants in each expression and then applying (3.28). We find that

$$\hat{A}^\dagger = 2|1\rangle\langle 0| + i|0\rangle\langle 1|$$

You Try It

Show that the adjoint of $\hat{B} = 3i|0\rangle\langle 0| + 2i|0\rangle\langle 1|$ is given by $\hat{B} = -3i|0\rangle\langle 0| - 2i|1\rangle\langle 0|$.

Considering the matrix representation of an operator, we compute its Hermitian adjoint in two steps:

- Compute the transpose of the matrix (swap rows and columns)
- Compute the complex conjugate of each element

For a general 2×2 matrix given by

$$A = \begin{pmatrix} a & b \\ c & d \end{pmatrix} \tag{3.30}$$

the Hermitian conjugate is

$$A^\dagger = \begin{pmatrix} a^* & c^* \\ b^* & d^* \end{pmatrix} \tag{3.31}$$

Definition: Hermitian Operator

An operator $\hat{A}$ is said to be *Hermitian* if

$$\hat{A} = \hat{A}^\dagger \tag{3.32}$$

Clearly, the operator used in Example 3.3, $\hat{A} = 2|0\rangle\langle 1| - i|1\rangle\langle 0|$, is not Hermitian. However, the Pauli operators are Hermitian. For example, the operator Y is written as $Y = -i|0\rangle\langle 1| + i|1\rangle\langle 0|$ using the computational basis. The adjoint of this expression is

$$Y^{\dagger} = (-i|0\rangle\langle 1| + i|1\rangle\langle 0|)^{\dagger} = i|1\rangle\langle 0| - i|0\rangle\langle 1| = Y \tag{3.33}$$

It turns out that in quantum mechanics, operators that represent physical observables are Hermitian.

The matrix representation of a Hermitian operator has real matrix elements along it's diagonal. In the space $\mathbb{C}^2$, given a matrix representation (3.30) it must be the case that a and d are real ($a = a^*$, $d = d^*$) and $c = b^*$.

Definition: Unitary Operator

The inverse of an operator A is denoted by A^{-1}. This operator satisfies $AA^{-1} = A^{-1}A = I$, where I is the identity operator. An operator is said to be unitary if its adjoint is equal to its inverse. Unitary operators are often denoted using the symbol U and we can state its definition as

$$UU^{\dagger} = U^{\dagger}U = I \tag{3.34}$$

Unitary operators are important because they describe the time evolution of a quantum state. The Pauli operators are both Hermitian *and* Unitary.

Definition: Normal Operator

An operator A is said to be *normal* if

$$AA^{\dagger} = A^{\dagger}A \tag{3.35}$$

Later in the chapter when we consider the commutator of two operators, we will see that this means a normal operator is one that commutes with its adjoint. Hermitian and unitary operators are normal.

EIGENVALUES AND EIGENVECTORS

A given vector is said to be an *eigenvector* of an operator A if the following equation is satisfied, where λ is a complex number:

$$A|\psi\rangle = \lambda|\psi\rangle$$

The number λ is called an *eigenvalue* of the operator A. For example, looking at (3.11) and Example 3.3, we see that the computational basis states are the eigenvectors of the Z operator.

A common problem in quantum mechanics is the following: given an operator, find its eigenvalues and eigenvectors. The first step in this process is to find the eigenvalues using what is known as the *characteristic equation.*

The Characteristic Equation

The characteristic equation for an operator A is found by solving the following equation:

$$\det|A - \lambda I| = 0 \tag{3.36}$$

where λ is an unknown variable, I is the identity matrix and *det* denotes the determinant of the matrix $A - \lambda I$. The values of λ that are the solutions to this equation are the *eigenvalues* of the operator A. The *determinant* of a 2×2 matrix (3.30) is given by

$$\det|A| = \det\begin{vmatrix} a & b \\ c & d \end{vmatrix} = ad - bc \tag{3.37}$$

Example 3.5

Find the eigenvalues of an operator with matrix representation

$$A = \begin{pmatrix} 2 & 1 \\ -1 & -1 \end{pmatrix}$$

Solution

First we construct the matrix $A - \lambda I$:

$$A - \lambda I = \begin{pmatrix} 2 & 1 \\ -1 & -1 \end{pmatrix} - \lambda \begin{pmatrix} 1 & 0 \\ 0 & 1 \end{pmatrix}$$

$$= \begin{pmatrix} 2 & 1 \\ -1 & -1 \end{pmatrix} - \begin{pmatrix} \lambda & 0 \\ 0 & \lambda \end{pmatrix} = \begin{pmatrix} 2-\lambda & 1 \\ -1 & -1-\lambda \end{pmatrix}$$

Then we compute the determinant

$$\det|A - \lambda I| = \det\begin{vmatrix} 2-\lambda & 1 \\ -1 & -1-\lambda \end{vmatrix} = (2-\lambda)(-1-\lambda) - (-1)(1)$$

$$= -2 + \lambda - 2\lambda + \lambda^2 + 1$$

Rearranging and combining terms, and setting them equal to zero, we obtain

$$\lambda^2 - \lambda - 1 = 0$$

The next step is to solve this equation for the unknown variable λ. We do so using the quadratic formula. This gives

$$\lambda_{1,2} = \frac{-b \pm \sqrt{b^2 - 4ac}}{2a} = \frac{1 \pm \sqrt{1 - 4(1)(-1)}}{2} = \frac{1 \pm \sqrt{5}}{2}$$

You Try It

Using the matrix representation of the Z operator (3.20), show that its eigenvalues are ± 1.

Once the eigenvalues are known, the eigenvectors can be found. This is done by writing out the eigenvalue equation

$$A|\psi\rangle = \lambda|\psi\rangle$$

for each eigenvalue λ and an eigenvector $|\psi\rangle$ with unknown components that we call a, b. This leads to a set of equations we can use to solve for a and b. Usually the equations allow us to relate the two variables but leave one, for example, a, undetermined. But in quantum mechanics there is an additional condition that must be satisfied, normalization. Normalization requires that

$$|a|^2 + |b|^2 = 1$$

So the procedure works as follows:

- Solve the characteristic equation to find the eigenvalues.
- For each eigenvalue, use the eigenvalue equation to generate relations among the components of the given eigenvector.
- Use the normalization condition to find the values of those components.

If each of the eigenvectors of an operator is associated with a unique eigenvalue, we say that they are *nondegenerate*. If two or more eigenvectors are *degenerate*, this means that they correspond to the same eigenvalue.

We illustrate the procedure for finding the eigenvalues and eigenvectors of a matrix with an example.

Example 3.6

Find the eigenvalues and eigenvectors for the "$\pi/8$" gate, which has the matrix representation

$$T = \begin{pmatrix} 1 & 0 \\ 0 & e^{i\pi/4} \end{pmatrix}$$

Solution

We begin by solving the characteristic equation

$$0 = \det|T - \lambda I| = \det\left|\begin{pmatrix} 1 & 0 \\ 0 & e^{i\pi/4} \end{pmatrix} - \lambda\begin{pmatrix} 1 & 0 \\ 0 & 1 \end{pmatrix}\right|$$

$$= \det\begin{vmatrix} 1-\lambda & 0 \\ 0 & e^{i\pi/4}-\lambda \end{vmatrix} = (1-\lambda)(e^{i\pi/4}-\lambda)$$

Since the characteristic equation comes out factored, we can set each term equal to zero to obtain the respective eigenvalue. These are

$$1 - \lambda = 0$$
$$\Rightarrow \lambda_1 = 1$$

and

$$e^{i\pi/4} - \lambda = 0$$
$$\Rightarrow \lambda_2 = e^{i\pi/4}$$

Now we find the eigenvectors that correspond to each eigenvalue. We label them $\{|\phi_1\rangle, |\phi_2\rangle\}$. For the first eigenvalue, which is $\lambda_1 = 1$, the eigenvalue equation is given by

$$T|\phi_1\rangle = |\phi_1\rangle$$

To find the eigenvector, we have to write the equation in matrix form. We label the unknown components of the eigenvector in the following way:

$$|\phi_1\rangle = \begin{pmatrix} a \\ b \end{pmatrix}$$

where the unknowns a, b can be complex numbers. We then have

$$T|\phi_1\rangle = \begin{pmatrix} 1 & 0 \\ 0 & e^{i\pi/4} \end{pmatrix}\begin{pmatrix} a \\ b \end{pmatrix} = \begin{pmatrix} a \\ b \end{pmatrix}$$

We carry out the matrix multiplication on the left side, obtaining

$$\begin{pmatrix} 1 & 0 \\ 0 & e^{i\pi/4} \end{pmatrix}\begin{pmatrix} a \\ b \end{pmatrix} = \begin{pmatrix} a \\ e^{i\pi/4}b \end{pmatrix}$$

Setting this equal to the column vector on the right side gives two equations

$$\begin{pmatrix} a \\ e^{i\pi/4}b \end{pmatrix} = \begin{pmatrix} a \\ b \end{pmatrix}$$
$$\Rightarrow a = a$$
$$e^{i\pi/4}b = b$$

Let's divide the second equation by b. This gives a nonsensical relationship

$$e^{i\pi/4} = 1$$
$$e^{i\pi/4} = \cos\left(\frac{\pi}{4}\right) + i\sin\left(\frac{\pi}{4}\right) = \frac{\sqrt{2}}{2} + i\frac{\sqrt{2}}{2} \neq 1$$

Since these terms are not equal, it must be the case that $b = 0$. That leaves the eigenvector as

$$|\phi_1\rangle = \begin{pmatrix} a \\ 0 \end{pmatrix}$$

Now, to find a, we apply the normalization condition. Stated in terms of inner products this condition is

$$\langle \phi_1 | \phi_1 \rangle = 1$$

Remember, to find the dual vector $\langle \phi_1 |$, we use

$$\text{if} \quad |\phi\rangle = \begin{pmatrix} a \\ b \end{pmatrix}, \quad \text{then} \quad \langle \phi | = \begin{pmatrix} a^* & b^* \end{pmatrix}$$

So for the current eigenvector this gives

$$1 = \langle \phi_1 | \phi_1 \rangle = \begin{pmatrix} a^* & 0 \end{pmatrix} \begin{pmatrix} a \\ 0 \end{pmatrix} = |a|^2, \Rightarrow a = 1$$

We have found that $a = 1$ and $b = 0$. Therefore

$$|\phi_1\rangle = \begin{pmatrix} 1 \\ 0 \end{pmatrix}$$

Now we tackle the second eigenvalue, $\lambda_2 = e^{i\pi/4}$. The eigenvector equation is

$$T|\phi_2\rangle = e^{i\pi/4}|\phi_2\rangle$$

Again, letting the eigenvector $|\phi_2\rangle$contain arbitrary components, we have

$$|\phi_2\rangle = \begin{pmatrix} a \\ b \end{pmatrix}$$

So we obtain

$$T|\phi_2\rangle = \begin{pmatrix} 1 & 0 \\ 0 & e^{i\pi/4} \end{pmatrix} \begin{pmatrix} a \\ b \end{pmatrix} = e^{i\pi/4} \begin{pmatrix} a \\ b \end{pmatrix}$$

Carrying out the matrix multiplication on the left leads to two equations:

$$a = e^{i\pi/4} a$$

$$e^{i\pi/4} b = e^{i\pi/4} b, \quad \Rightarrow b = b$$

The first equation is similar to what we obtained for the constant b for the first eigenvector. By the same reasoning we conclude that $a = 0$. We solve for b using normalization:

$$1 = \langle \phi_2 | \phi_2 \rangle = \begin{pmatrix} 0 & b^* \end{pmatrix} \begin{pmatrix} 0 \\ b \end{pmatrix} = |b|^2$$

$$\Rightarrow b = 1$$

So the second eigenvector is

$$|\phi_2\rangle = \begin{pmatrix} 0 \\ 1 \end{pmatrix}$$

It is always a good idea to double check your work. We check the relations we have found to make sure they satisfy the eigenvalue equations as expected:

$$T|\phi_1\rangle = \begin{pmatrix} 1 & 0 \\ 0 & e^{i\pi/4} \end{pmatrix}\begin{pmatrix} 1 \\ 0 \end{pmatrix} = \begin{pmatrix} 1^*1 + 0^*0 \\ 0^*1 + e^{i\pi/4}*0 \end{pmatrix} = \begin{pmatrix} 1 \\ 0 \end{pmatrix} = |\phi_1\rangle$$

$$T|\phi_2\rangle = \begin{pmatrix} 1 & 0 \\ 0 & e^{i\pi/4} \end{pmatrix}\begin{pmatrix} 0 \\ 1 \end{pmatrix} = \begin{pmatrix} 1^*0 + 0^*1 \\ 0^*0 + e^{i\pi/4}*1 \end{pmatrix} = \begin{pmatrix} 0 \\ e^{i\pi/4} \end{pmatrix} = e^{i\pi/4}\begin{pmatrix} 0 \\ 1 \end{pmatrix} = e^{i\pi/4}|\phi_2\rangle$$

An important item to note is that the eigenvectors of a Hermitian operator constitute an orthonormal basis set for the given vector space. The eigenvalues and eigenvectors of a Hermitian operator also satisfy the following important properties:

- The eigenvalues of a Hermitian operator are *real.*
- The eigenvectors of a Hermitian operator corresponding to different eigenvalues are orthogonal.

The eigenvalues and eigenvectors of a unitary operator satisfy the following:

- The eigenvalues of a unitary operator are complex numbers with modulus 1.
- A unitary operator with nondegenerate eigenvalues has mutually orthogonal eigenvectors.

SPECTRAL DECOMPOSITION

An operator A belonging to some vector space that is normal and has a diagonal matrix representation with respect to some basis of that vector space. This result is known as the spectral decomposition theorem. Suppose that an operator A satisfies the spectral decomposition theorem for some basis $|u_i\rangle$. This means that we can write the operator in the form

$$A = \sum_{i=1}^{n} a_i |u_i\rangle\langle u_i| \tag{3.38}$$

where a_i are the eigenvalues of the operator. In the computational basis the Z operator is diagonal. We have already seen the Z operator written in the form of (3.38) when we considered $Z = |0\rangle\langle 0| - |1\rangle\langle 1|$ in Example 3.1.

Example 3.7

Using the spectral decomposition theorem, write down the representation (3.38) for the operator

$$A = \begin{pmatrix} 0 & 0 & i \\ 0 & 1 & 0 \\ -i & 0 & 0 \end{pmatrix}$$

Solution

The eigenvectors of this matrix are (you try it)

$$|u_1\rangle = \frac{1}{\sqrt{2}}\begin{pmatrix} 1 \\ 0 \\ i \end{pmatrix}, \quad |u_2\rangle = \frac{1}{\sqrt{2}}\begin{pmatrix} 1 \\ 0 \\ -i \end{pmatrix}, \quad |u_3\rangle = \begin{pmatrix} 0 \\ 1 \\ 0 \end{pmatrix}$$

The eigenvalues corresponding to each eigenvector are $a_1 = -1$, $a_2 = a_3 = 1$. Since this matrix is Hermitian (check it) the eigenvectors of the matrix constitute an orthonormal basis. So we can write A as

$$A = \sum_i |u_i\rangle\langle u_i| = -|u_1\rangle\langle u_1| + |u_2\rangle\langle u_2| + |u_3\rangle\langle u_3|$$

THE TRACE OF AN OPERATOR

If an operator is in a matrix representation, the *trace* of the operator is the sum of the diagonal elements. For example,

$$A = \begin{pmatrix} a & b \\ c & d \end{pmatrix}, \quad Tr(A) = a + d$$

$$B = \begin{pmatrix} a & b & c \\ d & e & f \\ g & h & i \end{pmatrix}, \quad Tr(B) = a + e + i$$

If an operator is written down as an outer product, we take the trace by summing over inner products with the basis vectors. If we label a basis $|u_i\rangle$, then

$$Tr(A) = \sum_{i=1}^{n} \langle u_i|A|u_i\rangle$$

Example 3.8

An operator expressed in the $\{|0\rangle, |1\rangle\}$ basis is given by

$$A = 2i|0\rangle\langle 0| + 3|0\rangle\langle 1| - 2|1\rangle\langle 0| + 4|1\rangle\langle 1|$$

Find the trace.

Solution

We find the trace by computing

$$Tr(A) = \sum_{i=1}^{n} \langle \phi_i|A|\phi_i\rangle = \langle 0|A|0\rangle + \langle 1|A|1\rangle$$

Then we compute each term individually and add the results. We have

$$\begin{aligned}\langle 0|A|0\rangle &= \langle 0|(2i|0\rangle\langle 0| + 3|0\rangle\langle 1| - 2|1\rangle\langle 0| + 4|1\rangle\langle 1|)|0\rangle \\ &= 2i\langle 0|0\rangle\langle 0|0\rangle + 3\langle 0|0\rangle\langle 1|0\rangle - 2\langle 0|1\rangle\langle 0|0\rangle + 4\langle 0|1\rangle\langle 1|0\rangle\end{aligned}$$

Recall that the $\{|0\rangle, |1\rangle\}$ basis is orthonormal, so

$$\langle 0|1\rangle = \langle 1|0\rangle = 0$$

This expression reduces to

$$\langle 0|A|0\rangle = 2i\langle 0|0\rangle\langle 0|0\rangle$$

Also recall that

$$\langle 0|0\rangle = \langle 1|1\rangle = 1$$

So we find that

$$\langle 0|A|0\rangle = 2i\langle 0|0\rangle\langle 0|0\rangle = 2i$$

By similar reasoning for the other term we obtain

$$\begin{aligned}\langle 1|A|1\rangle &= \langle 1|(2i|0\rangle\langle 0| + 3|0\rangle\langle 1| - 2|1\rangle\langle 0| + 4|1\rangle\langle 1|)|1\rangle \\ &= 2i\langle 1|0\rangle\langle 0|1\rangle + 3\langle 1|0\rangle\langle 1|1\rangle - 2\langle 1|1\rangle\langle 0|1\rangle + 4\langle 1|1\rangle\langle 1|1\rangle \\ &= 4\langle 1|1\rangle\langle 1|1\rangle = 4\end{aligned}$$

Hence the trace is

$$Tr(A) = \langle 0|A|0\rangle + \langle 1|A|1\rangle = 2i + 4$$

Example 3.9

Find the trace of the Z operator.

Solution

Using the matrix representation of the Z operator in the computational basis:

$$Z = \begin{pmatrix} 1 & 0 \\ 0 & -1 \end{pmatrix}$$

Calculate the trace by summing the diagonal elements:

$$Tr(Z) = Tr\begin{pmatrix} 1 & 0 \\ 0 & -1 \end{pmatrix} = 1 + (-1) = 0$$

Important Properties of the Trace

The trace has some important properties that are good to know. These include the following:

- The trace is *cyclic*, meaning that $Tr(ABC) = Tr(CAB) = Tr(BCA)$.
- The trace of an outer product is the inner product $Tr(|\phi\rangle\langle\psi|) = \langle\phi|\phi\rangle$.
- By extension of the above it follows that $Tr(A|\psi\rangle\langle\phi|) = \langle\phi|A|\psi\rangle$.
- The trace is *basis independent*. Let $|u_i\rangle$ and $|v_i\rangle$ be two bases for some Hilbert space. Then $Tr(A) = \sum\langle u_i|A|u_i\rangle = \sum\langle v_i|A|v_i\rangle$.
- The trace of an operator is equal to the sum of its eigenvalues. If the eigenvalues of A are labeled by λ_i, then $Tr(A) = \sum_{i=1}^{n} \lambda_i$.
- The trace is linear, meaning that $Tr(\alpha A) = \alpha Tr(A)$, $Tr(A+B) = Tr(A) + Tr(B)$.

Example 3.10

Show that the trace of a matrix is equal to the sum of its eigenvalues for

$$X = \begin{pmatrix} 0 & 1 \\ 1 & 0 \end{pmatrix}, \quad T = \begin{pmatrix} 1 & 0 \\ 0 & e^{i\pi/4} \end{pmatrix}, \quad B = \begin{pmatrix} 1 & 0 & 2 \\ 0 & 3 & 4 \\ 1 & 0 & 2 \end{pmatrix}$$

Solution

For a matrix representation, the trace is the sum of the diagonal elements. We compute the trace of each matrix:

$$Tr(X) = Tr \begin{pmatrix} 0 & 1 \\ 1 & 0 \end{pmatrix} = 0 + 0 = 0,$$

$$Tr(T) = Tr \begin{pmatrix} 1 & 0 \\ 0 & e^{i\pi/4} \end{pmatrix} = 1 + e^{i\pi/4},$$

$$Tr(B) = Tr \begin{pmatrix} 1 & 0 & 2 \\ 0 & 3 & 4 \\ 1 & 0 & 2 \end{pmatrix} = 1 + 3 + 2 = 6$$

In the examples above we found that

$$\text{eigenvalues}(X) = 1, \quad -1 \Rightarrow$$

$$\sum \lambda_i = 1 - 1 = 0 = Tr(X)$$

$$\text{eigenvalues}(T) = 1, \quad e^{i\pi/4} \Rightarrow$$

$$\sum \lambda_i = 1 + e^{i\pi/4} = Tr(T)$$

In the exercises you will show that the eigenvalues of B are $\{0, 3, 3\}$ so that

$$\sum \lambda_i = 0 + 3 + 3 = 6 = Tr(B)$$

Example 3.11

Prove that

$$Tr(A|\phi\rangle\langle\psi|) = \langle\psi|A|\phi\rangle$$

Solution

Using an arbitrary basis, we have

$$Tr(A|\phi\rangle\langle\psi|) = \sum_{i=1}^{n} \langle u_i|A|\phi\rangle\langle\psi|u_i\rangle = \sum_{i=1}^{n} \langle\psi|u_i\rangle\langle u_i|A|\phi\rangle$$

$$= \langle\psi| \left(\sum_{i=1}^{n} |u_i\rangle\langle u_i| \right) A|\phi\rangle = \langle\psi|A|\phi\rangle$$

To obtain the result, we used the fact that inner products are just numbers to rearrange the terms in the third step. Then we used the completeness relation to write the sum over the basis vector outer products as the identity.

THE EXPECTATION VALUE OF AN OPERATOR

The *expectation value* of an operator is the *mean* or *average value* of that operator with respect to a given quantum state. In other words, we are asking the following question: If a quantum state $|\psi\rangle$ is prepared many times, and we measure a given operator A each time, what is the average of the measurement results?

This is the expectation value and we write this as

$$\langle A\rangle = \langle\psi|A|\psi\rangle \tag{3.39}$$

Example 3.12

A quantum system is in the state

$$|\psi\rangle = \frac{1}{\sqrt{3}}|0\rangle + \sqrt{\frac{2}{3}}|1\rangle$$

What is the average or expectation value of X in this state?

Solution

We wish to calculate $\langle X \rangle = \langle \psi | X | \psi \rangle$. First let's write down the action of X on the state, recalling that $X|0\rangle = |1\rangle$, $X|1\rangle = |0\rangle$. We obtain

$$X|\psi\rangle = X\left(\frac{1}{\sqrt{3}}|0\rangle + \sqrt{\frac{2}{3}}|1\rangle\right) = \frac{1}{\sqrt{3}}X|0\rangle + \sqrt{\frac{2}{3}}X|1\rangle = \frac{1}{\sqrt{3}}|1\rangle + \sqrt{\frac{2}{3}}|0\rangle$$

Using $\langle 0|1\rangle = \langle 1|0\rangle = 0$, we find that

$$\begin{aligned}\langle\psi|X|\psi\rangle &= \left(\frac{1}{\sqrt{3}}\langle 0| + \sqrt{\frac{2}{3}}\langle 1|\right)\left(\frac{1}{\sqrt{3}}|1\rangle + \sqrt{\frac{2}{3}}|0\rangle\right) \\ &= \frac{1}{3}\langle 0|1\rangle + \frac{\sqrt{2}}{3}\langle 0|0\rangle + \frac{\sqrt{2}}{3}\langle 1|1\rangle + \frac{2}{3}\langle 1|0\rangle \\ &= \frac{\sqrt{2}}{3}\langle 0|0\rangle + \frac{\sqrt{2}}{3}\langle 1|1\rangle \\ &= \frac{\sqrt{2}}{3} + \frac{\sqrt{2}}{3} = \frac{2\sqrt{2}}{3}\end{aligned}$$

It is important to recognize that the expectation value does not have to be a value that is ever actually measured. Remember, the eigenvalues of X are $+1$ and -1. No other value is ever measured for X, and the value we have found here for the average is never actually measured. What this exercise has told us is that if we prepare a large number of systems in the state $|\psi\rangle$ and measure X on each of those systems, and then average the results, the average will be $2\sqrt{2}/3$.

We can compute the expectation value of higher moments of an operator. It is common to compute

$$\langle A^2 \rangle$$

This allows us to calculate the standard deviation or *uncertainty* for an operator, which is defined as

$$\Delta A = \sqrt{\langle A^2 \rangle - \langle A \rangle^2}$$

In the next example we consider the *qutrit basis*, a basis for a three-level system of interest in quantum information theory.

Example 3.13

An operator acts on the qutrit basis states in the following way:

$$A|0\rangle = |1\rangle$$

$$A|1\rangle = \frac{|0\rangle + |1\rangle}{\sqrt{2}}$$

$$A|2\rangle = 0$$

Find, $\langle A \rangle$ for the state

$$|\psi\rangle = \frac{1}{2}|0\rangle - \frac{i}{2}|1\rangle + \frac{1}{\sqrt{2}}|2\rangle.$$

Solution

First we note that

$$A|\psi\rangle = \frac{1}{2}A|0\rangle - \frac{i}{2}A|1\rangle + \frac{1}{\sqrt{2}}A|2\rangle = \frac{1}{2}|1\rangle - \frac{i}{2}\frac{|0\rangle + |1\rangle}{\sqrt{2}} = -\frac{i}{2\sqrt{2}}|0\rangle + \left(\frac{\sqrt{2}-i}{2\sqrt{2}}\right)|1\rangle$$

So the expectation value is

$$\langle A \rangle = \langle\psi|A|\psi\rangle = \left(\frac{1}{2}\langle 0| + \frac{i}{2}\langle 1| + \frac{1}{\sqrt{2}}\langle 2|\right)\left(-\frac{i}{2\sqrt{2}}|0\rangle + \left(\frac{\sqrt{2}-i}{2\sqrt{2}}\right)|1\rangle\right)$$

$$= -\frac{i}{4\sqrt{2}}\langle 0|0\rangle + \left(\frac{\sqrt{2}-i}{4\sqrt{2}}\right)\langle 0|1\rangle + \frac{1}{4\sqrt{2}}\langle 1|0\rangle + \left(\frac{i\sqrt{2}+1}{2\sqrt{2}}\right)\langle 1|1\rangle$$

$$- \frac{i}{4}\langle 2|0\rangle + \left(\frac{\sqrt{2}-i}{4}\right)\langle 2|1\rangle$$

$$= -\frac{i}{4\sqrt{2}}\langle 0|0\rangle + \left(\frac{i\sqrt{2}+1}{2\sqrt{2}}\right)\langle 1|1\rangle$$

$$= -\frac{i}{4\sqrt{2}} + \frac{i\sqrt{2}+1}{2\sqrt{2}} = \frac{2+(2\sqrt{2}-1)i}{4\sqrt{2}}$$

FUNCTIONS OF OPERATORS

The function of an operator can be found by calculating its Taylor expansion:

$$f(A) = \sum_{n=0}^{\infty} a_n A^n \tag{3.40}$$

Something that is seen frequently is an operator in the argument of an exponential function, meaning e^{aA}. We can find out how this function acts by writing down its Taylor expansion:

$$e^{aA} = I + aA + \frac{a^2}{2!}A^2 + \cdots + \frac{a^n}{n!}A^n + \cdots \tag{3.41}$$

If an operator A is normal and has a spectral expansion given by $A = \sum_i a_i |u_i\rangle\langle u_i|$, then

$$f(A) = \sum_i f(a_i)|u_i\rangle\langle u_i| \tag{3.42}$$

If H is a Hermitian operator, then

$$U = e^{i\varepsilon H} \tag{3.43}$$

where ε is a scalar is a unitary operator. If the Hermitian operator $H = \sum_i \phi_i |u_i\rangle\langle u_i|$, we can use (3.42) and write

$$U = e^{i\varepsilon H} = \sum_i e^{i\varepsilon\phi_i}|u_i\rangle\langle u_i| \tag{3.44}$$

After writing out the Taylor expansion of (3.43) and taking only the first two terms, we obtain the *infinitesimal unitary transformation:*

$$U = I + i\varepsilon H \tag{3.45}$$

The operator H is called the generator of the transformation.

If the matrix representation of an operator is written in a basis such that the matrix is diagonal, then we can use (3.42) in a convenient fashion. Using the computational basis

$$Z = \begin{pmatrix} 1 & 0 \\ 0 & -1 \end{pmatrix}$$

Use (3.42) to obtain

$$e^Z = \begin{pmatrix} e & 0 \\ 0 & 1/e \end{pmatrix}$$

UNITARY TRANSFORMATIONS

A method known as a *unitary transformation* can be used to transform the matrix representation of an operator in one basis to a representation of that operator in another basis. For simplicity, we consider the two dimensional vector space $\mathbb{C}^2$. The *change of basis* matrix from a basis $|u_i\rangle$ to a basis $|v_i\rangle$ is given by

$$U = \begin{pmatrix} \langle v_1|u_1\rangle & \langle v_1|u_2\rangle \\ \langle v_2|u_1\rangle & \langle v_2|u_2\rangle \end{pmatrix} \tag{3.46}$$

We write a state vector $|\psi\rangle$ given in the $|u_i\rangle$ basis in terms of the new $|v_i\rangle$ basis as

$$|\psi'\rangle = U|\psi\rangle \tag{3.47}$$

Here $|\psi'\rangle$ is the *same* vector— but expressed in terms of the $|v_i\rangle$ basis. Now suppose that we are given an operator in the $|u_i\rangle$ basis. To write it in terms of the $|v_i\rangle$ basis, we use

$$A' = UAU^\dagger \tag{3.48}$$

Example 3.14

Find the change of basis matrix to go from the computational basis $\{|0\rangle, |1\rangle\}$ to the $\{|\pm\rangle\}$ basis. Then use it to write

$$|\psi\rangle = \frac{1}{\sqrt{3}}|0\rangle + \sqrt{\frac{2}{3}}|1\rangle$$

and the operator $T = \begin{pmatrix} 1 & 0 \\ 0 & e^{i\pi/4} \end{pmatrix}$ in the $\{|\pm\rangle\}$ basis.

Solution

From (3.46) we see that the change of basis matrix in this case is

$$U = \begin{pmatrix} \langle +|0\rangle & \langle +|1\rangle \\ \langle -|0\rangle & \langle -|1\rangle \end{pmatrix} \tag{3.49}$$

Using the column vector representations of these basis states, we have

$$\langle +|0\rangle = (1/\sqrt{2})\,(1 \quad 1)\begin{pmatrix} 1 \\ 0 \end{pmatrix} = \frac{1}{\sqrt{2}} \tag{3.50}$$

$$\langle +|1\rangle = (1/\sqrt{2})\,(1 \quad 1)\begin{pmatrix} 0 \\ 1 \end{pmatrix} = \frac{1}{\sqrt{2}} \tag{3.51}$$

$$\langle -|0\rangle = (1/\sqrt{2})\,(1 \quad -1)\begin{pmatrix} 1 \\ 0 \end{pmatrix} = \frac{1}{\sqrt{2}} \tag{3.52}$$

$$\langle -|1\rangle = (1/\sqrt{2})\,(1 \quad -1)\begin{pmatrix} 0 \\ 1 \end{pmatrix} = -\frac{1}{\sqrt{2}} \tag{3.53}$$

Hence

$$U = \frac{1}{\sqrt{2}}\begin{pmatrix} 1 & 1 \\ 1 & -1 \end{pmatrix}$$

It is easy to check that $U = U^\dagger$ (you try it). Using (3.47), we can rewrite $|\psi\rangle = \frac{1}{\sqrt{3}}|0\rangle + \sqrt{\frac{2}{3}}|1\rangle$ in the $\{|\pm\rangle\}$ basis. In the computational basis, in column vector form, the state is

$$|\psi\rangle = \frac{1}{\sqrt{3}}|0\rangle + \sqrt{\frac{2}{3}}|1\rangle = \frac{1}{\sqrt{3}}\begin{pmatrix} 1 \\ \sqrt{2} \end{pmatrix}$$

Then

$$|\psi\rangle = \frac{1}{\sqrt{2}}\begin{pmatrix} 1 & 1 \\ 1 & -1 \end{pmatrix}\frac{1}{\sqrt{3}}\begin{pmatrix} 1 \\ \sqrt{2} \end{pmatrix} = \frac{1}{\sqrt{6}}\begin{pmatrix} 1+\sqrt{2} \\ 1-\sqrt{2} \end{pmatrix}$$

in the $\{|\pm\rangle\}$ basis. Now let's find the representation of the T operator. First we find that

$$TU^{\dagger} = \frac{1}{\sqrt{2}}\begin{pmatrix} 1 & 0 \\ 0 & e^{i\pi/4} \end{pmatrix}\begin{pmatrix} 1 & 1 \\ 1 & -1 \end{pmatrix} = \frac{1}{\sqrt{2}}\begin{pmatrix} 1 & 1 \\ e^{i\pi/4} & -e^{i\pi/4} \end{pmatrix}$$

So the representation of the operator in the new basis is

$$\begin{aligned} UTU^{\dagger} &= \frac{1}{\sqrt{2}}\begin{pmatrix} 1 & 1 \\ 1 & -1 \end{pmatrix}\frac{1}{\sqrt{2}}\begin{pmatrix} 1 & 1 \\ e^{i\pi/4} & -e^{i\pi/4} \end{pmatrix} \\ &= \frac{1}{2}\begin{pmatrix} 1+e^{i\pi/4} & 1-e^{i\pi/4} \\ 1-e^{i\pi/4} & 1+e^{i\pi/4} \end{pmatrix} \end{aligned}$$

PROJECTION OPERATORS

A *projection operator* is an operator that can be formed by writing the outer product (3.12) using a single ket. That is, given a state $|\psi\rangle$, the operator

$$P = |\psi\rangle\langle\psi| \tag{3.54}$$

is a projection operator. A projection operator is Hermitian. If the state $|\psi\rangle$ is normalized, then a projection operator is equal to its own square:

$$P^2 = P \tag{3.55}$$

If P_1 and P_2 are projection operators that commute, meaning that $P_1P_2 = P_2P_1$, then their product is also a projection operator. Suppose that a given vector space has n dimensions and a basis given by $|1\rangle, |2\rangle, \ldots, |n\rangle$ and that $m > n$. The operator given by

$$P = \sum_{i=1}^{m} |i\rangle\langle i| \tag{3.56}$$

projects onto the subspace spanned by the set $|1\rangle, |2\rangle, \ldots, |m\rangle$. The spectral decomposition theorem (3.38) allows us to write an operator A in terms of projection operators. Recall that we wrote

$$A = \sum_{i=1}^{n} a_i |u_i\rangle\langle u_i|$$

The projection operator $P_i = |u_i\rangle\langle u_i|$ projects onto the subspace defined by the eigenvalue a_i. The eigenvalue a_i represents a given measurement result for the operator A. As we will see later, the projection operators represent a type of measurement described by quantum theory.

The notion of projection operators allows us to rewrite the spectral decomposition of an operator A in terms of projection operators:

$$A = \sum_{i=1}^{n} a_i P_i \tag{3.57}$$

Since the basis states satisfy the completeness relation, we can say that the projection operators in the spectral decomposition of A do as well:

$$\sum_i P_i = I \tag{3.58}$$

Example 3.15

Describe the projection operators formed by the computational basis states and show they satisfy the completeness relation.

Solution

The computational basis states are $|0\rangle$ and $|1\rangle$. We can form two projection operators from these states:

$$P_0 = |0\rangle\langle 0| = \begin{pmatrix} 1 \\ 0 \end{pmatrix} \begin{pmatrix} 1 & 0 \end{pmatrix} = \begin{pmatrix} 1 & 0 \\ 0 & 0 \end{pmatrix}$$

$$P_1 = |1\rangle\langle 1| = \begin{pmatrix} 0 \\ 1 \end{pmatrix} \begin{pmatrix} 0 & 1 \end{pmatrix} = \begin{pmatrix} 0 & 0 \\ 0 & 1 \end{pmatrix}$$

We can verify the completeness relation directly:

$$P_0 + P_1 = |0\rangle\langle 0| + |1\rangle\langle 1| = \begin{pmatrix} 1 & 0 \\ 0 & 0 \end{pmatrix} + \begin{pmatrix} 0 & 0 \\ 0 & 1 \end{pmatrix} = \begin{pmatrix} 1 & 0 \\ 0 & 1 \end{pmatrix} = I$$

You Try It

Using the basis states

$$|+\rangle = \frac{1}{\sqrt{2}} \begin{pmatrix} 1 \\ 1 \end{pmatrix}, \quad |-\rangle = \frac{1}{\sqrt{2}} \begin{pmatrix} 1 \\ -1 \end{pmatrix}$$

construct the projection operators P_+ and P_-, and show that they satisfy the completeness relation.

We can write the projection operators P_+ and P_- in terms of the computational basis. Given that

$$|+\rangle = \frac{|0\rangle + |1\rangle}{\sqrt{2}}$$

you can show that

$$P_+ = |+\rangle\langle+| = \frac{1}{2}\left(|0\rangle\langle 0| + |0\rangle\langle 1| + |1\rangle\langle 0| + |1\rangle\langle 1|\right) \tag{3.59}$$

And given that

$$|-\rangle = \frac{|0\rangle - |1\rangle}{\sqrt{2}} \tag{3.60}$$

you can show that

$$P_- = |-\rangle\langle-| = \frac{1}{2}\left(|0\rangle\langle 0| - |0\rangle\langle 1| - |1\rangle\langle 0| + |1\rangle\langle 1|\right) \tag{3.61}$$

These expressions allow us to write any operator in terms of different projection operators. For example, we have seen that the Z operator is

$$Z = |0\rangle\langle 0| - |1\rangle\langle 1| = P_0 - P_1$$

This is just another example of the spectral decomposition theorem. But let's think about the projection operators $P_\pm$ now. Using (3.59) and (3.61), we can see they satisfy the closure relation:

$$P_+ + P_- = |0\rangle\langle 0| + |1\rangle\langle 1| = I$$

In Chapter 6 we will see the role that projection operators play within the context of measurement. For now let's suppose that a system is prepared in the state

$$|\psi\rangle = \sum_{i=1}^{n} c_i |u_i\rangle$$

where the $|u_i\rangle$ constitute an orthonormal basis. To find the probability of finding the ith outcome if a measurement is made, we can use the projection operators

$$P_i = |u_i\rangle\langle u_i|$$

Then the probability of finding the ith outcome when a measurement is made on a system prepared in the state $|\psi\rangle$ is

$$\Pr(i) = |P_i|\psi\rangle|^2 = \langle\psi|P_i^\dagger P_i|\psi\rangle = \langle\psi|P_i^2|\psi\rangle = \langle\psi|P_i|\psi\rangle \tag{3.62}$$

To obtain result (3.62), we used the fact that projection operators are Hermitian so $P_i^\dagger = P_i$ together with $P_i^2 = P_i$. Now looking at the expansion of the state $|\psi\rangle$ in terms of the basis $|u_i\rangle$, we have the probability of the ith outcome as

$$\begin{aligned}\Pr(i) = \langle\psi|P_i|\psi\rangle &= \left(\sum_{j=1}^{n} c_j^*\langle u_j|\right)(|u_i\rangle\langle u_i|)\left(\sum_{k=1}^{n} c_k|u_k\rangle\right) \\ &= \sum_{j=1}^{n} c_j^*\langle u_j|u_i\rangle \sum_{k=1}^{n} c_k\langle u_i|u_k\rangle = \sum_{j=1}^{n} c_j^*\delta_{ij}\sum_{k=1}^{n} c_k\delta_{ik} \\ &= c_i^* c_i = |c_i|^2\end{aligned} \tag{3.63}$$

This gives us what we expect, that the probability of the ith outcome is equal to the modulus squared of the coefficient c_i.

Example 3.16

A qubit is in the state

$$|\psi\rangle = \frac{1}{\sqrt{3}}|0\rangle + \sqrt{\frac{2}{3}}|1\rangle$$

Using the projection operators P_0 and P_1, indicate the probability of finding $|0\rangle$ and the probability of finding $|1\rangle$ when a measurement is made.

Solution

In Example 3.15 we saw that $P_0 = |0\rangle\langle 0|$ and $P_1 = |1\rangle\langle 1|$. The respective probabilities that we seek are given by $\Pr(0) = \langle\psi|P_0|\psi\rangle$ and $\Pr(1) = \langle\psi|P_1|\psi\rangle$. Now

$$\begin{aligned}P_0|\psi\rangle &= (|0\rangle\langle 0|)\left(\frac{1}{\sqrt{3}}|0\rangle + \sqrt{\frac{2}{3}}|1\rangle\right) = \frac{1}{\sqrt{3}}|0\rangle\langle 0|0\rangle + \sqrt{\frac{2}{3}}|0\rangle\langle 0|1\rangle \\ &= \frac{1}{\sqrt{3}}|0\rangle\end{aligned}$$

Hence

$$\Pr(0) = \langle\psi|P_0|\psi\rangle = \left(\frac{1}{\sqrt{3}}\langle 0| + \sqrt{\frac{2}{3}}\langle 1|\right)\frac{1}{\sqrt{3}}|0\rangle = \frac{1}{3}\langle 0|0\rangle + \frac{\sqrt{2}}{3}\langle 1|0\rangle = \frac{1}{3}$$

In the second case

$$\begin{aligned}P_1|\psi\rangle &= (|1\rangle\langle 1|)\left(\frac{1}{\sqrt{3}}|0\rangle + \sqrt{\frac{2}{3}}|1\rangle\right) = \frac{1}{\sqrt{3}}|1\rangle\langle 1|0\rangle + \sqrt{\frac{2}{3}}|1\rangle\langle 1|1\rangle \\ &= \sqrt{\frac{2}{3}}|1\rangle\end{aligned}$$

So the probability of finding $|1\rangle$ is

$$\Pr(1) = \langle\psi|P_1|\psi\rangle = \left(\frac{1}{\sqrt{3}}\langle 0| + \sqrt{\frac{2}{3}}\langle 1|\right)\sqrt{\frac{2}{3}}|1\rangle = \frac{2}{3}\langle 1|1\rangle = \frac{2}{3}$$

POSITIVE OPERATORS

Consider the vector space $\mathbb{C}^n$. An operator A is said to be *positive semidefinite* if

$$\langle\psi|A|\psi\rangle \geq 0 \tag{3.64}$$

for all $|\psi\rangle \in \mathbb{C}^n$. When we study measurement theory in Chapter 6, we will see that a specific type of operator known as a *positive operator valued measure* or POVM is of fundamental importance. A POVM consists of a set of operators

$$\{E_1, E_2, \ldots, E_n\}$$

where each E_i is positive semidefinite. A set of POVM operators satisfies

$$\sum_i E_i = I \tag{3.65}$$

We will have more to say about POVM's in Chapter 6.

COMMUTATOR ALGEBRA

The *commutator* of two operators A and B is defined as

$$[A, B] = AB - BA \tag{3.66}$$

When $[A, B] = 0$, we say that the operators A and B *commute.* This allows us to interchange the order of the operators in expressions since $AB = BA$. However, in general, $[A, B] \neq 0$, so the ordering of operators is important. We sometimes say that if two operators do not commute, then they are *incompatible.* The commutator is antisymmetric, meaning that

$$[A, B] = -[B, A] \tag{3.67}$$

The commutator is linear, means that

$$[A, B + C] = [A, B] + [A, C] \tag{3.68}$$

We can use the following rule for distributivity:

$$[A, BC] = [A, B]C + B[A, C] \tag{3.69}$$

The most famous commutator relation is that between the position operator X and the momentum operator P:

$$[X, P] = i\hbar I \tag{3.70}$$

(note that X is not the Pauli operator σ_1 in this context). Since we can write operators in terms of their matrix representations, we can calculate the commutator of two operators using matrix multiplication.

Example 3.17

Show that $[\sigma_1, \sigma_2] = 2i\sigma_3$.

Solution

Writing the matrix representations of the Pauli operators in the computational basis, we have

$$\sigma_1 = \begin{pmatrix} 0 & 1 \\ 1 & 0 \end{pmatrix}, \quad \sigma_2 = \begin{pmatrix} 0 & -i \\ i & 0 \end{pmatrix}, \quad \sigma_3 = \begin{pmatrix} 1 & 0 \\ 0 & -1 \end{pmatrix}$$

Now

$$\sigma_1\sigma_2 = \begin{pmatrix} 0 & 1 \\ 1 & 0 \end{pmatrix}\begin{pmatrix} 0 & -i \\ i & 0 \end{pmatrix} = \begin{pmatrix} i & 0 \\ 0 & -i \end{pmatrix}$$

and

$$\sigma_2\sigma_1 = \begin{pmatrix} 0 & -i \\ i & 0 \end{pmatrix}\begin{pmatrix} 0 & 1 \\ 1 & 0 \end{pmatrix} = \begin{pmatrix} -i & 0 \\ 0 & i \end{pmatrix}$$

So the commutator is

$$[\sigma_1, \sigma_2] = \sigma_1\sigma_2 - \sigma_2\sigma_1 = \begin{pmatrix} i & 0 \\ 0 & -i \end{pmatrix} - \begin{pmatrix} -i & 0 \\ 0 & i \end{pmatrix} = \begin{pmatrix} 2i & 0 \\ 0 & -2i \end{pmatrix} = 2i\begin{pmatrix} 1 & 0 \\ 0 & -1 \end{pmatrix} = 2i\sigma_3$$

Earlier we defined a normal operator A as one that satisfies $AA^\dagger = A^\dagger A$. Now that we know the definition of the commutator, we can state that a normal operator is one that commutes with its adjoint. That is, an operator A is normal if

$$[A, A^\dagger] = 0 \tag{3.71}$$

If two operators commute, they possess a set of *common eigenvectors*. Let A and B be two operators such that $[A, B] = 0$. Moreover suppose that $|u_i\rangle$ is a nondegenerate eigenvector of A with eigenvalue a_n, that is, $A|u_i\rangle = a_i|u_i\rangle$. Assume A and B are Hermitian. Then

$$\langle u_i|[A, B]|u_j\rangle = \langle u_i|(AB - BA)|u_j\rangle = (a_i - a_j)\langle u_i|B|u_j\rangle \tag{3.72}$$

Since $[A, B]=0$, we have $\langle u_i|[A, B]|u_j\rangle = \langle u_i|0|u_j\rangle = 0$, meaning that $(a_i - a_j)\langle u_i|B|u_j\rangle = 0$. This implies that if $i \neq j$, $\langle u_i|B|u_j\rangle = 0$. In other words,

$$\langle u_i|B|u_j\rangle \propto \delta_{ij} \tag{3.73}$$

This means that the eigenvectors of A, which we have denoted by $|u_i\rangle$, are also eigenvectors of B. The simultaneous eigenvector of A and B can be written as $|u_n^{(a)}, u_m^{(b)}\rangle$, where

$$\begin{aligned} A|u_n^{(a)}, u_m^{(b)}\rangle &= a_n|u_n^{(a)}, u_m^{(b)}\rangle \\ B|u_n^{(a)}, u_m^{(b)}\rangle &= b_m|u_n^{(a)}, u_m^{(b)}\rangle \end{aligned} \tag{3.74}$$

Now let's consider the exponential of two operators and their commutator. If two operators A and B commute, then

$$e^A e^B = e^B e^A = e^{A+B} \tag{3.75}$$

The more restrictive condition is where $[A, B] \neq 0$, but A and B each commute with $[A, B]$ (e.g., meaning that $[A, [A, B]] = 0$. In that case

$$e^A e^B = e^{A+B} e^{1/2[A,B]} \tag{3.76}$$

The anticommutator of two operators A and B is

$$\{A, B\} = AB + BA \tag{3.77}$$

THE HEISENBERG UNCERTAINTY PRINCIPLE

The expectation or mean value of an operator A was defined in (3.39) as $\langle A\rangle = \langle\psi|A|\psi\rangle$. We can compute higher operator moments such as the mean value of A^2, which is given by

$$\langle A^2\rangle = \langle\psi|A^2|\psi\rangle \tag{3.78}$$

The *uncertainty* ΔA, which is a statistical measure of the spread of measurements about the mean, is given by

$$\Delta A = \sqrt{\langle A^2\rangle - \langle A\rangle^2} \tag{3.79}$$

Then, for two operators A and B, it can be shown that the product of their uncertainties satisfies

$$\Delta A \Delta B \geq \frac{1}{2}|\langle[A, B]\rangle| \tag{3.80}$$

This is a generalization of the famous *Heisenberg uncertainty principle.* What this tells us is that there is a limit to the precision with which we can know the values of two incompatible observables simultaneously. Equation (3.80) sets a lower bound on the precision that we can obtain in our measurements. If $[A, B] \neq 0$, if we make the uncertainty in A smaller (i.e., ΔA gets smaller) in order to satisfy (3.80) the uncertainty in B must grow. The most famous uncertainty relation is that between position and momentum

$$\Delta x \Delta p \geq \frac{\hbar}{2}$$

POLAR DECOMPOSITION AND SINGULAR VALUES

If the matrix representation of an operator A is nonsingular, then the polar decomposition theorem tells us that we can decompose the operator A into a unitary operator U and a positive semidefinite Hermitian operator P in the following way:

$$A = UP \tag{3.81}$$

This form of the polar decomposition of A is called the *left polar decomposition* of A. The operator P is given by

$$P = \sqrt{A^\dagger A} \tag{3.82}$$

It is also possible to derive the *right polar decomposition* of A. If we let $Q = \sqrt{AA^\dagger}$, then

$$A = QU \tag{3.83}$$

We can determine the operator U from

$$U = AP^{-1} = A(\sqrt{A^\dagger A})^{-1} \tag{3.84}$$

We call this the polar decomposition of A because we find an analogue to the polar representation of a complex number in the following expression:

$$\det A = \det U \det P = re^{i\theta} \tag{3.85}$$

Example 3.18

Write the polar decomposition of the matrix

$$A = \begin{pmatrix} a & -b \\ b & a \end{pmatrix}$$

Solution

Defining $r = \sqrt{a^2 + b^2}$ and $\theta = \tan^{-1}(b/a)$, the polar decomposition of this matrix is

$$A = \begin{pmatrix} a & -b \\ b & a \end{pmatrix} = \begin{pmatrix} r\cos\theta & -r\sin\theta \\ r\sin\theta & r\cos\theta \end{pmatrix} = \begin{pmatrix} \cos\theta & -\sin\theta \\ \sin\theta & \cos\theta \end{pmatrix} \begin{pmatrix} r & 0 \\ 0 & r \end{pmatrix}$$

The singular value decomposition of a matrix A is given by

$$A = UDV \tag{3.86}$$

where D is a diagonal matrix consisting of the singular values of the matrix A.

THE POSTULATES OF QUANTUM MECHANICS

Having examined quantum states (qubits) and observables (operators), we close the chapter with a look at how to put these together into the framework of a workable physical theory. This is done by listing the "postulates" of quantum mechanics. These postulates are a set of axioms that define how the theory operates.

Postulate 1: The State of a System

The state of a quantum system is a vector $|\psi(t)\rangle$ in a Hilbert space (we emphasize it can change with time). The state $|\psi(t)\rangle$ contains all information that we can obtain about the system. We work with normalized states such that $\langle\psi|\psi\rangle = 1$, which we call state vectors. A qubit is a state vector in a complex two-dimensional vector space $|\psi\rangle = \alpha|0\rangle + \beta|1\rangle$ normalized such that $|\alpha|^2 + |\beta|^2 = 1$.

Postulate 2: Observable Quantities Represented by Operators

To every dynamical variable A that is a physically measurable quantity, there corresponds an operator A. The operator A is Hermitian and its eigenvectors form a complete orthonormal basis of the vector space.

Postulate 3: Measurements

The possible results of measurement of a dynamical variable A are the eigenvalues a_n of the operator A corresponding to that variable. Using spectral decomposition, we can write the operator A in terms of it's eigenvalues and corresponding projection operators $P_n = |u_n\rangle\langle u_n|$ as $A = \sum_n a_n P_n$. The probability of obtaining measurement result a_n is given by

$$\Pr(a_n) = \langle\psi|P_n|\psi\rangle = Tr(P_n|\psi\rangle\langle\psi|) \tag{3.87}$$

The probability amplitude $c_n = \langle u_n|\psi\rangle$ gives us the probability of obtaining measurement result a_n as

$$\Pr(a_n) = \frac{|\langle u_n|\psi\rangle|^2}{\langle\psi|\psi\rangle} = \frac{|c_n|^2}{\langle\psi|\psi\rangle} \tag{3.88}$$

(Recall that if the state is normalized, then $\langle\psi|\psi\rangle = 1$.) A measurement result a_n causes the collapse of the wavefunction, meaning that the system is left in state $|u_n\rangle$. We can write the postmeasurement state of the system in terms of the projection operator $P_n = |u_n\rangle\langle u_n|$ as

$$|\psi\rangle \overset{measurement}{\longrightarrow} \frac{P_n|\psi\rangle}{\sqrt{\langle\psi|P_n|\psi\rangle}} \tag{3.89}$$

Postulate 4: Time Evolution of the System

The time evolution of a closed (i.e., physically isolated) quantum system is governed by the *Schrödinger equation.* This equation is given by

$$i\hbar\frac{\partial}{\partial t}|\psi\rangle = H|\psi\rangle \tag{3.90}$$

where H is an operator called the *Hamiltonian* of the system. This operator corresponds to the total energy of the system. The possible energies the system can have are the eigenvalues of the H operator. The state of the system at a later time t is given by

$$|\psi(t)\rangle = e^{-iHt/\hbar}|\psi(0)\rangle \tag{3.91}$$

Therefore the time evolution of a quantum state is governed by the unitary operator

$$U = e^{-iHt/\hbar} \tag{3.92}$$

where again H is the Hamiltonian operator describing the total energy of the system.

EXERCISES

3.1. *Verify that the outer product representations of Xand Yare given by* $X = |0\rangle\langle 1| + |1\rangle\langle 0|$ *and* $Y = -i|0\rangle\langle 1| + i|1\rangle\langle 0|$ *by letting them act on the state* $|\psi\rangle = \alpha|0\rangle + \beta|1\rangle$ *and comparing with (3.9) and (3.10).*

3.2. *Show that the matrix representation of the X operator with respect to the computational basis is*

$$X = \begin{pmatrix} 0 & 1 \\ 1 & 0 \end{pmatrix}$$

3.3. *Consider the basis states given by*

$$|+\rangle = \frac{|0\rangle + |1\rangle}{\sqrt{2}}, \quad |-\rangle = \frac{|0\rangle - |1\rangle}{\sqrt{2}}$$

Show that the matrix representation of the X operator with respect to this basis is

$$X = \begin{pmatrix} 1 & 0 \\ 0 & -1 \end{pmatrix}$$

3.4. *Consider the space* $\mathbb{C}^3$ *with the basis* $\{|1\rangle, |2\rangle, |3\rangle\}$. *An operator* $\hat{A}$ *is given by* $\hat{A} = i|1\rangle\langle 1| + \frac{\sqrt{3}}{2}|1\rangle\langle 2| + 2|2\rangle\langle 1| - |2\rangle\langle 3|$. *Write down the adjoint of this operator* $\hat{A}^\dagger$.

3.5. *Find the eigenvalues and eigenvectors of the X operator.*

3.6. *Show that the Y operator is traceless.*

3.7. *Find the eigenvalues of*

$$B = \begin{pmatrix} 1 & 0 & 2 \\ 0 & 3 & 4 \\ 1 & 0 & 2 \end{pmatrix}.$$

3.8. *Prove the following relations involving the trace operation:*

$$Tr(A + B) = Tr(A) + Tr(B)$$

$$Tr(\lambda A) = \lambda Tr(A)$$

$$Tr(AB) = Tr(BA)$$

3.9. *Show that* $X = |0\rangle\langle 1| + |1\rangle\langle 0| = P_+ - P_-$.

3.10. *A three-state system is in the state*

$$|\psi\rangle = \frac{1}{2}|0\rangle + \frac{1}{2}|1\rangle - \frac{i}{\sqrt{2}}|2\rangle$$

Write down the necessary projection operators and calculate the probabilities Pr(0), Pr(1), and Pr(2).

3.11. *In Example 3.17 we showed that* $[\sigma_1, \sigma_2] = 2i\sigma_3$. *Following the same procedure, show that* $[\sigma_2, \sigma_3] = 2i\sigma_1$ *and* $[\sigma_3, \sigma_1] = 2i\sigma_2$.

3.12. *Show that* $\{\sigma_i, \sigma_j\} = 0$ *when* $i \neq j$.

4

TENSOR PRODUCTS

In quantum mechanics we don't always work with single particles in isolation. In many cases, some of which are seen in the context of quantum information processing, it is necessary to work with multiparticle states. Mathematically, to understand multiparticle systems in quantum mechanics, it is necessary to be able to construct a Hilbert space H that is a composite of the independent Hilbert spaces that are associated with each individual particle. The machinery required to do this goes by the name of the *Kronecker* or *tensor product*. We consider the two-particle case.

Suppose that H_1 and H_2 are two Hilbert spaces of dimension N_1 and N_2. We can put these two Hilbert spaces together to construct a larger Hilbert space. We denote this larger space by H and use the tensor product operation symbol $\otimes$. So we write

$$H = H_1 \otimes H_2 \tag{4.1}$$

The dimension of the larger Hilbert space is the product of the dimensions of H_1 and H_2. Once again, we assume that $\dim(H_1) = N_1$ and $\dim(H_2) = N_2$. Then

$$\dim(H) = N_1 N_2 \tag{4.2}$$

Next we start getting down to business and learn how to represent state vectors in the composite Hilbert space.

REPRESENTING COMPOSITE STATES IN QUANTUM MECHANICS

A state vector belonging to H is the tensor product of state vectors belonging to H_1 and H_2. We will show how to represent such vectors explicitly in a moment. For now we will just present some notation, sticking to the more abstract Dirac notation. Let $|\phi\rangle \in H_1$ and $|\chi\rangle \in H_2$ be two vectors that belong to the Hilbert spaces used to construct H. We can construct a vector $|\psi\rangle \in H$ using the tensor product in the following way:

$$|\psi\rangle = |\phi\rangle \otimes |\chi\rangle \tag{4.3}$$

The tensor product of two vectors is linear. That is,

$$\begin{aligned} |\phi\rangle \otimes [|\chi_1\rangle + |\chi_2\rangle] &= |\phi\rangle \otimes |\chi_1\rangle + |\phi\rangle \otimes |\chi_2\rangle \\ [|\phi_1\rangle + |\phi_2\rangle] \otimes |\chi\rangle &= |\phi_1\rangle \otimes |\chi\rangle + |\phi_2\rangle \otimes |\chi\rangle \end{aligned} \tag{4.4}$$

Moreover the tensor product is linear with respect to scalars

$$|\phi\rangle \otimes (\alpha|\chi\rangle) = \alpha|\phi\rangle \otimes |\chi\rangle \tag{4.5}$$

and vice versa. To construct a basis for the larger Hilbert space, we simply form the tensor products of basis vectors from the spaces H_1 and H_2. Let us denote the basis of H_1 by $|u_i\rangle$ and the basis of H_2 by $|v_i\rangle$. Then it follows that we can construct a basis $|w_i\rangle$ for $H = H_1 \otimes H_2$ using

$$|w_i\rangle = |u_i\rangle \otimes |v_i\rangle \tag{4.6}$$

Note that the order of the tensor product is not relevant, meaning

$$|\phi\rangle \otimes |\chi\rangle = |\chi\rangle \otimes |\phi\rangle$$

It is often cumbersome to write the $\otimes$ symbol. Therefore you should be aware that the tensor product $|\phi\rangle \otimes |\chi\rangle$ is often written more simply as $|\phi\rangle|\chi\rangle$, or even as $|\phi\chi\rangle$.

Example 4.1

Let H_1 and H_2 be two Hilbert spaces for qubits. Describe the basis of $H = H_1 \otimes H_2$.

Solution

The basis for each of the qubits is $\{|0\rangle, |1\rangle\}$. The basis of $H = H_1 \otimes H_2$ is formed by writing all possible products of the basis states for H_1 and H_2. The basis vectors are

$$|w_1\rangle = |0\rangle \otimes |0\rangle$$

$$|w_2\rangle = |0\rangle \otimes |1\rangle$$

$$|w_3\rangle = |1\rangle \otimes |0\rangle$$

$$|w_4\rangle = |1\rangle \otimes |1\rangle$$

Now consider the expansion of an arbitrary vector from H_i and an arbitrary vector from H_2 in terms of the basis states in each of the respective Hilbert spaces. That is, if we again denote the basis of H_1 by $|u_i\rangle$ and the basis of H_2 by $|v_i\rangle$, then

$$|\phi\rangle = \sum_i \alpha_i |u_i\rangle \quad \text{and} \quad |\chi\rangle = \sum_i \beta_i |v_i\rangle$$

To expand a vector $|\psi\rangle = |\phi\rangle \otimes |\chi\rangle$ that belongs to the larger Hilbert space H, we simply sum up the products of the individual terms, that is,

$$|\psi\rangle = \sum_{i,j} \alpha_i \beta_j |u_i\rangle \otimes |v_j\rangle \tag{4.7}$$

This result can be summarized by saying that if $|\psi\rangle = |\phi\rangle \otimes |\chi\rangle$ is a vector formed by a tensorproduct, then the components of $|\psi\rangle$ are found by multiplying the components of the two vectors $|\phi\rangle$ and $|\chi\rangle$ used to form the tensor product.

Example 4.2

Let $|\phi\rangle \in H_1$ with basis vectors $\{|x\rangle, |y\rangle\}$ and expansion

$$|\phi\rangle = a_x |x\rangle + a_y |y\rangle$$

And let $|\chi\rangle \in H_2$ with basis vectors $\{|u\rangle, |v\rangle\}$ and expansion

$$|\chi\rangle = b_u |u\rangle + b_v |v\rangle$$

Describe the vector $|\psi\rangle = |\phi\rangle \otimes |\chi\rangle$.

Solution

Using (7), we can write down the expansion of $|\psi\rangle = |\phi\rangle \otimes |\chi\rangle$ in terms of the basis vectors and given expansion coefficients. We simply write down all possible combinations of products between the vectors and then add them up. There are four possible combinations in this case:

$$a_x |x\rangle \otimes b_u |u\rangle$$
$$a_x |x\rangle \otimes b_v |v\rangle$$
$$a_y |y\rangle \otimes b_u |u\rangle$$
$$a_y |y\rangle \otimes b_v |v\rangle$$

Adding them up and using the linearity properties, we obtain

$$|\psi\rangle = a_x b_u |x\rangle \otimes |u\rangle + a_x b_v |x\rangle \otimes |v\rangle + a_y b_u |y\rangle \otimes |u\rangle + a_y b_v |y\rangle \otimes |v\rangle$$

COMPUTING INNER PRODUCTS

The next item of business is to determine how to compute the inner product of two vectors belonging to the larger Hilbert space H. This is actually quite simple—we just take the inner products of the vectors belonging to H_1 and H_2 and multiply them together. That is, suppose

$$|\psi_1\rangle = |\phi_1\rangle \otimes |\chi_1\rangle$$
$$|\psi_2\rangle = |\phi_2\rangle \otimes |\chi_2\rangle$$

Then

$$\langle\psi_1|\psi_2\rangle = \langle\phi_1| \otimes \langle\chi_1|)(|\phi_2\rangle \otimes |\chi_2\rangle) = \langle\phi_1|\phi_2\rangle\langle\chi_1|\chi_2\rangle \tag{4.8}$$

In many cases we want to construct a basis for $\mathbb{C}^4$ out of $\mathbb{C}^2$. The next example shows how to do this.

Example 4.3

Use the $|+\rangle$, $|-\rangle$ states to construct a basis of $\mathbb{C}^4$, and verify that the basis is orthonormal.

Solution

The basis states for $H = \mathbb{C}^4$ can be constructed by using $|+\rangle, |-\rangle$ as the basis for H_1 and H_2. Following Example 4.1, we have

$$|w_1\rangle = |+\rangle|+\rangle$$
$$|w_2\rangle = |+\rangle|-\rangle$$
$$|w_3\rangle = |-\rangle|+\rangle$$
$$|w_4\rangle = |-\rangle|-\rangle$$

If the basis is orthonormal, then we must have $\langle w_1|w_1\rangle = \langle w_2|w_2\rangle = \langle w_3|w_3\rangle = \langle w_4|w_4\rangle = 1$ and all other inner products equal to zero. We consider $\langle w_1|w_1\rangle$, $\langle w_2|w_2\rangle$, and $\langle w_1|w_2\rangle$, and apply (4.8).

$$\langle w_1|w_1\rangle = (\langle +|\langle +|)(|+\rangle|+\rangle) = \langle +|+\rangle\langle +|+\rangle = (1)(1) = 1$$

$$\langle w_2|w_2\rangle = (\langle +|\langle -|)(|+\rangle|-\rangle) = \langle +|+\rangle\langle -|-\rangle = (1)(1) = 1$$

$$\langle w_1|w_2\rangle = (\langle +|\langle +|)(|+\rangle|+\rangle) = \langle +|+\rangle\langle +|-\rangle = (1)(0) = 0$$

$$\langle w_2|w_1\rangle = (\langle +|\langle -|)(|+\rangle|-\rangle) = \langle +|+\rangle\langle -|+\rangle = (1)(0) = 0$$

You Try It

Show that $\langle w_3|w_3\rangle = \langle w_4|w_4\rangle = 1$ and $\langle w_2|w_3\rangle = \langle w_3|w_3 w_2\rangle = 0$.

Example 4.4

Given that $\langle a|b\rangle = 4$ and $\langle c|d\rangle = 7$, calculate $\langle\psi|\phi\rangle$, where

$$|\psi\rangle = |a\rangle \otimes |c\rangle \text{ and } |\phi\rangle = |b\rangle \otimes |d\rangle.$$

Solution

Applying (4.8), we find that

$$\langle\psi|\phi\rangle = ((\langle a| \otimes \langle c|)(|b\rangle \otimes |d\rangle) = \langle a|b\rangle\langle c|d\rangle = (4)(7) = 28$$

You Try It

Given that $\langle a|b\rangle = 1$ and $\langle c|d\rangle = -2$, calculate $\langle\psi|\phi\rangle$, where

$$|\psi\rangle = |a\rangle \otimes |c\rangle \text{ and } |\phi\rangle = |b\rangle \otimes |d\rangle.$$

We will see later not all states can be written as straightforward products like $|\phi\rangle|\chi\rangle$. States that can are referred to as *product states*.

Example 4.5

If $|\psi\rangle = \frac{1}{2}(|0\rangle|0\rangle - |0\rangle|1\rangle + |1\rangle|0\rangle - |1\rangle|1\rangle)$, could it be written as a product state?

Solution

Yes it can. Let

$$|\phi\rangle = \frac{|0\rangle + |1\rangle}{\sqrt{2}} \quad \text{and} \quad |\chi\rangle = \frac{|0\rangle - |1\rangle}{\sqrt{2}}$$

Then

$$|\psi\rangle = |\phi\rangle \otimes |\chi\rangle = \left(\frac{|0\rangle + |1\rangle}{\sqrt{2}}\right) \otimes \left(\frac{|0\rangle - |1\rangle}{\sqrt{2}}\right) = \frac{1}{2}(|0\rangle|0\rangle - |0\rangle|1\rangle + |1\rangle|0\rangle - |1\rangle|1\rangle)$$

You Try It

Write $|\psi\rangle = \frac{1}{2}(|0\rangle|0\rangle + |0\rangle|1\rangle + |1\rangle|0\rangle + |1\rangle|1\rangle)$ as a tensor product of two states.

TENSOR PRODUCTS OF COLUMN VECTORS

In this section we describe how to calculate tensor products when state vectors are written in a matrix representation, meaning as column vectors. We are primarily concerned with going from $\mathbb{C}^2 \to \mathbb{C}^4$, so there is only one procedure for us to learn. Let

$$|\phi\rangle = \begin{pmatrix} a \\ b \end{pmatrix} \quad \text{and} \quad |\chi\rangle = \begin{pmatrix} c \\ d \end{pmatrix}$$

Then the tensor product is given by

$$|\phi\rangle \otimes |\chi\rangle = \begin{pmatrix} a \\ b \end{pmatrix} \otimes \begin{pmatrix} c \\ d \end{pmatrix} = \begin{pmatrix} ac \\ ad \\ bc \\ bd \end{pmatrix} \tag{4.9}$$

Example 4.6

Calculate the tensor product of

$$|a\rangle = \frac{1}{\sqrt{2}}\begin{pmatrix} 1 \\ -1 \end{pmatrix} \quad \text{and} \quad |b\rangle = \frac{1}{\sqrt{3}}\begin{pmatrix} \sqrt{2} \\ 1 \end{pmatrix}$$

Solution

Remember, the tensor product is linear with respect to scalars. So

$$|a\rangle \otimes |b\rangle = \frac{1}{\sqrt{2}}\begin{pmatrix} 1 \\ -1 \end{pmatrix} \otimes \frac{1}{\sqrt{3}}\begin{pmatrix} \sqrt{2} \\ 1 \end{pmatrix} = \left(\frac{1}{\sqrt{2}}\right)\left(\frac{1}{\sqrt{3}}\right)\left[\begin{pmatrix} 1 \\ -1 \end{pmatrix} \otimes \begin{pmatrix} \sqrt{2} \\ 1 \end{pmatrix}\right]$$

Now we apply (4.9):

$$|a\rangle \otimes |b\rangle = \left(\frac{1}{\sqrt{2}}\right)\left(\frac{1}{\sqrt{3}}\right)\left[\begin{pmatrix} 1 \\ -1 \end{pmatrix} \otimes \begin{pmatrix} \sqrt{2} \\ 1 \end{pmatrix}\right] = \left(\frac{1}{\sqrt{6}}\right)\begin{pmatrix} \sqrt{2} \\ 1 \\ -\sqrt{2} \\ -1 \end{pmatrix}$$

You Try It

Calculate the tensor product of

$$|a\rangle = \frac{1}{\sqrt{2}}\begin{pmatrix} 1 \\ 1 \end{pmatrix} \quad \text{and} \quad |b\rangle = \begin{pmatrix} 2 \\ 3 \end{pmatrix}$$

OPERATORS AND TENSOR PRODUCTS

Operators act on tensor products in the following manner: Once again, let $|\phi\rangle \in H_1$ and $|\chi\rangle \in H_2$ be two vectors that belong to the Hilbert spaces used to construct H. Now let A be an operator that acts on vectors $|\phi\rangle \in H_1$, and let B be an operator that acts on vectors $|\chi\rangle \in H_2$. We can create an operator $A \otimes B$ that acts on the vectors $|\rangle \in H$ as follows:

$$(A \otimes B)|\psi\rangle = (A \otimes B)(|\phi\rangle \otimes |\chi\rangle) = (A|\phi\rangle) \otimes (B|\chi\rangle) \tag{4.10}$$

Example 4.7

Suppose $|\psi\rangle = |a\rangle \otimes |b\rangle$ and $A|a\rangle = a|a\rangle,\ B|b\rangle = b|b\rangle$. What is $A \otimes B|\psi\rangle$?

Solution

Using (4.10), we just apply each operator to the vector from its respective space. First we write $|\psi\rangle$ as the tensor product:

$$A \otimes B|\psi\rangle = (A \otimes B)(|a\rangle \otimes |b\rangle)$$

Then we use (4.10) to distribute the operators:

$$A \otimes B|\psi\rangle = (A \otimes B)(|a\rangle \otimes |b\rangle) = A|a\rangle \otimes B|b\rangle$$

Next we use $A|a\rangle = a|a\rangle,\ B|b\rangle = b|b\rangle$ to write

$$A|a\rangle \otimes B|b\rangle = a|a\rangle \otimes b|b\rangle$$

Now recall (4.5), which tells us that $|\phi\rangle \otimes (\alpha|\chi\rangle) = \alpha|\phi\rangle \otimes |\chi\rangle$, so we can pull the scalars to the outside:

$$a|a\rangle \otimes b|b\rangle = ab(|a\rangle \otimes |b\rangle) = ab|\psi\rangle$$

We have shown that

$$A \otimes B|\psi\rangle = ab|\psi\rangle$$

You Try It

Given that X—$0\rangle = |1\rangle$ and Z—$1\rangle = -|1\rangle$, calculate $X \otimes Z|\psi\rangle$ where $|\psi\rangle = |0\rangle \otimes |1\rangle$.

Tensor products of operators are linear in the usual way

$$(A \otimes B)\left(\sum_i c_i|\phi_i\rangle \otimes |\chi_i\rangle\right) = \sum_i c_i(A \otimes B)|\phi_i\rangle \otimes |\chi_i\rangle = \sum_i c_i A|\phi_i\rangle \otimes B|\chi_i\rangle \tag{4.11}$$

Example 4.8

Find $X \otimes Z|\psi\rangle$, where

$$|\psi\rangle = \frac{|0\rangle|0\rangle - |1\rangle|1\rangle}{\sqrt{2}}$$

Solution

First we write

$$X \otimes Z|\psi\rangle = (X \otimes Z)\left(\frac{|0\rangle|0\rangle - |1\rangle|1\rangle}{\sqrt{2}}\right)$$

Using (4.11), we distribute the operator over the sum

$$(X \otimes Z)\left(\frac{|0\rangle|0\rangle - |1\rangle|1\rangle}{\sqrt{2}}\right) = \frac{1}{\sqrt{2}}(X \otimes Z)|0\rangle|0\rangle - \frac{1}{\sqrt{2}}(X \otimes Z)|1\rangle|1\rangle$$

Next we apply (4.10) to allow each component operator to act on the state vector in its respective space:

$$\frac{1}{\sqrt{2}}(X \otimes Z)|0\rangle|0\rangle - \frac{1}{\sqrt{2}}(X \otimes Z)|1\rangle|1\rangle = \frac{1}{\sqrt{2}}(X|0\rangle)(Z|0\rangle) - \frac{1}{\sqrt{2}}(X|1\rangle)(Z|1\rangle)$$

Since $X|0\rangle = |1\rangle$, $X|1\rangle = |0\rangle$ and $Z|0\rangle = |0\rangle$, $Z|1\rangle = -|1\rangle$, this becomes

$$\frac{1}{\sqrt{2}}(X|0\rangle)(Z|0\rangle) - \frac{1}{\sqrt{2}}(X|1\rangle)(Z|1\rangle) = \frac{1}{\sqrt{2}}|1\rangle|0\rangle + \frac{1}{\sqrt{2}}|0\rangle|1\rangle$$

So

$$X \otimes Z|\psi\rangle = \frac{|1\rangle|0\rangle + |0\rangle|1\rangle}{\sqrt{2}}$$

The tensor product of two operators $A \otimes B$ satisfies the following properties:

- If A and B are Hermitian, then $A \otimes B$ is Hermitian.
- If A and B are projection operators, then $A \otimes B$ is a projection operator.
- If A and B are unitary, then $A \otimes B$ is unitary.
- If A and B are positive, then $A \otimes B$ is positive.

Example 4.9

Suppose that A is a projection operator in H_1 where $A = |0\rangle\langle 0|$ and B is a projection operator in H_2 where $B = |1\rangle\langle 1|$. Find $A \otimes B|\psi\rangle$ where

$$|\psi\rangle = \frac{|01\rangle + |10\rangle}{\sqrt{2}}$$

Solution

Using what we know about the action of tensor products of operators, we write

$$A \otimes B|\psi\rangle = A \otimes B\left(\frac{|01\rangle + |10\rangle}{\sqrt{2}}\right) = \frac{1}{\sqrt{2}}[(A|0\rangle)(B|1\rangle) + (A|1\rangle)(B|0\rangle)]$$

Now

$$A|0\rangle = (|0\rangle\langle 0|)|0\rangle = |0\rangle\langle 0|0\rangle = |0\rangle$$

$$A|1\rangle = (|0\rangle\langle 0|)|1\rangle = |0\rangle\langle 0|1\rangle = 0$$

$$B|0\rangle = (|1\rangle\langle 1|)|0\rangle = |0\rangle\langle 1|0\rangle = 0$$

$$B|1\rangle = (|1\rangle\langle 1|)|1\rangle = |1\rangle\langle 1|1\rangle = |1\rangle$$

Therefore we find that

$$A \otimes B|\psi\rangle = \frac{1}{\sqrt{2}}|0\rangle|1\rangle$$

Example 4.10

Show that if A and B are Hermitian, then $A \otimes B$ is Hermitian.

Solution

Let's define two tensor product states

$$|\psi\rangle = |\alpha\rangle \otimes |\beta\rangle$$

$$|\phi\rangle = |\mu\rangle \otimes |\nu\rangle$$

We will also say that $C = A \otimes B$. To show that the tensor product of the two operators is Hermitian, we need to show that

$$\langle\phi|C|\psi\rangle^\dagger = \langle\psi|C^\dagger|\phi\rangle = \langle\psi|C|\phi\rangle$$

Working out the first term obtains

$$C|\psi\rangle = C|\alpha\rangle \otimes |\beta\rangle = A \otimes B|\alpha\rangle \otimes |\beta\rangle = (A|\alpha\rangle)(B|\beta\rangle)$$

Now recall how we compute inner products of tensor product vectors (4.8), We have

$$\langle\phi|C|\psi\rangle = \langle\phi|(A|\alpha\rangle B|\beta\rangle) = (\langle\mu|\langle\nu|)(A|\alpha\rangle B|\beta\rangle) = \langle\mu|A|\alpha\rangle\langle\nu|B|\beta\rangle$$

The Hermitian conjugate of this expression is

$$\langle\phi|C|\psi\rangle^\dagger = (\langle\mu|A|\alpha\rangle\langle\nu|B|\beta\rangle)^\dagger = \langle\beta|B^\dagger|\nu\rangle\langle\alpha|A^\dagger|\mu\rangle$$

But A and B are Hermitian and the inner products are numbers we can just move around. So we write

$$\langle\phi|C|\psi\rangle^\dagger = \langle\alpha|A|\mu\rangle\langle\beta|B|\nu\rangle$$

Now let's apply (4.8) in reverse to write

$$\langle\phi|C|\psi\rangle^\dagger = (\langle\alpha|\langle\beta|)A\otimes B(|\mu\rangle|\nu\rangle) = \langle\psi|(A\otimes B)|\phi\rangle = \langle\psi|C|\phi\rangle$$

Therefore we conclude that $A\otimes B$ is Hermitian.

You Try It

Show that if A and B are unitary, then $A\otimes B$ is unitary.
We can also form a tensor product operator that acts only on H_1 while doing nothing to vectors from H_2. This is done by forming the tensor product $A\otimes I$, where I is the identity operator acting in H_2. Similarly $I\otimes B$ does nothing to vectors from H_1 but acts on vectors from H_2.

Example 4.11

Suppose

$$|\psi\rangle = \frac{|00\rangle - |11\rangle}{\sqrt{2}}$$

Describe the action of $X\otimes I$ on the state.

Solution

First remember that —00⟩ is just a shorthand notation for —0⟩⊗|0⟩ and similarly for —11⟩. The action of $X\otimes I$ is as follows:

$$X\otimes I|\psi\rangle = X\otimes I\left(\frac{|00\rangle - |11\rangle}{\sqrt{2}}\right) = \frac{1}{\sqrt{2}}[(X|0\rangle)|0\rangle - (X|1\rangle)|1\rangle]$$

$$= \frac{|10\rangle - |01\rangle}{\sqrt{2}}$$

You Try It

If

$$|\psi\rangle = \frac{|00\rangle + |11\rangle}{\sqrt{2}}$$

show that

$$Z \otimes I|\psi\rangle = \frac{|00\rangle - |11\rangle}{\sqrt{2}}$$

TENSOR PRODUCTS OF MATRICES

We conclude the chapter by looking at how to compute tensor products of matrices, something we will need to do frequently later. We're going to keep things simple and focus on taking tensor products of operators in two-dimensional Hilbert spaces to produce an operator that acts on a four-dimensional Hilbert space.

Let

$$A = \begin{pmatrix} a_{11} & a_{12} \\ a_{21} & a_{22} \end{pmatrix}, \quad B = \begin{pmatrix} b_{11} & b_{12} \\ b_{21} & b_{22} \end{pmatrix}$$

be the matrix representations of two operators A and B. The matrix representation of the tensor product $A \otimes B$ is

$$A \otimes B = \begin{pmatrix} a_{11}B & a_{12}B \\ a_{21}B & a_{22}B \end{pmatrix} = \begin{pmatrix} a_{11}b_{11} & a_{11}b_{12} & a_{12}b_{11} & a_{12}b_{12} \\ a_{11}b_{21} & a_{11}b_{22} & a_{12}b_{21} & a_{12}b_{22} \\ a_{21}b_{11} & a_{21}b_{12} & a_{22}b_{11} & a_{22}b_{12} \\ a_{21}b_{21} & a_{21}b_{22} & a_{22}b_{21} & a_{22}b_{22} \end{pmatrix} \tag{4.12}$$

Example 4.12

Find the tensor product of the Pauli matrices X and Z.

Solution

First let's write down the Pauli matrices:

$$X = \begin{pmatrix} 0 & 1 \\ 1 & 0 \end{pmatrix}, \quad Z = \begin{pmatrix} 1 & 0 \\ 0 & -1 \end{pmatrix}$$

Now we apply (4.12):

$$X \otimes Z = \begin{pmatrix} (0)Z & (1)Z \\ (1)Z & (0)Z \end{pmatrix} = \begin{pmatrix} 0 & 0 & 1 & 0 \\ 0 & 0 & 0 & -1 \\ 1 & 0 & 0 & 0 \\ 0 & -1 & 0 & 0 \end{pmatrix}$$

You Try It

Calculate the matrix representation of $Z \otimes X$, showing that $X \otimes Z \neg Z \otimes X$.

EXERCISES

4.1. *Consider the basis in Example 4.1 Show that it is orthonormal.*

4.2. *Returning to Example 4.1, show that* $\langle w_3 | w_4 \rangle = \langle w_4 | w_4 w_3 \rangle = 0$.

4.3. *Given that* $\langle a|b \rangle = 1/2$ *and* $\langle c|d \rangle = 3/4$, *calculate* $\langle \psi | \phi \rangle$, *where* $|\psi\rangle = |a\rangle \otimes |c\rangle$ *and* $|\phi\rangle = |b\rangle \otimes |d\rangle$.

4.4. *Calculate the tensor product of*

$$|\psi\rangle = \frac{1}{\sqrt{2}}\begin{pmatrix} 1 \\ 1 \end{pmatrix} \quad \textit{and} \quad |\phi\rangle = \frac{1}{2}\begin{pmatrix} 1 \\ \sqrt{3} \end{pmatrix}$$

4.5. *Can* $|\psi\rangle = \frac{1}{2}(|0\rangle|0\rangle - |0\rangle|1\rangle - |1\rangle|0\rangle + |1\rangle|1\rangle)$ *be written as a product state?*

4.6. *Can*

$$|\psi\rangle = \frac{|0\rangle|0\rangle + |1\rangle|1\rangle}{\sqrt{2}}$$

be written as a product state?

4.7. *Find* $X \otimes Y|\psi\rangle$, *where*

$$|\psi\rangle = \frac{|0\rangle|1\rangle - |1\rangle|0\rangle}{\sqrt{2}}$$

4.8. *Show that* $(A \otimes B)^\dagger = A^\dagger \otimes B^\dagger$.

4.9. *If*

$$|\psi\rangle = \frac{|00\rangle + |11\rangle}{\sqrt{2}}$$

find $I \otimes Y|\psi\rangle$.

4.10. *Calculate the matrix representation of* $X \otimes Y$.

5

THE DENSITY OPERATOR

In many cases of practical interest, rather than considering a single quantum system, we need to study a large number or collection of systems called an *ensemble*. Furthermore, rather than being in a single state, members of the ensemble can be found in one of two or more different quantum states. There is a given probability that a member of the ensemble is found in each of these states. We make this more concrete with a simple example.

Consider a two-dimensional Hilbert space with basis vectors $\{|x\rangle, |y\rangle\}$. We prepare a large number N of systems, where each member of the system can be in one of two state vectors

$$|a\rangle = \alpha|x\rangle + \beta|y\rangle$$
$$|b\rangle = \gamma|x\rangle + \delta|y\rangle$$

These states are normalized, so $|\alpha|^2 + |\beta|^2 = |\gamma|^2 + |\delta|^2 = 1$, and the usual rules of quantum mechanics apply. For a system in state $|a\rangle$, if a measurement is made, then there is a probability $|\alpha|^2$ of finding $|x\rangle$ while there is a probability $|\beta|^2$ of finding $|y\rangle$, and similarly for state $|b\rangle$.

Now suppose that we prepare n_a of these systems in state $|a\rangle$ and n_b of the systems in state $|b\rangle$. Since we have N total systems, then

$$n_a + n_b = N$$

Quantum Computing Explained, by David McMahon

If we divide by N,

$$\frac{n_a}{N} + \frac{n_b}{N} = 1$$

This relation tells us that if we randomly select a member of the ensemble, the probability that it is found in state $|a\rangle$ is given by $p = n_a/N$. Probabilities must sum to one. Therefore the probability of finding a member of the ensemble in state $|b\rangle$ is given by $1 - n_a/N = 1 - p$.

So we see that with a collection of systems where each member of the ensemble is prepared in one of two or more states, the use of probability is operating on two different levels: "The Born rule tells us that the probabilities given by squared amplitudes must sum to of".

- At the level of a single quantum system, where the Born rule gives us the probability of obtaining a given a measurement result
- On the ensemble level, where if we draw a member of the ensemble, there is a certain probability that the system was prepared in one state vector or another

At the ensemble level the use of probability is acting in a *classical* way. That is, it is a simple reflection of incomplete information. So, at the ensemble level, we have a simple statistical mixture.

The question at hand is how do we describe a system like where there are classical probabilities of finding each member of the system in different states? We need to calculate the usual quantities that we would for any quantum system, such as the expectation values of operators and the probabilities of obtaining different measurement results. However, in the case of a statistical mixture of states, we need to weight the calculated quantities by the probabilities of finding different states. The proper way to do this is with the *density operator.*

THE DENSITY OPERATOR FOR A PURE STATE

We can start learning the mathematical machinery of the density operator by looking at a single state. Consider a system that is in some known state $|\psi\rangle$. To review, if there exists some orthonormal basis $|u_i\rangle$, then we can expand the state in that basis:

$$|\psi\rangle = c_1|u_1\rangle + c_2|u_2\rangle + \cdots + c_n|u_n\rangle$$

Then, using the Born rule, we know that the probability of finding the system in state $|u_i\rangle$ upon measurement is given by $|c_i|^2$. When a system is in a definite state like this, we say the system is in a *pure state*. In light of the discussion above, we seek a different way to describe quantum states that can be generalized to a statistical mixture. This can be done with an operator called the density operator, which is denoted by the symbol ρ. The density operator is an average operator that will allow us to describe a statistical mixture. In the case of a pure state this is done by constructing an outer product from the state. To see this, we begin by finding the expectation value or *average* of some operator A. We write

$$\langle A \rangle = \langle \psi | A | \psi \rangle$$

Expanding the state in some orthonormal basis as $|\psi\rangle = c_1|u_1\rangle + c_2|u_2\rangle + \cdots + c_n|u_n\rangle$, this becomes

$$\langle A \rangle = \langle \psi | A | \psi \rangle = \left(c_1^*\langle u_1| + c_2^*\langle u_2| + \cdots + c_n^*\langle u_n|\right) A \left(c_1|u_1\rangle + c_2|u_2\rangle + \cdots + c_n|u_n\rangle\right)$$

$$= \sum_{k,l=1}^{n} c_k^* c_l \langle u_k | A | u_l \rangle$$

$$= \sum_{k,l=1}^{n} c_k^* c_l A_{kl}$$

Recall how the coefficients in the expansion of a state vector are defined. That is,

$$c_m = \langle u_m | \psi \rangle$$

Taking the complex conjugate, we obtain $c_m^* = \langle \psi | u_m \rangle$. This allows us to write

$$c_k^* c_l = \langle \psi | u_k \rangle \langle u_l | \psi \rangle = \langle u_l | \psi \rangle \langle \psi | u_k \rangle$$

We were able to switch the order of the terms because $c_m = \langle u_m | \psi \rangle$ is just a complex number. Notice now that we have a *projection operator* sandwiched in between the basis vectors. That is,

$$c_k^* c_l = \langle u_l | \psi \rangle \langle \psi | u_k \rangle = \langle u_l | (|\psi\rangle\langle\psi|) | u_k \rangle$$

We call this the density operator and denote it by $\rho = |\psi\rangle\langle\psi|$. So the expectation or average value of an operator A with respect to a state $|\psi\rangle$ can be written as

$$\langle A \rangle = \sum_{k,l=1}^{n} c_k^* c_l A_{kl} = \sum_{k,l=1}^{n} \langle u_l | (|\psi\rangle\langle\psi|) | u_k \rangle A_{kl} = \sum_{k,l=1}^{n} \langle u_l | \rho | u_k \rangle A_{kl}$$

Let's set aside the definition we've just come up with.

Definition: Density Operator for a Pure State

The density operator for a state $|\psi\rangle$ is given by

$$\rho = |\psi\rangle\langle\psi| \tag{5.1}$$

Let's write down the expectation value of A again, and then do a little manipulation that will allow us to write this in terms of a trace operation. Then the expectation value or average of the operator A can be written as

$$\begin{aligned}\langle A\rangle &= \sum_{k,l=1}^{n} c_k^* c_l \langle u_k|A|u_l\rangle \\ &= \sum_{k,l=1}^{n} \langle u_l|\psi\rangle\langle\psi|u_k\rangle\langle u_k|A|u_l\rangle \\ &= \sum_{k,l=1}^{n} \langle u_l|\rho|u_k\rangle\langle u_k|A|u_l\rangle\end{aligned}$$

To write this expectation value in terms of a trace, we can use the completeness relation. Recall that $\sum_k |u_k\rangle\langle u_k| = 1$ for an orthnormal basis $|u_k\rangle$. This leads us to our next definition.

Definition: Using the Density Operator to Find the Expectation Value

The expectation value of an operator A can be written in terms of the density operator as

$$\begin{aligned}\langle A\rangle &= \sum_{k,l=1}^{n} \langle u_l|\rho|u_k\rangle\langle u_k|A|u_l\rangle \\ &= \sum_{l=1}^{n} \langle u_l|\rho\left(\sum_{k=1}^{n} |u_k\rangle\langle u_k|\right)A|u_l\rangle \\ &= \sum_{l=1}^{n} \langle u_l|\rho A|u_l\rangle \\ &= Tr(\rho A)\end{aligned} \tag{5.2}$$

What happens if we take the trace of the density operator by itself? We can work in reverse and expand the state $|\psi\rangle$ in terms of a basis. Here we assume that the state is normalized. Then we will use the fact that $\sum_j |c_j|^2 = 1$ to write down the trace of the density operator. We have

$$Tr(\rho) = \sum_j \langle u_j|\rho|u_j\rangle = \sum_j \langle u_j|\psi\rangle\langle\psi|u_j\rangle = \sum_j c_j c_j^* = \sum_j |c_j|^2 = 1 \tag{5.3}$$

This expression can be summarized by saying that *due to the conservation of probability*, the trace of the density operator is always 1.

Example 5.1

A system is in the state $|\psi\rangle = \frac{1}{\sqrt{3}}|u_1\rangle + i\sqrt{\frac{2}{3}}|u_2\rangle$, where the $|u_k\rangle$ constitute an orthonormal basis. Write down the density operator, and show it has unit trace.

Solution

First we write down the bra corresponding to the given state. Remember, we must take the conjugate of any complex numbers. We obtain

$$\langle\psi| = \frac{1}{\sqrt{3}}\langle u_1| - i\sqrt{\frac{2}{3}}\langle u_2|$$

Then, using (5.1), we have

$$\rho = |\psi\rangle\langle\psi| = \left(\frac{1}{\sqrt{3}}|u_1\rangle + i\sqrt{\frac{2}{3}}|u_2\rangle\right)\left(\frac{1}{\sqrt{3}}\langle u_1| - i\sqrt{\frac{2}{3}}\langle u_2|\right)$$

$$= \frac{1}{3}|u_1\rangle\langle u_1| - i\frac{\sqrt{2}}{3}|u_1\rangle\langle u_2| + i\frac{\sqrt{2}}{3}|u_2\rangle\langle u_1| + \frac{2}{3}|u_2\rangle\langle u_2|$$

The trace is

$$Tr(\rho) = \sum_{i=1}^{2}\langle u_i|\rho|u_i\rangle = \langle u_1|\rho|u_1\rangle + \langle u_2|\rho|u_2\rangle$$

$$= \frac{1}{3}\langle u_1|u_1\rangle\langle u_1|u_1\rangle - i\frac{\sqrt{2}}{3}\langle u_1|u_1\rangle\langle u_2|u_1\rangle + i\frac{\sqrt{2}}{3}\langle u_1|u_2\rangle\langle u_1|u_1\rangle + \frac{2}{3}\langle u_1|u_2\rangle\langle u_2|u_1\rangle$$

$$+ \frac{1}{3}\langle u_2|u_1\rangle\langle u_1|u_2\rangle - i\frac{\sqrt{2}}{3}\langle u_2|u_1\rangle\langle u_2|u_2\rangle + i\frac{\sqrt{2}}{3}\langle u_2|u_2\rangle\langle u_1|u_2\rangle + \frac{2}{3}\langle u_2|u_2\rangle\langle u_2|u_2\rangle$$

$$= \frac{1}{3} + \frac{2}{3} = 1$$

Mixed terms like $\langle u_1|u_2\rangle$ drop out because the basis is orthonormal, so $\langle u_i|u_j\rangle = \delta_{ij}$.

You Try It

Show that if $|\psi\rangle = \frac{1}{2}|u_1\rangle + \frac{1}{\sqrt{2}}|u_2\rangle + \frac{1}{2}|u_3\rangle$, then

$$\rho = \frac{1}{4}|u_1\rangle\langle u_1| + \frac{1}{2}|u_2\rangle\langle u_2| + \frac{1}{4}|u_3\rangle\langle u_3|$$

$$+ \frac{1}{2\sqrt{2}}(|u_1\rangle\langle u_2| + |u_2\rangle\langle u_1| + |u_2\rangle\langle u_3| + |u_3\rangle\langle u_2|) + \frac{1}{4}(|u_1\rangle\langle u_3| + |u_3\rangle\langle u_1|)$$

Then show that the trace of the density operator is 1.

Now let's look at the matrix representation of the density operator. In this case we refer to the *density matrix*. For the density operator in Example 5.1, where we had

$$\rho = \frac{1}{3}|u_1\rangle\langle u_1| - i\frac{\sqrt{2}}{3}|u_1\rangle\langle u_2| + i\frac{\sqrt{2}}{3}|u_2\rangle\langle u_1| + \frac{2}{3}|u_2\rangle\langle u_2|$$

the matrix representation is given by

$$[\rho] = \begin{pmatrix}\langle u_1|\rho|u_1\rangle & \langle u_1|\rho|u_2\rangle \\ \langle u_2|\rho|u_1\rangle & \langle u_2|\rho|u_2\rangle\end{pmatrix} = \begin{pmatrix}1/3 & -i\sqrt{2}/3 \\ i\sqrt{2}/3 & 2/3\end{pmatrix}$$

There are a couple of things to notice:

- The sum of the diagonal elements of the matrix is one. This shouldn't be surprising, since we showed earlier that $Tr(\rho) = 1$ because probabilities sum to one.
- The values along the diagonal are the probabilities of finding the system in the respective states. For example, looking at the state given in Example 5.1, you can see that the probability of finding the system in the state $|u_1\rangle$ is given by $\langle u_1|\rho|u_1\rangle$, which is 1/3. More generally, the matrix element ρ_{nn} is the average probability of finding the system in the state $|u_n\rangle$.

The off-diagonal elements of the density matrix are called *coherences*. These terms are a representation of the interference effects among different states, in this case between $|u_m\rangle$ and $|u_n\rangle$. The nondiagonal components of the density matrix ρ_{mn} represent the averages of these cross terms. Since we have a pure state in this example, these terms are nonzero. If you see a density matrix with nonzero off-diagonal terms, the state is coherent or has some interference effects, which isn't the case for a statistical mixture. However, keep in mind that a matrix representation of an operator can be taken with respect to a different basis, so we can find a representation of the density matrix that is diagonal. As we will see below, there is a better way to determine if the state is pure or mixed.

Time Evolution of the Density Operator

The time evolution of the density operator can be found readily by application of the Schrödinger equation. Recall that

$$i\hbar\frac{d}{dt}|\psi\rangle = H|\psi\rangle$$

Since $H = H^\dagger$, we can also write

$$-i\hbar\frac{d}{dt}\langle\psi| = \langle\psi|H$$

Writing the density operator as $\rho = |\psi\rangle\langle\psi|$, and applying the product rule for derivatives, we have

$$\frac{d\rho}{dt} = \frac{d}{dt}(|\psi\rangle\langle\psi|) = \left(\frac{d}{dt}|\psi\rangle\right)\langle\psi| + |\psi\rangle\left(\frac{d}{dt}\langle\psi|\right)$$

Now we use our result from the Schrödinger equation. This becomes

$$\frac{d\rho}{dt} = \left(\frac{H}{i\hbar}|\psi\rangle\right)\langle\psi| + |\psi\rangle\left(\langle\psi|\frac{H}{-i\hbar}\right) = \frac{H}{i\hbar}\rho - \rho\frac{H}{i\hbar} = \frac{1}{i\hbar}[H, \rho]$$

Definition: Time Evolution of the Density Operator

$$i\hbar\frac{d\rho}{dt} = [H, \rho] \tag{5.4}$$

For a closed system the time evolution of the density operator can also be written in terms of a unitary operator U. If we let $\rho(t)$ represent the density operator at some later time t and $\rho(t_o)$ represent the density operator at an initial time t_o, then

$$\rho(t) = U\rho(t_o)U^\dagger \tag{5.5}$$

Finally, looking at the definition $\rho = |\psi\rangle\langle\psi|$, it is obvious that the density operator is Hermitian, meaning $\rho = \rho^\dagger$. In the case of pure states, since

$$\rho^2 = (|\psi\rangle\langle\psi|)(|\psi\rangle\langle\psi|) = |\psi\rangle(\langle\psi|\psi\rangle)\langle\psi| = |\psi\rangle\langle\psi| = \rho$$

we have the following definition. If a system is in a pure state $|\psi\rangle$ with density operator $\rho = |\psi\rangle\langle\psi|$, then

$$Tr(\rho^2) = 1 \quad \text{(pure state only)} \tag{5.6}$$

THE DENSITY OPERATOR FOR A MIXED STATE

We now turn to the reason we considered density operators in the first place—we need a way to describe a statistical mixture of states. What we are looking for is a density matrix that describes an ensemble. This can be done by following these three steps:

- Construct a density operator for each individual state that can be found in the ensemble.
- Weight it by the probability of finding that state in the ensemble.
- Sum up the possibilities.

To see how this procedure works, we return to the example mentioned at the beginning of the chapter. In that case, members of the ensemble could be found in one of two states

$$|a\rangle = \alpha|x\rangle + \beta|y\rangle$$
$$|b\rangle = \gamma|x\rangle + \delta|y\rangle$$

The density operators for each of these states are

$$\rho_a = |a\rangle\langle a|$$
$$\rho_b = |b\rangle\langle b|$$

Or, in terms of the basis states these are

$$\begin{aligned}\rho_a &= |a\rangle\langle a| = (\alpha|x\rangle + \beta|y\rangle)(\alpha^*\langle x| + \beta^*\langle y|) \\ &= |\alpha|^2|x\rangle\langle x| + \alpha\beta^*|x\rangle\langle y| + \alpha^*\beta|y\rangle\langle x| + |\beta|^2|y\rangle\langle y| \\ \rho_b &= |b\rangle\langle b| = (\gamma|x\rangle + \delta|y\rangle)(\gamma^*\langle x| + \delta^*\langle y|) \\ &= |\gamma|^2|x\rangle\langle x| + \gamma\delta^*|x\rangle\langle y| + \gamma^*\delta|y\rangle\langle x| + |\delta|^2|y\rangle\langle y|\end{aligned}$$

The probability of a member of the ensemble being in state $|a\rangle$ is p while the probability of a member of the ensemble being in state $|b\rangle$ is $1-p$. In terms of the states $|a\rangle$ and $|b\rangle$ the density operator for the ensemble is

$$\rho = p\rho_a + (1-p)\rho_b = p|a\rangle\langle a| + (1-p)|b\rangle\langle b|$$

If we wish, we can write this in terms of the basis states $\{|x\rangle,|y\rangle\}$ as well, but we'll save that algebraic mess for worked examples. The main point is to see how to weigh the density operator for each pure state in the ensemble by the respective probability, and then add them up.

In general, suppose that there are n possible states. For a state $|\psi_i\rangle$, using (1) the density operator is written as $\rho_i = |\psi_i\rangle\langle\psi_i|$. Denote the probability that a member of the ensemble has been prepared in the state $|\psi_i\rangle$ as p_i. Then the density operator for the entire system is

$$\rho = \sum_{i=1}^{n} p_i\rho_i = \sum_{i=1}^{n} p_i|\psi_i\rangle\langle\psi_i| \tag{5.7}$$

Now that we know what the density operator is, let's write down the key properties.

KEY PROPERTIES OF A DENSITY OPERATOR

An operator ρ is a density operator if and only if it satisfies the following three requirements:

- The density operator is Hermitian, meaning $\rho = \rho^\dagger$.
- $Tr(\rho) = 1$.
- ρ is a positive operator, meaning $\langle u|\rho|u\rangle \geq 0$ for any state vector $|u\rangle$.

Recall that an operator is positive if and only if it is Hermitian and has nonnegative eigenvalues.

Example 5.2

Consider the state

$$|a\rangle = \begin{pmatrix} e^{-i\phi}\sin\theta \\ \cos\theta \end{pmatrix}$$

Is $\rho = |a\rangle\langle a|$ a density operator?

Solution

In the $\{|0\rangle, |1\rangle\}$ basis, the state is written as

$$|a\rangle = \begin{pmatrix} e^{-i\phi}\sin\theta \\ \cos\theta \end{pmatrix} = e^{-i\phi}\sin\theta\begin{pmatrix}1\\0\end{pmatrix} + \cos\theta\begin{pmatrix}0\\1\end{pmatrix} = e^{-i\phi}\sin\theta|0\rangle + \cos\theta|1\rangle$$

The dual vector is

$$\langle a| = e^{i\phi}\sin\theta\langle 0| + \cos\theta\langle 1|$$

So we have

$$\begin{aligned}\rho = |a\rangle\langle a| &= (e^{-i\phi}\sin\theta|0\rangle + \cos\theta|1\rangle)(e^{i\phi}\sin\theta\langle 0| + \cos\theta\langle 1|)\\ &= \sin^2\theta|0\rangle\langle 0| + e^{-i\phi}\sin\theta\cos\theta|0\rangle\langle 1| + e^{i\phi}\sin\theta\cos\theta|1\rangle\langle 0| + \cos^2\theta|1\rangle\langle 1|\end{aligned}$$

The matrix representation of this density operator is

$$\begin{pmatrix} \langle 0|\rho|0\rangle & \langle 0|\rho|1\rangle \\ \langle 1|\rho|0\rangle & \langle 1|\rho|1\rangle \end{pmatrix} = \begin{pmatrix} \sin^2\theta & e^{-i\phi}\sin\theta\cos\theta \\ e^{i\phi}\sin\theta\cos\theta & \cos^2\theta \end{pmatrix}$$

First, we check to see if the matrix is Hermitian. The transpose of the matrix is

$$\rho^T = \begin{pmatrix} \sin^2\theta & e^{-i\phi}\sin\theta\cos\theta \\ e^{i\phi}\sin\theta\cos\theta & \cos^2\theta \end{pmatrix}^T = \begin{pmatrix} \sin^2\theta & e^{i\phi}\sin\theta\cos\theta \\ e^{-i\phi}\sin\theta\cos\theta & \cos^2\theta \end{pmatrix}$$

Taking the complex conjugate gives

$$\rho^\dagger = (\rho^T)^* = \begin{pmatrix} \sin^2\theta & e^{i\phi}\sin\theta\cos\theta \\ e^{-i\phi}\sin\theta\cos\theta & \cos^2\theta \end{pmatrix}^* = \begin{pmatrix} \sin^2\theta & e^{-i\phi}\sin\theta\cos\theta \\ e^{i\phi}\sin\theta\cos\theta & \cos^2\theta \end{pmatrix}$$

Since $\rho = \rho^\dagger$, the matrix is Hermitian. Second, we see that $Tr(\rho) = \sin^2\theta + \cos^2\theta = 1$. The trace of a density matrix is always unity. Finally, we consider an arbitrary state

$$|\psi\rangle = \begin{pmatrix} a \\ b \end{pmatrix}$$

Then

$$\langle\psi|\rho|\psi\rangle = |a|^2 \sin^2\theta + ab^* e^{i\phi} \sin\theta\cos\theta + a^* b e^{-i\phi} \sin\theta\cos\theta + |b|^2 \cos^2\theta$$

For complex numbers z and w, we can write

$$(z+w)(z^*+w^*) = zz^* + wz^* + w^*z + ww^* = |z+w|^2$$

It is also true that the modulus of any complex number satisfies $|\varsigma|^2 \geq 0$, and also $|z+w|^2 \geq 0$. So we make the following definitions:

$$z = ae^{-i\phi}\sin\theta, \quad \Rightarrow zz^* = |a|^2\sin^2\theta$$

$$w = b\cos\theta, \quad \Rightarrow ww^* = |b|^2\cos^2\theta$$

We see that we can identify

$$ab^* e^{i\phi}\sin\theta\cos\theta = wz^*$$

$$a^* b e^{-i\phi}\sin\theta\cos\theta = zw^*$$

Now recall that for any complex number z, $|z|^2 \geq 0$. So

$$\langle\psi|\rho|\psi\rangle = |ae^{-i\phi}\sin\theta + b\cos\theta|^2 \geq 0$$

since $|z+w|^2 \geq 0$. Hence the operator is positive. Since ρ is Hermitian, has unit trace, and is a positive operator, it qualifies as a density operator. We can also verify that the density operator is positive by examining its eigenvalues. It can be shown that the eigenvalues of

$$\rho = \begin{pmatrix} \sin^2\theta & e^{-i\phi}\sin\theta\cos\theta \\ e^{i\phi}\sin\theta\cos\theta & \cos^2\theta \end{pmatrix}$$

are given by $\lambda_{1,2} = \{1, 0\}$. Since both eigenvalues are nonnegative and the matrix is Hermitian, we conclude that the operator is positive.

We have already seen how the trace condition is a reflection of the conservation of probability in the pure state case. We can quickly work this out in the case of the statistical mixture by recalling that given a scalar a and an operator A, $Tr(aA) = aTr(A)$. Recall also that the trace turns outer products into inner products, meaning $Tr(|\psi\rangle\langle\phi|) = \langle\psi|\phi\rangle$. Remember, the trace is linear, so $Tr(A+B) = Tr(A) + Tr(B)$ and $Tr(\alpha A) = \alpha Tr(A)$ if α is a scalar. Putting these facts together to compute the trace of (5.7), we have

$$Tr(\rho) = Tr\left(\sum_{i=1}^{n} p_i|\psi_i\rangle\langle\psi_i|\right) = \sum_{i=1}^{n} p_i Tr(|\psi_i\rangle\langle\psi_i|) = \sum_{i=1}^{n} p_i\langle\psi_i|\psi_i\rangle = \sum_{i=1}^{n} p_i = 1$$

To get this result, we made the reasonable assumption that the states are normalized so that $\langle\psi_i|\psi_i\rangle = 1$.

Now let's show that in the general case the density operator is a positive operator. We consider an arbitrary state vector $|\phi\rangle$ and consider $\langle\phi|\rho|\phi\rangle$. Using (5.7), we obtain

$$\langle\phi|\rho|\phi\rangle = \sum_{i=1}^{n} p_i \langle\phi|\psi_i\rangle\langle\psi_i|\phi\rangle = \sum_{i=1}^{n} p_i |\langle\phi|\psi_i\rangle|^2$$

Note that the numbers p_i are probabilities—so they all satisfy $0 \le p_i \le 1$. Recall that the inner product satisfies $|\langle\phi|\psi_i\rangle|^2 \ge 0$. Therefore we have found that $\langle\phi|\rho|\phi\rangle \ge 0$ for an arbitrary state vector $|\phi\rangle$. We conclude that ρ is a positive operator.

Since ρ is a positive operator, the first property we stated for density operators—that ρ is Hermitian—follows automatically. This property tells us that the eigenvalues $\lambda_i \ge 0$ and that we can use spectral decomposition to write the density operator in a diagonal representation $\rho = \sum_i \lambda_i |u_i\rangle\langle u_i|$.

Expectation Values

The result that we found in the pure state case for calculating the expectation value of a given operator holds true in the general case of a statistical mixture. That is, the expectation value of an operator with respect to a statistical mixture of states can be calculated using

$$\langle A\rangle = Tr(\rho A) \tag{5.8}$$

Probability of Obtaining a Given Measurement Result

Given a projection operator $P_n = |u_n\rangle\langle u_n|$ corresponding to measurement result a_n, the probability of obtaining a_n can be calculated using the density operator as follows:

$$p(a_n) = \langle u_n|\rho|u_n\rangle = Tr(|u_n\rangle\langle u_n|\rho) = Tr(P_n\rho) \tag{5.9}$$

In terms of the more general measurement operator formalism, the probability of obtaining measurement result m associated with measurement operator M_m is

$$P(m) = Tr(M_m^{\dagger} M_m \rho) \tag{5.10}$$

After a measurement described by a projection operator, the system is in the state described by

$$\rho \to \frac{P_n \rho P_n}{Tr(P_n \rho)} \tag{5.11}$$

In terms of general measurement operators, the state of the system after measurement is

$$\frac{M_m \rho M_m^{\dagger}}{Tr(M_m^{\dagger} M_m \rho)} \tag{5.12}$$

Example 5.3

Does the matrix

$$\rho = \begin{pmatrix} \frac{1}{4} & \frac{1-i}{4} \\ \frac{1-i}{4} & \frac{3}{4} \end{pmatrix}$$

represent a density operator?

Solution

We see that the matrix has unit trace

$$Tr(\rho) = \frac{1}{4} + \frac{3}{4} = 1$$

So it looks like it might be a valid density operator. However,

$$\rho^\dagger = \begin{pmatrix} \frac{1}{4} & \frac{1+i}{4} \\ \frac{1+i}{4} & \frac{3}{4} \end{pmatrix} \neq \rho$$

Since the matrix is not Hermitian, it cannot represent a density operator.

You Try It

Does $\rho = \begin{pmatrix} 1 & 0 \\ 0 & 1 \end{pmatrix}$ represent a density operator? If not why not?

Example 5.4

A system is found to be in the state

$$|\psi\rangle = \frac{1}{\sqrt{5}}|0\rangle + \frac{2}{\sqrt{5}}|1\rangle$$

(a) Write down the density operator for this state.

(b) Write down the matrix representation of the density operator in the $\{|0\rangle, |1\rangle\}$ basis. Verify that $Tr(\rho) = 1$, and show this is a pure state.

(c) A measurement of Z is made. Calculate the probability that the system is found in the state $|0\rangle$ and the probability that the system is found in the state $|1\rangle$.

(d) Find $\langle X \rangle$.

Solution

(a) To write down the density operator, first we construct the dual vector $\langle\psi|$. This can be done by inspection

$$\langle\psi| = \frac{1}{\sqrt{5}}\langle 0| + \frac{2}{\sqrt{5}}\langle 1|$$

The density operator is

$$\rho = |\psi\rangle\langle\psi| = \left(\frac{1}{\sqrt{5}}|0\rangle + \frac{2}{\sqrt{5}}|1\rangle\right)\left(\frac{1}{\sqrt{5}}\langle 0| + \frac{2}{\sqrt{5}}\langle 1|\right)$$
$$= \frac{1}{5}|0\rangle\langle 0| + \frac{2}{5}|0\rangle\langle 1| + \frac{2}{5}|1\rangle\langle 0| + \frac{4}{5}|1\rangle\langle 1|$$

(b) The matrix representation of the density operator in the $\{|0\rangle, |1\rangle\}$ basis is found by writing,

$$[\rho] = \begin{pmatrix} \langle 0|\rho|0\rangle & \langle 0|\rho|1\rangle \\ \langle 1|\rho|0\rangle & \langle 1|\rho|1\rangle \end{pmatrix}$$

For the state given in this problem, we have

$$\langle 0|\rho|0\rangle = \frac{1}{5}$$
$$\langle 0|\rho|1\rangle = \frac{2}{5} = \langle 1|\rho|0\rangle$$
$$\langle 1|\rho|1\rangle = \frac{4}{5}$$

So in the $\{|0\rangle, |1\rangle\}$ basis the density matrix is

$$\rho = \begin{pmatrix} \frac{1}{5} & \frac{2}{5} \\ \frac{2}{5} & \frac{4}{5} \end{pmatrix}$$

The trace is just the sum of the diagonal elements. In this case

$$Tr(\rho) = \frac{1}{5} + \frac{4}{5} = 1$$

To determine whether or not this is a pure state, we need to determine if $Tr(\rho^2) = 1$. Now

$$\rho^2 = \begin{pmatrix} \frac{1}{5} & \frac{2}{5} \\ \frac{2}{5} & \frac{4}{5} \end{pmatrix}\begin{pmatrix} \frac{1}{5} & \frac{2}{5} \\ \frac{2}{5} & \frac{4}{5} \end{pmatrix} = \begin{pmatrix} \frac{1}{25} + \frac{4}{25} & \frac{2}{25} + \frac{8}{25} \\ \frac{2}{25} + \frac{8}{25} & \frac{4}{25} + \frac{16}{25} \end{pmatrix}$$
$$= \begin{pmatrix} \frac{5}{25} & \frac{10}{25} \\ \frac{10}{25} & \frac{20}{25} \end{pmatrix} = \begin{pmatrix} \frac{1}{5} & \frac{2}{5} \\ \frac{2}{5} & \frac{4}{5} \end{pmatrix} = \rho$$

Since $\rho^2 = \rho$, it follows that $Tr(\rho^2) = 1$ and this is a pure state.

(c) In this simple example we can see by inspection that the probability that the system is in state $|0\rangle$ is 1/5 while the probability that the system is found in state $|1\rangle$ is 4/5. Let's see if we can verify this using the density operator formalism. First, we write down the projection operators in matrix form. The measurement operator that corresponds to the measurement result $|0\rangle$ is $P_0 = |0\rangle\langle 0|$ while the measurement operator $P_1 = |1\rangle\langle 1|$. The matrix representation in the given basis is

$$P_0 = \begin{pmatrix} \langle 0|P_0|0\rangle & \langle 0|P_0|1\rangle \\ \langle 1|P_0|0\rangle & \langle 1|P_0|1\rangle \end{pmatrix} = \begin{pmatrix} 1 & 0 \\ 0 & 0 \end{pmatrix} \quad \text{and} \quad P_1 = \begin{pmatrix} \langle 0|P_1|0\rangle & \langle 0|P_1|1\rangle \\ \langle 1|P_1|0\rangle & \langle 1|P_1|1\rangle \end{pmatrix} = \begin{pmatrix} 0 & 0 \\ 0 & 1 \end{pmatrix}$$

The probability of finding the system in state $|0\rangle$ is

$$p(0) = Tr(P_0\rho) = Tr\left[\begin{pmatrix} 1 & 0 \\ 0 & 0 \end{pmatrix}\begin{pmatrix} \frac{1}{5} & \frac{2}{5} \\ \frac{2}{5} & \frac{4}{5} \end{pmatrix}\right] = Tr\begin{pmatrix} \frac{1}{5} & \frac{2}{5} \\ 0 & 0 \end{pmatrix} = \frac{1}{5}$$

The probability of finding the system in state $|1\rangle$ is

$$p(1) = Tr(P_1\rho) = Tr\left[\begin{pmatrix} 0 & 0 \\ 0 & 1 \end{pmatrix}\begin{pmatrix} \frac{1}{5} & \frac{2}{5} \\ \frac{2}{5} & \frac{4}{5} \end{pmatrix}\right] = Tr\begin{pmatrix} 0 & 0 \\ \frac{2}{5} & \frac{4}{5} \end{pmatrix} = \frac{4}{5}$$

(d) We can find the expectation value of X by calculating $Tr(X\rho)$. First, let's do the matrix multiplication:

$$X\rho = \begin{pmatrix} 0 & 1 \\ 1 & 0 \end{pmatrix}\begin{pmatrix} \frac{1}{5} & \frac{2}{5} \\ \frac{2}{5} & \frac{4}{5} \end{pmatrix} = \begin{pmatrix} \frac{2}{5} & \frac{4}{5} \\ \frac{1}{5} & \frac{2}{5} \end{pmatrix}$$

The trace is the sum of the diagonal elements of this matrix

$$\langle X\rangle = Tr(X\rho) = Tr\begin{pmatrix} \frac{2}{5} & \frac{4}{5} \\ \frac{1}{5} & \frac{2}{5} \end{pmatrix} = \frac{2}{5} + \frac{2}{5} = \frac{4}{5}$$

You Try It

If $|\psi\rangle = \frac{2}{3}|0\rangle + \frac{\sqrt{5}}{3}|1\rangle$, show that the state is normalized, and then show that the density matrix for this state is

$$\rho = \begin{pmatrix} \frac{4}{9} & \frac{2\sqrt{5}}{9} \\ \frac{2\sqrt{5}}{9} & \frac{5}{9} \end{pmatrix}$$

You Try It

Show that for the state given in Example 5.2, $\langle Z \rangle = -3/5$.

You Try It

Continue with the state in Example 5.2. Recall the $\{|+\rangle, |-\rangle\}$ basis where

$$|+\rangle = \frac{1}{\sqrt{2}}\begin{pmatrix} 1 \\ 1 \end{pmatrix}, \quad |-\rangle = \frac{1}{\sqrt{2}}\begin{pmatrix} 1 \\ -1 \end{pmatrix}$$

(a) Show that the matrix representation of the projection operator onto the $|-\rangle$ state is given by

$$P_- = \frac{1}{2}\begin{pmatrix} 1 & -1 \\ -1 & 1 \end{pmatrix}$$

in the $\{|0\rangle, |1\rangle\}$ basis.

(b) For the state given in Example 5.2, show that the probability of finding the system in the $|-\rangle$ state is 1/10.

CHARACTERIZING MIXED STATES

Recall that coherence is the capability of different components of a state to interfere with one another. To emphasize that coherence is not present in a statistical mixture, some authors refer to a plain old statistical mixture as an *incoherent mixture*. In a statistical mixture there will not be coherences, but in a pure state or a state in a linear superposition there will be coherences. One indication of whether or not a state is a pure state or a mixed state is the presence of off-diagonal terms in the matrix representation of the density operator. In summary:

- A mixed state is a classical statistical mixture of two or more states. The state has no coherences. Therefore the off-diagonal terms of the density operator are zero, that is, $\rho_{mn} = 0$ when $m \neq n$.
- A pure state will have nonzero off-diagonal terms.

Recall from Chapter 3, that we can represent an operator as a matrix by computing the components of that operator with respect to a given basis. We can choose to represent the operator using one basis or another valid basis for the space, so the matrix representation will not always be the same. We also indicated it possible to find a diagonal representation of the density operator. Therefore a stronger criterion must be used to determine whether or not the density operator represents a mixed state rather than considering the off-diagonal terms of the matrix.

Recall that for a pure state the density operator is a projection operator, meaning $\rho^2 = \rho$. We also showed that since $Tr(\rho) = 1$ for any density operator, this means that $Tr(\rho^2) = 1$ for a pure state. The same is not true in the case of a mixed state, so we can use this fact to construct a definitive test to determine whether or not a density operator represents a mixed state or a pure state. Summarizing:

- $Tr(\rho^2) < 1$ for a mixed state.
- $Tr(\rho^2) = 1$ for a pure state.

Example 5.5

Recalling the basis states

$$|+\rangle = \frac{1}{\sqrt{2}}(|0\rangle + |1\rangle) \quad \text{and} \quad |-\rangle = \frac{1}{\sqrt{2}}(|0\rangle - |1\rangle)$$

Suppose that a statistical mixture has 75% in the $|+\rangle$ state and 25% in the $|-\rangle$ state. A member of the ensemble is drawn. What are the probabilities of finding it in the $|0\rangle$ state and $|1\rangle$ state, respectively? Show that it is a mixed state and that this calculation doesn't depend on the basis we use. Contrast the statistical mixture with 75% of the systems in the $|+\rangle$ state and 25% of the systems in the $|-\rangle$ state, with the pure state described by

$$|\psi\rangle = \sqrt{\frac{3}{4}}|+\rangle + \sqrt{\frac{1}{4}}|-\rangle$$

Write down the probability that a measurement of the pure state $|\psi\rangle$ finds the system in $|0\rangle$.

Solution

The density operator for the ensemble is given by

$$\rho = \frac{3}{4}|+\rangle\langle+| + \frac{1}{4}|-\rangle\langle-|$$

Now

$$|+\rangle\langle+| = \left(\frac{1}{\sqrt{2}}(|0\rangle + |1\rangle)\right)\left(\frac{1}{\sqrt{2}}(\langle 0| + \langle 1|)\right) = \frac{1}{2}(|0\rangle\langle 0| + |0\rangle\langle 1| + |1\rangle\langle 0| + |1\rangle\langle 1|)$$

$$|-\rangle\langle-| = \left(\frac{1}{\sqrt{2}}(|0\rangle - |1\rangle)\right)\left(\frac{1}{\sqrt{2}}(\langle 0| - \langle 1|)\right) = \frac{1}{2}(|0\rangle\langle 0| - |0\rangle\langle 1| - |1\rangle\langle 0| + |1\rangle\langle 1|)$$

So we find that

$$\rho = \frac{3}{4}|+\rangle\langle+| + \frac{1}{4}|-\rangle\langle-|$$
$$= \left(\frac{3}{4}\right)\left(\frac{1}{2}\right)(|0\rangle\langle 0| + |0\rangle\langle 1| + |1\rangle\langle 0| + |1\rangle\langle 1|)$$

$$+\left(\frac{1}{4}\right)\left(\frac{1}{2}\right)((|0\rangle\langle 0| - |0\rangle\langle 1| - |1\rangle\langle 0| + |1\rangle\langle 1|))$$
$$= \frac{1}{2}|0\rangle\langle 0| + \frac{1}{4}|0\rangle\langle 1| + \frac{1}{4}|1\rangle\langle 0| + \frac{1}{2}|1\rangle\langle 1|$$

The respective probabilities are

$$\begin{aligned}
p(0) &= Tr(\rho|0\rangle\langle 0|) \\
&= \langle 0|\rho|0\rangle \\
&= \langle 0|\left(\frac{1}{2}|0\rangle\langle 0| + \frac{1}{4}|0\rangle\langle 1| + \frac{1}{4}|1\rangle\langle 0| + \frac{1}{2}|1\rangle\langle 1|\right)|0\rangle = \frac{1}{2} \\
p(1) &= Tr(\rho|1\rangle\langle 1|) \\
&= \langle 1|\rho|1\rangle \\
&= \langle 1|\left(\frac{1}{2}|0\rangle\langle 0| + \frac{1}{4}|0\rangle\langle 1| + \frac{1}{4}|1\rangle\langle 0| + \frac{1}{2}|1\rangle\langle 1|\right)|1\rangle = \frac{1}{2}
\end{aligned}$$

In the $\{|+\rangle, |-\rangle\}$ basis the density matrix is

$$\rho = \begin{pmatrix} \frac{3}{4} & 0 \\ 0 & \frac{1}{4} \end{pmatrix}$$

Notice that the trace of the density matrix is one, as it should be. Also note that the off-diagonal terms are zero. Squaring, we find that

$$\rho^2 = \begin{pmatrix} \frac{3}{4} & 0 \\ 0 & \frac{1}{4} \end{pmatrix}\begin{pmatrix} \frac{3}{4} & 0 \\ 0 & \frac{1}{4} \end{pmatrix} = \begin{pmatrix} \frac{9}{16} & 0 \\ 0 & \frac{1}{16} \end{pmatrix}$$
$$\Rightarrow Tr(\rho^2) = \frac{9}{16} + \frac{1}{16} = \frac{10}{16} = \frac{5}{8}$$

So we see that even though the off-diagonal terms of the matrix were zero, $Tr(\rho^2) < 1$, telling us that this *is a mixed state*. Let's look at the matrix in the $\{|0\rangle, |1\rangle\}$ basis. We find that

$$\rho = \begin{pmatrix} \frac{1}{2} & \frac{1}{4} \\ \frac{1}{4} & \frac{1}{2} \end{pmatrix}$$

So

$$\rho^2 = \begin{pmatrix} \frac{1}{2} & \frac{1}{4} \\ \frac{1}{4} & \frac{1}{2} \end{pmatrix}\begin{pmatrix} \frac{1}{2} & \frac{1}{4} \\ \frac{1}{4} & \frac{1}{2} \end{pmatrix} = \begin{pmatrix} \frac{5}{16} & \frac{1}{4} \\ \frac{1}{4} & \frac{5}{16} \end{pmatrix}$$
$$\Rightarrow Tr(\rho^2) = \frac{5}{16} + \frac{5}{16} = \frac{10}{16} = \frac{5}{8}$$

Once again, notice that $Tr(\rho^2) = 5/8 < 1$, confirming that this is a mixed state.

For the last part of this problem, recall that for the statistical mixture the probability of obtaining measurement result 0 was $p_m(0) = 1/2$, where we used the subscript m to remind ourselves this is the probability given the mixed state. Now we consider the pure state:

$$|\psi\rangle = \sqrt{\frac{3}{4}}|+\rangle + \sqrt{\frac{1}{4}}|-\rangle$$

This state has a superficial resemblance to the statistical mixture with 75% of the systems in the $|+\rangle$ state and 25% of the systems in the $|-\rangle$ state, since a measurement in the $\{|+\rangle, |-\rangle\}$ basis gives $|+\rangle$ with a probability $p(+) = 3/4 = 0.75$ and $|-\rangle$ with a probability $p(-) = 1/4 = 0.25$. For example, look at the density operator for the mixed state:

$$\rho = \frac{3}{4}|+\rangle\langle+| + \frac{1}{4}|-\rangle\langle-|$$

The probability of finding $|+\rangle$ is

$$p_m(+) = Tr(|+\rangle\langle+|\rho) = \langle+|\rho|+\rangle = \langle+|\left(\frac{3}{4}|+\rangle\langle+| + \frac{1}{4}|-\rangle\langle-|\right)|+\rangle = \frac{3}{4}$$

However, it turns out that if we instead consider measurements with respect to the $\{|0\rangle, |1\rangle\}$ basis, we will find dramatically different results. Let's rewrite the given pure state in that basis. We find that

$$\begin{aligned}
|\psi\rangle &= \sqrt{\frac{3}{4}}|+\rangle + \sqrt{\frac{1}{4}}|-\rangle = \sqrt{\frac{3}{4}}\left(\frac{|0\rangle + |1\rangle}{\sqrt{2}}\right) + \sqrt{\frac{1}{4}}\left(\frac{|0\rangle - |1\rangle}{\sqrt{2}}\right) \\
&= \left(\sqrt{\frac{3}{4}}\frac{1}{\sqrt{2}} + \sqrt{\frac{1}{4}}\frac{1}{\sqrt{2}}\right)|0\rangle + \left(\sqrt{\frac{3}{4}}\frac{1}{\sqrt{2}} - \sqrt{\frac{1}{4}}\frac{1}{\sqrt{2}}\right)|1\rangle \\
&= \left(\frac{\sqrt{3}+1}{2\sqrt{2}}\right)|0\rangle + \left(\frac{\sqrt{3}-1}{2\sqrt{2}}\right)|1\rangle
\end{aligned}$$

Therefore the probability that a measurement finds $|\psi\rangle$ in the state $|0\rangle$ is

$$p(0) = \left(\frac{\sqrt{3}+1}{2\sqrt{2}}\right)^2 = \frac{2+\sqrt{2}}{4} \approx 0.85$$

So we see that even though $p(+) = p_m(+)$, $p(0) \neq p_m(0)$.

Example 5.6

Consider the ensemble in the previous example, where

$$\rho = \frac{3}{4}|+\rangle\langle+| + \frac{1}{4}|-\rangle\langle-|$$

A measurement in the $\{|0\rangle, |1\rangle\}$ basis finds the result $|0\rangle$. Use the density operator formalism to show that the state of the system after measurement is in fact $|0\rangle\langle 0|$.

Solution

The state of the system after measurement is $\rho \to P_n \rho P_n / Tr(P_n \rho)$. Writing the density operator in the $\{|0\rangle, |1\rangle\}$ basis obtains

$$\rho = \frac{1}{2}|0\rangle\langle 0| + \frac{1}{4}|0\rangle\langle 1| + \frac{1}{4}|1\rangle\langle 0| + \frac{1}{2}|1\rangle\langle 1|$$

The projection operator for the $|0\rangle$ state is $P_0 = |0\rangle\langle 0|$. So we see that

$$\rho P_0 = \left(\frac{1}{2}|0\rangle\langle 0| + \frac{1}{4}|0\rangle\langle 1| + \frac{1}{4}|1\rangle\langle 0| + \frac{1}{2}|1\rangle\langle 1|\right)|0\rangle\langle 0| = \frac{1}{2}|0\rangle\langle 0| + \frac{1}{4}|1\rangle\langle 0|$$

Hence

$$P_0 \rho P_0 = |0\rangle\langle 0|\left(\frac{1}{2}|0\rangle\langle 0| + \frac{1}{4}|1\rangle\langle 0|\right) = \frac{1}{2}|0\rangle\langle 0|$$

Now

$$Tr(\rho P_0) = Tr(\rho|0\rangle\langle 0|) = \langle 0|\rho|0\rangle$$
$$= \langle 0|\left(\frac{1}{2}|0\rangle\langle 0| + \frac{1}{4}|0\rangle\langle 1| + \frac{1}{4}|1\rangle\langle 0| + \frac{1}{2}|1\rangle\langle 1|\right)|0\rangle = \frac{1}{2}$$

Therefore the state of the system after measurement is

$$\rho \to \frac{P_0 \rho P_0}{Tr(P_0 \rho)} = \frac{(1/2)|0\rangle\langle 0|}{(1/2)} = |0\rangle\langle 0|$$

You Try It

An ensemble is described by the density operator $\rho = \frac{3}{8}|+\rangle\langle +| + \frac{5}{8}|-\rangle\langle -|$. Write down the density operator in the $\{|0\rangle, |1\rangle\}$ basis. Find the probability that if a member of the ensemble is drawn, measurement determines it to be in state $|1\rangle$. Show that if measurement obtains the state $|1\rangle$, the density operator is $|1\rangle\langle 1|$.

Example 5.7

Suppose

$$|a\rangle = \frac{1}{\sqrt{3}}|+\rangle + \sqrt{\frac{2}{3}}|-\rangle \quad \text{and} \quad |b\rangle = \frac{2}{3}|+\rangle - \frac{\sqrt{5}}{9}|-\rangle$$

with 75% of the systems prepared in state $|a\rangle$ and 25% of the systems prepared in state $|b\rangle$.

(a) Write down the density operators ρ_a and ρ_b.
(b) Compute the density operator for the ensemble.
(c) A measurement is made. What are the probabilities of finding $|+\rangle$ and $|-\rangle$?
(d) Instead of question (c), what are the probabilities of finding $|0\rangle$ and $|1\rangle$?

Solution

(a) We write the density operators in the $\{|+\rangle, |-\rangle\}$ basis. We find that

$$\rho_a = |a\rangle\langle a| = \left(\frac{1}{\sqrt{3}}|+\rangle + \sqrt{\frac{2}{3}}|-\rangle\right)\left(\frac{1}{\sqrt{3}}\langle +| + \sqrt{\frac{2}{3}}\langle -|\right)$$

$$= \frac{1}{3}|+\rangle\langle +| + \frac{\sqrt{2}}{3}|+\rangle\langle -| + \frac{\sqrt{2}}{3}|-\rangle\langle +| + \frac{2}{3}|-\rangle\langle -|$$

$$\rho_b = |b\rangle\langle b| = \left(\frac{2}{3}|+\rangle - \frac{\sqrt{5}}{3}|-\rangle\right)\left(\frac{2}{3}\langle +| - \frac{\sqrt{5}}{3}\langle -|\right)$$

$$= \frac{4}{9}|+\rangle\langle +| - \frac{2\sqrt{5}}{9}|+\rangle\langle -| - \frac{2\sqrt{5}}{9}|-\rangle\langle +| + \frac{5}{9}|-\rangle\langle -|$$

The density matrices, in the $\{|+\rangle, |-\rangle\}$ basis are given by

$$\rho_a = \begin{pmatrix} \frac{1}{3} & \frac{\sqrt{2}}{3} \\ \frac{\sqrt{2}}{3} & \frac{2}{3} \end{pmatrix}, \quad \rho_b = \begin{pmatrix} \frac{4}{9} & \frac{-2\sqrt{5}}{9} \\ \frac{-2\sqrt{5}}{9} & \frac{5}{9} \end{pmatrix}$$

Notice that $Tr(\rho_a) = Tr(\rho_b) = 1$, which must be the case for a density operator.

(b) The density operator for the ensemble is given by

$$\rho = (3/4)\rho_a + (1/4)\,\rho_b$$

$$= (3/4)\left[\frac{1}{3}|+\rangle\langle +| + \frac{\sqrt{2}}{3}|+\rangle\langle -| + \frac{\sqrt{2}}{3}|-\rangle\langle +| + \frac{2}{3}|-\rangle\langle -|\right]$$

$$+ (1/4)\left[\frac{4}{9}|+\rangle\langle +| - \frac{2\sqrt{5}}{9}|+\rangle\langle -| - \frac{2\sqrt{5}}{9}|-\rangle\langle +| + \frac{5}{9}|-\rangle\langle -|\right]$$

$$= \frac{13}{36}|+\rangle\langle +| + \frac{(9\sqrt{2} - 2\sqrt{5})}{36}(|+\rangle\langle -| + |-\rangle\langle +|) + \frac{23}{36}|-\rangle\langle -|$$

Notice that the trace is still unity

$$Tr(\rho) = \langle +|\rho|+\rangle + \langle -|\rho|-\rangle = \frac{13}{36} + \frac{23}{36} = \frac{36}{36} = 1$$

(c) If we pull a member of the ensemble, the probability that a measurement finds it in the $|+\rangle$ state is

$$p(+) = Tr(\rho|+\rangle\langle +|)$$

$$= \langle +|\left(\frac{13}{36}|+\rangle\langle +| + \frac{(9\sqrt{2} - 2\sqrt{5})}{36}(|+\rangle\langle -| + |-\rangle\langle +|) + \frac{23}{36}|-\rangle\langle -|\right)|+\rangle = \frac{13}{36} \approx 0.36$$

The probability that measurement finds a member of the ensemble in the $|-\rangle$ state is

$$p(-) = Tr(\rho|-\rangle\langle-|)$$

$$= \langle -|\left(\frac{13}{36}|+\rangle\langle+| + \frac{(9\sqrt{2}-2\sqrt{5})}{36}(|+\rangle\langle-| + |-\rangle\langle+|) + \frac{23}{36}|-\rangle\langle-|\right)|-\rangle = \frac{23}{36} \approx 0.64$$

Notice that these probabilities sum to one, as they should.

Example 5.8

Prove that for a mixed state, $Tr(\rho^2) < 1$.

Solution

First recall that we can write the density operator in terms of the spectral decomposition

$$\rho = \sum_i \lambda_i |u_i\rangle\langle u_i|$$

The $\{|u_i\rangle\}$ constitute an orthonormal basis, and so $\langle u_i|u_j\rangle = \delta_{ij}$. The trace of a density operator is 1, Therefore

$$Tr(\rho) = Tr\left(\sum_i \lambda_i |u_i\rangle\langle u_i|\right) = \sum_j \langle u_j|\left(\sum_i \lambda_i |u_i\rangle\langle u_i|\right)|u_j\rangle$$

$$= \sum_{i,j} \lambda_i \langle u_j|u_i\rangle\langle u_i|u_j\rangle$$

$$= \sum_{i,j} \lambda_i \langle u_j|u_i\rangle\delta_{ij} = \sum_i \lambda_i \langle u_i|u_i\rangle = \sum_i \lambda_i$$

So we have $\sum_i \lambda_i = 1$, which can only be true if each of the eigenvalues of the density operator satisfies $\lambda_i < 1$. This result implies that $\sum_i \lambda_i^2 < 1$. To see this better, note that if $\lambda_i < 1$, then $\lambda_i^2 < \lambda_i$.

With this mixed state relation of Example 5.8 in mind, we consider the square of the density operator. We write it as

$$Tr(\rho^2) = \sum_i \langle u_i|\left(\sum_j \lambda_j |u_j\rangle\langle u_j|\right)\left(\sum_k \lambda_k |u_k\rangle\langle u_k|\right)|u_i\rangle$$

$$= \sum_{i,j,k} \lambda_j \lambda_k \langle u_i|u_j\rangle\langle u_j|u_k\rangle\langle u_k|u_i\rangle$$

$$= \sum_{i,j,k} \lambda_j \lambda_k \langle u_i|u_j\rangle\langle u_j|u_k\rangle\delta_{ki}$$

$$= \sum_{i,j} \lambda_j \lambda_i \langle u_i | u_j \rangle \langle u_j | u_i \rangle$$

$$= \sum_{i,j} \lambda_j \lambda_i \langle u_i | u_j \rangle \delta_{ij}$$

$$= \sum_{i,j} \lambda_i \lambda_i \langle u_i | u_i \rangle$$

$$= \sum_{i,j} \lambda_i \lambda_i$$

$$= \sum_{i,j} \lambda_i^2 < 1$$

Example 5.9

Suppose

$$\rho = \begin{pmatrix} \frac{1}{3} & \frac{i}{3} \\ \frac{-i}{3} & \frac{2}{3} \end{pmatrix}$$

(a) Show that ρ is Hermitian and has positive eigenvalues that satisfy $0 \leq \lambda_i \leq 1$, and $\sum \lambda_i = 1$.

(b) Is this a mixed state?

(c) Find $\langle X \rangle$ for this state.

Solution

(a) To determine if this matrix is a representation of a valid density operator, we will check to see if it is Hermitian, if it has unit trace, and if it is a positive operator.

First notice that ρ is Hermitian. So we take the transpose and exchange rows and columns:

$$\rho^T = \begin{pmatrix} \frac{1}{3} & \frac{-i}{3} \\ \frac{i}{3} & \frac{2}{3} \end{pmatrix}$$

Next we take the complex conjugate to give $\rho^\dagger$:

$$\rho^\dagger = \begin{pmatrix} \frac{1}{3} & \frac{i}{3} \\ \frac{-i}{3} & \frac{2}{3} \end{pmatrix}$$

We have at least $\rho = \rho^{\dagger}$. Now the trace is

$$Tr(\rho) = \frac{1}{3} + \frac{2}{3} = 1$$

Since the trace is unity, it looks like this matrix might be a representation of a density operator. All we have to confirm is that the operator is a positive. Begin by finding the eigenvalues. Remember, we do this by solving the characteristic equation $\det|\rho - \lambda I| = 0$. We have

$$\det|\rho - \lambda I| = \det\left|\begin{pmatrix} \frac{1}{3} & \frac{i}{3} \\ \frac{-i}{3} & \frac{2}{3} \end{pmatrix} - \begin{pmatrix} \lambda & 0 \\ 0 & \lambda \end{pmatrix}\right| = \det\begin{vmatrix} \frac{1}{3} - \lambda & \frac{i}{3} \\ \frac{-i}{3} & \frac{2}{3} - \lambda \end{vmatrix}$$

$$= \left(\frac{1}{3} - \lambda\right)\left(\frac{2}{3} - \lambda\right) - \left(\frac{-i}{3}\right)\left(\frac{i}{3}\right)$$

$$= \lambda^2 - \lambda + \frac{1}{9}$$

Therefore the eigenvalues are

$$\lambda_{1,2} = \frac{1}{2} \pm \frac{\sqrt{5}}{6}$$

Because we have

$$\lambda_1 = \frac{1}{2} + \frac{\sqrt{5}}{6} \approx 0.87$$

$$\lambda_2 = \frac{1}{2} - \frac{\sqrt{5}}{6} \approx 0.13$$

the eigenvalues are real and nonnegative, and both are less than one. It is obvious that

$$\sum_{i=1}^{2} \lambda_i = \left(\frac{1}{2} + \frac{\sqrt{5}}{6}\right) + \left(\frac{1}{2} - \frac{\sqrt{5}}{6}\right) = \frac{1}{2} + \frac{1}{2} = 1$$

(b) Let's square the matrix:

$$\rho^2 = \begin{pmatrix} \frac{1}{3} & \frac{i}{3} \\ \frac{-i}{3} & \frac{2}{3} \end{pmatrix}\begin{pmatrix} \frac{1}{3} & \frac{i}{3} \\ \frac{-i}{3} & \frac{2}{3} \end{pmatrix} = \begin{pmatrix} \frac{2}{9} & \frac{i}{3} \\ \frac{-i}{3} & \frac{5}{9} \end{pmatrix}$$

The trace of this quantity is

$$Tr(\rho^2) = \frac{2}{9} + \frac{5}{9} = \frac{7}{9} < 1$$

Therefore this is a mixed state.

(c) We can find the expectation value using $\langle X \rangle = Tr(X\rho)$. Now

$$X\rho = \begin{pmatrix} 0 & 1 \\ 1 & 0 \end{pmatrix} \begin{pmatrix} \frac{1}{3} & \frac{i}{3} \\ \frac{-i}{3} & \frac{2}{3} \end{pmatrix} = \begin{pmatrix} \frac{-i}{3} & \frac{2}{3} \\ \frac{1}{3} & \frac{i}{3} \end{pmatrix}$$

The trace of this matrix is zero, so we see that $\langle X \rangle = 0$.

You Try It

Suppose

$$\rho = \begin{pmatrix} \frac{3}{5} & \frac{1}{5} \\ \frac{1}{5} & \frac{2}{5} \end{pmatrix}$$

Show that this is a valid density operator. You should find that the eigenvalues are $\lambda_{1,2} = 5 \pm \sqrt{5}/10$. Show that this is a mixed state and that $\langle Z \rangle = 1/5$.

Probability of Finding an Element of the Ensemble in a Given State

Once again, we ask the question: If an individual member is drawn from the ensemble and a measurement is made, what are the probabilities of obtaining each possible measurement result? In this case we write down the answer using the density matrix.

We can compute the density operator from the states that make up the ensemble using (5.7). Then it is convenient to write the density matrix in a given basis. If we choose the compute the density matrix in the $\{|0\rangle, |1\rangle\}$ basis, the matrix will be of the form

$$\rho = \begin{pmatrix} \langle 0|\rho|0\rangle & \langle 0|\rho|1\rangle \\ \langle 1|\rho|0\rangle & \langle 1|\rho|1\rangle \end{pmatrix}$$

If a measurement on a member of the ensemble is made, then the probability that the particle is found to be in the state $|0\rangle$ is given by

$\langle 0|\rho|0\rangle$ (probability a member of the ensemble is found to be in the state $|0\rangle$)

If a measurement on a member of the ensemble is made, then the probability the particle is found to be in the state $|1\rangle$ is

$\langle 1|\rho|1\rangle$ (probability a member of the ensemble is found to be in the state $|1\rangle$)

Notice this statement is equivalent to the earlier definition where we used the trace together with the given projection operator compute the various probabilities.

Example 5.10

Consider an ensemble in which 40% of the systems are known to be prepared in the state

$$|\psi\rangle = \frac{1}{\sqrt{3}}|0\rangle + \sqrt{\frac{2}{3}}|1\rangle$$

and 60% of the systems are prepared in the state

$$|\phi\rangle = \frac{1}{2}|0\rangle + \frac{\sqrt{3}}{2}|1\rangle$$

(a) Find the density operators for each of these states, and show they are pure states. If measurements are made on systems in each of these states, what are the probabilities they are found to be in states $|0\rangle$ and state $|1\rangle$, respectively?

(b) Determine the density operator for the ensemble.

(c) Show that $Tr(\rho) = 1$.

(d) A measurement of Z is made on a member drawn from the ensemble. What are the probabilities it is found to be in state $|0\rangle$ and state $|1\rangle$, respectively?

Solution

(a) By looking at $|\psi\rangle$ and using the Born rule, we see that the probability of finding the system in state $|0\rangle$ is 1/3, while the probability of finding the system in the state $|1\rangle$ is 2/3. If the system is prepared in state $|\phi\rangle$, the probability that measurement finds the system in state $|0\rangle$ is 1/4, and the probability that the system is found in state $|1\rangle$ is 3/4.

The density operators for each individual state are given by $\rho_\psi = |\psi\rangle\langle\psi|$ and $\rho_\phi = |\phi\rangle\langle\phi|$. We obtain

$$\rho_\psi = |\psi\rangle\langle\psi| = \left(\frac{1}{\sqrt{3}}|0\rangle + \sqrt{\frac{2}{3}}|1\rangle\right)\left(\frac{1}{\sqrt{3}}\langle 0| + \sqrt{\frac{2}{3}}\langle 1|\right)$$

$$= \frac{1}{3}|0\rangle\langle 0| + \frac{\sqrt{2}}{3}|0\rangle\langle 1| + \frac{\sqrt{2}}{3}|1\rangle\langle 0| + \frac{2}{3}|1\rangle\langle 1|$$

$$\rho_\phi = |\phi\rangle\langle\phi| = \left(\frac{1}{2}|0\rangle + \frac{\sqrt{3}}{2}|1\rangle\right)\left(\frac{1}{2}\langle 0| + \frac{\sqrt{3}}{2}\langle 1|\right)$$

$$= \frac{1}{4}|0\rangle\langle 0| + \frac{\sqrt{3}}{4}|0\rangle\langle 1| + \frac{\sqrt{3}}{4}|1\rangle\langle 0| + \frac{3}{4}|1\rangle\langle 1|$$

The matrix representations are given by

$$\rho_\psi = \begin{pmatrix} \langle 0|\rho_\psi|0\rangle & \langle 0|\rho_\psi|1\rangle \\ \langle 1|\rho_\psi|0\rangle & \langle 1|\rho_\psi|1\rangle \end{pmatrix} = \begin{pmatrix} \frac{1}{3} & \frac{\sqrt{2}}{3} \\ \frac{\sqrt{2}}{3} & \frac{2}{3} \end{pmatrix}$$

and

$$\rho_\phi = \begin{pmatrix} \langle 0|\rho_\phi|0\rangle & \langle 0|\rho_\phi|1\rangle \\ \langle 1|\rho_\phi|0\rangle & \langle 1|\rho_\phi|1\rangle \end{pmatrix} = \begin{pmatrix} \frac{1}{4} & \frac{\sqrt{3}}{4} \\ \frac{\sqrt{3}}{4} & \frac{3}{4} \end{pmatrix}$$

To determine whether or not these matrices are pure states, we square each density matrix and then compute the trace. We calculate the first explicitly:

$$\rho_\psi^2 = \begin{pmatrix} \frac{1}{3} & \frac{\sqrt{2}}{3} \\ \frac{\sqrt{2}}{3} & \frac{2}{3} \end{pmatrix}\begin{pmatrix} \frac{1}{3} & \frac{\sqrt{2}}{3} \\ \frac{\sqrt{2}}{3} & \frac{2}{3} \end{pmatrix} = \begin{pmatrix} \frac{1}{3} & \frac{\sqrt{2}}{3} \\ \frac{\sqrt{2}}{3} & \frac{2}{3} \end{pmatrix}$$

Since $\rho_\psi = \rho_\psi^2$ and $Tr(\rho_\psi) = 1$, this is a pure state. It is easy to show that ρ_ϕ is also a pure state.

(b) We use (5.7), which we restate here:

$$\rho = \sum_{i=1}^{n} p_i \rho_i = \sum_{i=1}^{n} p_i |\psi_i\rangle\langle\psi_i|$$

The probabilities given in the problem are

$$p_\psi = 40\% = \frac{2}{5}$$

$$p_\phi = 60\% = \frac{3}{5}$$

The density matrix for the ensemble is

$$\rho = p_\psi \rho_\psi + p_\phi \rho = \frac{2}{5}\begin{pmatrix} \frac{1}{3} & \frac{\sqrt{2}}{3} \\ \frac{\sqrt{2}}{3} & \frac{2}{3} \end{pmatrix} + \frac{3}{5}\begin{pmatrix} \frac{1}{4} & \frac{\sqrt{3}}{4} \\ \frac{\sqrt{3}}{4} & \frac{3}{4} \end{pmatrix}$$

$$= \begin{pmatrix} \frac{17}{60} & \frac{(8\sqrt{2}+9\sqrt{3})}{60} \\ \frac{(8\sqrt{2}+9\sqrt{3})}{60} & \frac{43}{60} \end{pmatrix}$$

(c) Notice that $Tr(\rho) = 17/60 + 43/60 = 60/60 = 1$, which must be the case for a density matrix.

(d) To find the respective probabilities, recall that we can write the matrix representation in the $\{|0\rangle, |1\rangle\}$ basis:

$$\rho = \begin{pmatrix} \langle 0|\rho|0\rangle & \langle 0|\rho|1\rangle \\ \langle 1|\rho|0\rangle & \langle 1|\rho|1\rangle \end{pmatrix}$$

The probability of finding a member of the ensemble in the state $|0\rangle$ is given by $\langle 0|\rho|0\rangle$. Looking at the density matrix calculated in part (c), we see that the probability is $17/60 \approx 0.28$.

Similarly, the probability of finding a member of the ensemble in the state $|1\rangle$ is given by $\langle 1|\rho|1\rangle$. Looking at the density matrix calculated in part (c), we see that the probability is $43/60 \approx 0.72$.

Completely Mixed States

A completely mixed state can be thought of as the opposite end along a continuum of density operators, with a pure state on the other side. In a completely mixed state the probability for the system to be in each given state is identical. In that case the density operator is a constant multiple of the identity matrix. If the state space has n dimensions, then

$$\rho = \frac{1}{n} I \tag{5.13}$$

Since $I^2 = I$, we have $\rho^2 = \frac{1}{n^2} I$. Furthermore, in n dimensions, $Tr(I) = n$. So for a completely mixed state we have

$$Tr(\rho^2) = Tr\left(\frac{1}{n^2} I\right) = \frac{1}{n^2} Tr(I) = \frac{1}{n} \tag{5.14}$$

In most if not all cases of interest to us, $n = 2$. For $n = 2$ the lower bound, given by a completely mixed state, is $Tr(\rho^2) = \frac{1}{2}$ while the upper bound for a pure state is given by $Tr(\rho^2) = 1$.

THE PARTIAL TRACE AND THE REDUCED DENSITY OPERATOR

A very important application of the density operator is in the characterization of composite systems—systems that are made up of two or more individual subsystems. Think entanglement.

The density operator is a very useful tool for characterizing and working with states of subsystems. In particular, we are going to start by thinking about a composite system where Alice has one part of the system and Bob has another part of the system and they fly off in opposite directions. The complete state of the system contains information about both subsystems, but obviously Bob, who is a long way from Alice, can't know about her half of the system unless she happened to call him up and tell him about it. We need a way to take the density operator for the entire system and distill it down into a density operator that just represents what Bob alone sees. This can be done by calculating the partial trace, which will give us Bob's very own density operator. We call that density operator the *reduced density operator*.

A typical example is imagining that Alice and Bob each share one member of an entangled EPR pair. Let's suppose that the system is in one of the Bell states

$$|\beta_{10}\rangle = \frac{|0_A\rangle|0_B\rangle - |1_A\rangle|1_B\rangle}{\sqrt{2}}$$

We construct the density operator for a composite system the same way we have been doing for individual systems. The density operator for this particular state is

$$\begin{aligned}\rho &= |\beta_{10}\rangle\langle\beta_{10}| \\ &= \left(\frac{|0_A\rangle|0_B\rangle - |1_A\rangle|1_B\rangle}{\sqrt{2}}\right)\left(\frac{\langle 0_A|\langle 0_B| - \langle 1_A|\langle 1_B|}{\sqrt{2}}\right) \\ &= \frac{|0_A\rangle|0_B\rangle\langle 0_A|\langle 0_B| - |0_A\rangle|0_B\rangle\langle 1_A|\langle 1_B| - |1_A\rangle|1_B\rangle\langle 0_A|\langle 0_B| + |1_A\rangle|1_B\rangle\langle 1_A|\langle 1_B|}{2}\end{aligned} \tag{5.15}$$

Because idea behind the partial trace is to obtain the density operator for one of the composite systems alone, we want a mathematical tool that describes quantities that are observable in that composite system. Basically with ρ as we have written it here, we have a description of the complete system. Now, if Alice and Bob fly off in opposite directions, neither one of them alone has access to the complete system. We need a tool that will tell us what Alice alone sees and what Bob alone sees. To implement the partial trace, we compute the trace by summing over the basis states of one party alone. Let's say for the sake of example that we're in Bob's shoes. Then we need to trace over Alice's basis states. We have

$$\rho_B = Tr_A(\rho) = Tr_A(|\beta_{10}\rangle\langle\beta_{10}|) = \langle 0_A|(|\beta_{10}\rangle\langle\beta_{10}|)|0_A\rangle + \langle 1_A|(|\beta_{10}\rangle\langle\beta_{10}|)|1_A\rangle$$

Using our expression for ρ, we have

$$\begin{aligned}&\langle 0_A|(|\beta_{10}\rangle\langle\beta_{10}|)|0_A\rangle \\ &\quad= \langle 0_A|\left(\frac{|0_A\rangle|0_B\rangle\langle 0_A|\langle 0_B| - |0_A\rangle|0_B\rangle\langle 1_A|\langle 1_B| - |1_A\rangle|1_B\rangle\langle 0_A|\langle 0_B| + |1_A\rangle|1_B\rangle\langle 1_A|\langle 1_B|}{2}\right)|0_A\rangle \\ &\quad= \frac{1}{2}\left(\frac{\begin{array}{c}\langle 0_A|0_A\rangle|0_B\rangle\langle 0_B|\langle 0_A|0_A\rangle - \langle 0_A|0_A\rangle|0_B\rangle\langle 1_B|\langle 0_A|1_A\rangle - \langle 0_A|1_A\rangle|1_B\rangle\langle 0_B|\langle 0_A|0_A\rangle \\ + \langle 0_A|1_A\rangle|1_B\rangle\langle 1_B|\langle 1_A|0_A\rangle\end{array}}{2}\right) \\ &\quad= \frac{|0_B\rangle\langle 0_B|}{2}\end{aligned}$$

and

$$\begin{aligned}&\langle 1_A|(|\beta_{10}\rangle\langle\beta_{10}|)|1_A\rangle \\ &\quad= \langle 1_A|\left(\frac{|0_A\rangle|0_B\rangle\langle 0_A|\langle 0_B| - |0_A\rangle|0_B\rangle\langle 1_A|\langle 1_B| - |1_A\rangle|1_B\rangle\langle 0_A|\langle 0_B| + |1_A\rangle|1_B\rangle\langle 1_A|\langle 1_B|}{2}\right)|1_A\rangle \\ &\quad= \frac{1}{2}\left(\frac{\begin{array}{c}\langle 1_A|0_A\rangle|0_B\rangle\langle 0_B|\langle 0_A|1_A\rangle - \langle 1_A|0_A\rangle|0_B\rangle\langle 1_B|\langle 0_A|1_A\rangle - \langle 1_A|1_A\rangle|1_B\rangle\langle 0_B|\langle 0_A|1_A\rangle \\ + \langle 1_A|1_A\rangle|1_B\rangle\langle 1_B|\langle 1_A|1_A\rangle\end{array}}{2}\right) \\ &\quad= \frac{|1_B\rangle\langle 1_B|}{2}\end{aligned}$$

So the density operator for Bob is

$$\rho_B = Tr_A(\rho) = Tr_A(|\beta_{10}\rangle\langle\beta_{10}|) = \langle 0_A|(|\beta_{10}\rangle\langle\beta_{10}|)|0_A\rangle + \langle 1_A|(|\beta_{10}\rangle\langle\beta_{10}|)|1_A\rangle$$
$$= \frac{|0\rangle\langle 0| + |1\rangle\langle 1|}{2}$$

We dropped the subscripts because this is Bob's state alone. The matrix representation with respect to Bob's $\{|0\rangle, |1\rangle\}$ basis is

$$\rho_B = \frac{1}{2}\begin{pmatrix} 1 & 0 \\ 0 & 1 \end{pmatrix}$$

Notice that $Tr(\rho_B) = 1/2 + 1/2 = 1$ (remember, the trace of a density matrix is always 1). Now let's square it. This is easy because we just have the identity matrix:

$$\rho_B = \frac{1}{2}\begin{pmatrix} 1 & 0 \\ 0 & 1 \end{pmatrix} = \frac{I}{2}, \Rightarrow \rho_B^2 = \frac{I^2}{4} = \frac{1}{4}\begin{pmatrix} 1 & 0 \\ 0 & 1 \end{pmatrix}$$

Then we have

$$Tr(\rho_B^2) = \frac{1}{4} + \frac{1}{4} = \frac{1}{2} < 1$$

That is, Bob has a *completely mixed state*.

You Try It

Compute the density matrix for Alice by taking the partial trace with respect to Bob's basis. Then show that Alice has the same completely mixed state in her possession.

What about the state of the joint system? The matrix representation of ρ as given in (15) is

$$[\rho] = \begin{pmatrix} \langle 00|\rho|00\rangle & \langle 00|\rho|01\rangle & \langle 00|\rho|10\rangle & \langle 00|\rho|11\rangle \\ \langle 01|\rho|00\rangle & \langle 01|\rho|01\rangle & \langle 01|\rho|10\rangle & \langle 01|\rho|11\rangle \\ \langle 10|\rho|00\rangle & \langle 10|\rho|01\rangle & \langle 10|\rho|10\rangle & \langle 10|\rho|11\rangle \\ \langle 11|\rho|00\rangle & \langle 11|\rho|01\rangle & \langle 11|\rho|10\rangle & \langle 11|\rho|11\rangle \end{pmatrix}$$
$$= \begin{pmatrix} \frac{1}{2} & 0 & 0 & \frac{-1}{2} \\ 0 & 0 & 0 & 0 \\ 0 & 0 & 0 & 0 \\ \frac{-1}{2} & 0 & 0 & \frac{1}{2} \end{pmatrix}$$

It can be easily verified that

$$\rho^2 = \begin{pmatrix} \frac{1}{2} & 0 & 0 & \frac{-1}{2} \\ 0 & 0 & 0 & 0 \\ 0 & 0 & 0 & 0 \\ \frac{-1}{2} & 0 & 0 & \frac{1}{2} \end{pmatrix} \begin{pmatrix} \frac{1}{2} & 0 & 0 & \frac{-1}{2} \\ 0 & 0 & 0 & 0 \\ 0 & 0 & 0 & 0 \\ \frac{-1}{2} & 0 & 0 & \frac{1}{2} \end{pmatrix} = \begin{pmatrix} \frac{1}{2} & 0 & 0 & \frac{-1}{2} \\ 0 & 0 & 0 & 0 \\ 0 & 0 & 0 & 0 \\ \frac{-1}{2} & 0 & 0 & \frac{1}{2} \end{pmatrix}$$

So we have $Tr(\rho^2) = 1$. That is, the joint system described by the state $|\beta_{10}\rangle$ is a pure state, while Alice and Bob alone see completely mixed states. We will see more of joint systems when we study entanglement in detail.

Example 5.11

Suppose that

$$|A\rangle = \frac{|0\rangle - i|1\rangle}{\sqrt{2}}, \quad |B\rangle = \sqrt{\frac{2}{3}}|0\rangle + \frac{1}{\sqrt{3}}|1\rangle$$

(a) Write down the product state $|A\rangle|B\rangle$.

(b) Compute the density operator. Is this a pure state?

Solution

(a) The product state is

$$|A\rangle \otimes |B\rangle = \left(\frac{|0\rangle - i|1\rangle}{\sqrt{2}}\right) \otimes \left(\sqrt{\frac{2}{3}}|0\rangle + \frac{1}{\sqrt{3}}|1\rangle\right)$$

$$= \frac{1}{\sqrt{3}}|00\rangle + \frac{1}{\sqrt{6}}|01\rangle - \frac{i}{\sqrt{3}}|10\rangle - \frac{i}{\sqrt{6}}|11\rangle$$

(b) The density operator is

$$\rho = \left(\frac{1}{\sqrt{3}}|00\rangle + \frac{1}{\sqrt{6}}|01\rangle - \frac{i}{\sqrt{3}}|10\rangle - \frac{i}{\sqrt{6}}|11\rangle\right)\left(\frac{1}{\sqrt{3}}\langle 00| + \frac{1}{\sqrt{6}}\langle 01| + \frac{i}{\sqrt{3}}\langle 10| + \frac{i}{\sqrt{6}}\langle 11|\right)$$

$$= \frac{1}{3}|00\rangle\langle 00| + \frac{1}{\sqrt{18}}|00\rangle\langle 01| + \frac{i}{3}|00\rangle\langle 10| + \frac{i}{\sqrt{18}}|00\rangle\langle 11| + \frac{1}{\sqrt{18}}|01\rangle\langle 00|$$

$$+ \frac{1}{6}|01\rangle\langle 01| + \frac{i}{\sqrt{18}}|01\rangle\langle 10| + \frac{i}{6}|01\rangle\langle 11|$$

$$-\frac{i}{3}|10\rangle\langle 00| - \frac{i}{\sqrt{18}}|10\rangle\langle 01| + \frac{1}{3}|10\rangle\langle 10| + \frac{1}{\sqrt{18}}|10\rangle\langle 11| - \frac{i}{\sqrt{18}}|11\rangle\langle 00|$$
$$-\frac{i}{6}|11\rangle\langle 01| + \frac{1}{\sqrt{18}}|11\rangle\langle 10| + \frac{1}{6}|11\rangle\langle 11|$$

The matrix representation is

$$\rho = \begin{pmatrix} \frac{1}{3} & \frac{1}{\sqrt{18}} & \frac{i}{3} & \frac{i}{\sqrt{18}} \\ \frac{1}{\sqrt{18}} & \frac{1}{6} & \frac{i}{\sqrt{18}} & \frac{i}{6} \\ \frac{-i}{3} & \frac{-i}{\sqrt{18}} & \frac{1}{3} & \frac{1}{\sqrt{18}} \\ \frac{-i}{\sqrt{18}} & \frac{-i}{6} & \frac{1}{\sqrt{18}} & \frac{1}{6} \end{pmatrix}$$

Notice that the trace is unity

$$Tr(\rho) = \frac{1}{3} + \frac{1}{6} + \frac{1}{3} + \frac{1}{6} = 1$$

Squaring, we have

$$\rho^2 = \begin{pmatrix} \frac{1}{3} & \frac{1}{3\sqrt{2}} & \frac{i}{3} & \frac{i}{3\sqrt{2}} \\ \frac{1}{3\sqrt{2}} & \frac{1}{6} & \frac{i}{3\sqrt{2}} & \frac{i}{6} \\ \frac{-i}{3} & \frac{-i}{3\sqrt{2}} & \frac{1}{3} & \frac{1}{3\sqrt{2}} \\ \frac{-i}{3\sqrt{2}} & \frac{-i}{6} & \frac{1}{3\sqrt{2}} & \frac{1}{6} \end{pmatrix}$$

Of course, this matrix is just the original density matrix, so

$$Tr(\rho^2) = \frac{1}{3} + \frac{1}{6} + \frac{1}{3} + \frac{1}{6} = 1$$

We see that the product state is a pure state.

THE DENSITY OPERATOR AND THE BLOCH VECTOR

In the second chapter we encountered the Bloch sphere, which provided a graphical representation of two-level quantum states. We will now see how to relate the density operator of a given system to the Bloch sphere.

The density operator of a system in a two-dimensional Hilbert space can be decomposed in the following way. First we define

$$\vec{\sigma} = \sigma_x\hat{x} + \sigma_y\hat{y} + \sigma_z\hat{z}$$

Then a density operator can be decomposed as

$$\rho = \frac{1}{2}(I + \vec{S} \cdot \vec{\sigma})$$

$\vec{S}$ is called the *Bloch vector*. The magnitude of the Bloch vector satisfies $|\vec{S}| \leq 1$, with equality for pure states. Otherwise, if $|\vec{S}| < 1$, then the state is a mixed state. The components of the Bloch vector are calculated by considering the expectation values of the operators X, Y, and Z. That is,

$$\vec{S} = S_x\hat{x} + S_y\hat{y} + S_z\hat{z} = \langle X \rangle \hat{x} + \langle Y \rangle \hat{y} + \langle Z \rangle \hat{z}$$

Using what we have learned about density operators, the components of the Bloch vector are given by

$$S_x = Tr(\rho X), \quad S_y = Tr(\rho Y), \quad S_z = Tr(\rho Z)$$

The Bloch vector provides us with another way to determine whether a state is pure or mixed, as the next example shows.

Example 5.12

Consider the following matrix:

$$\rho = \begin{pmatrix} \frac{5}{8} & \frac{i}{4} \\ \frac{-i}{4} & \frac{3}{8} \end{pmatrix}$$

(a) Is this a valid density operator?

(b) Does this represent a pure state or a mixed state?

Solution

(a) First it is easy to verify that the matrix is Hermitian. The transpose is

$$\rho^T = \begin{pmatrix} \frac{5}{8} & \frac{-i}{4} \\ \frac{i}{4} & \frac{3}{8} \end{pmatrix}$$

Taking the complex conjugate, we observe that $\rho = \rho^{dag}$, so the matrix is Hermitian. Next we see that

$$Tr(\rho) = \frac{5}{8} + \frac{3}{8} = 1$$

as required for density operators. Finally, we check the eigenvalues. A simple calculation shows that the eigenvalues of the matrix are

$$\lambda_{1,2} = \frac{4 \pm \sqrt{5}}{8}$$

Both eigenvalues satisfy $0 < \lambda_{1,2} < 1$, so the matrix represents a positive operator. We conclude that this is a valid density matrix.

(b) We compute the components of the Bloch vector:

$$S_x = Tr(X\rho) = Tr\left[\begin{pmatrix} 0 & 1 \\ 1 & 0 \end{pmatrix}\begin{pmatrix} \frac{5}{8} & \frac{i}{4} \\ \frac{-i}{4} & \frac{3}{8} \end{pmatrix}\right] = Tr\begin{pmatrix} \frac{-i}{4} & \frac{3}{8} \\ \frac{5}{8} & \frac{i}{4} \end{pmatrix} = 0$$

$$S_y = Tr(Y\rho) = Tr\left[\begin{pmatrix} 0 & -i \\ i & 0 \end{pmatrix}\begin{pmatrix} \frac{5}{8} & \frac{i}{4} \\ \frac{-i}{4} & \frac{3}{8} \end{pmatrix}\right] = Tr\begin{pmatrix} \frac{-1}{4} & \frac{-i3}{8} \\ \frac{i5}{8} & \frac{-1}{4} \end{pmatrix} = \frac{-1}{2}$$

$$S_z = Tr(Z\rho) = Tr\left[\begin{pmatrix} 1 & 0 \\ 0 & -1 \end{pmatrix}\begin{pmatrix} \frac{5}{8} & \frac{i}{4} \\ \frac{-i}{4} & \frac{3}{8} \end{pmatrix}\right] = Tr\begin{pmatrix} \frac{5}{8} & \frac{i}{4} \\ \frac{i}{4} & \frac{-3}{8} \end{pmatrix} = \frac{1}{4}$$

The magnitude of the Bloch vector is

$$\begin{aligned} |\vec{S}| &= \sqrt{S_x^2 + S_y^2 + S_z^2} \\ &= \sqrt{\left(\frac{-1+}{2}\right)^2 + \left(\frac{1}{4}\right)^2} \\ &= \sqrt{\frac{1}{4} + \frac{1}{16}} \\ &= \sqrt{\frac{5}{16}} \\ &= \frac{\sqrt{5}}{4} \approx 0.56 < 1 \end{aligned}$$

Since $|\vec{S}| < 1$, we conclude that this density matrix represents a mixed state.

EXERCISES

5.1. *Consider the following state vector:*

$$|\psi\rangle = \sqrt{\frac{5}{6}}|0\rangle + \frac{1}{\sqrt{6}}|1\rangle$$

(A) Is the state normalized?
(B) What is the probability that the system is found to be in state $|0\rangle$ if Z is measured?
(C) Write down the density operator.
(D) Find the density matrix in the $\{|0\rangle, |1\rangle\}$ basis, and show that $Tr(\rho) = 1$.

5.2. *Consider the state*

$$|\psi\rangle = \begin{pmatrix} \cos\theta \\ i\sin\theta \end{pmatrix}$$

Is this state normalized? Is $\rho = |\psi\rangle\langle\psi|$ a density operator?

5.3. *Let*

$$|\psi\rangle = \sqrt{\frac{3}{7}}|0\rangle + \frac{2}{\sqrt{7}}|1\rangle$$

(A) Write down the density matrix in the $\{|0\rangle, |1\rangle\}$ basis.
(B) Determine whether or not this is a pure state.
(C) Write down the density matrix in the $\{|+\rangle, |-\rangle\}$ basis, show that $Tr(\rho) = 1$ still holds, and determine if you still obtain the same result as in part (b).

5.4. *Suppose that a system is in the state*

$$|\psi\rangle = \sqrt{\frac{2}{3}}|0\rangle + \frac{1}{\sqrt{3}}|1\rangle$$

(A) Compute $Tr(\rho)$and $Tr(\rho^2)$. Is this a mixed state?
(B) Find $\langle X\rangle$ for this state.

5.5. *Suppose that*

$$\rho = \begin{pmatrix} \frac{1}{3} & \frac{i}{4} \\ \frac{-i}{4} & \frac{2}{3} \end{pmatrix}$$

(A) Is this a valid density matrix? If not, why not?
(B) If this is a valid density matrix, does it represent a pure state or a mixed state?

5.6. *For the density matrix given by*

$$\rho = \frac{1}{5}\begin{pmatrix} 3 & 1-i \\ 1+i & 2 \end{pmatrix}$$

(A) Is this a mixed state?
(B) Find $\langle X\rangle$, $\langle Y\rangle$, and $\langle Z\rangle$ for this state.

5.7. *Consider an ensemble in which 25% of the systems are known to be prepared in the state*

$$|\psi\rangle = \frac{2}{\sqrt{5}}|0\rangle + \frac{1}{\sqrt{5}}|1\rangle$$

and 75% of the systems are prepared in the state

$$|\phi\rangle = \frac{1}{\sqrt{2}}|0\rangle + \frac{1}{\sqrt{2}}|1\rangle$$

(A) *Find the density operators for each of these states, and show they are pure states. If measurements are made on systems in each of these states, what are the probabilities they are found to be in states $|0\rangle$ and state $|1\rangle$, respectively?*
(B) *Determine the density operator for the ensemble.*
(C) *Show that $Tr(\rho) = 1$.*
(D) *A measurement of Z is made on a member drawn from the ensemble. What are the probabilities it is found to be in state $|0\rangle$ and state $|1\rangle$, respectively?*

5.8. *Suppose that we have an ensemble with 60% of the states prepared as*

$$|a\rangle = \sqrt{\tfrac{2}{5}}|+\rangle - \sqrt{\tfrac{3}{5}}|-\rangle$$

and 40% of the states are prepared as

$$|b\rangle = \sqrt{\frac{5}{8}}|+\rangle + \sqrt{\frac{3}{8}}|-\rangle$$

A member is drawn from the ensemble. What is the probability that measurement finds it in the $|0\rangle$ state?

5.9. *Suppose that Alice and Bob share the entangled state*

$$|\psi\rangle = \frac{|00\rangle + |11\rangle}{\sqrt{2}}$$

(A) *Write down the density operator for this state.*
(B) *Compute the density matrix. Verify that $Tr(\rho) = 1$, and determine if this is a pure state.*
(C) *Find the density matrix that represents the reduced density operator as seen by Alice.*
(D) *Show that the reduced density operator as seen by Alice is a completely mixed state.*

5.10. *Consider the following matrix:*

$$\rho = \begin{pmatrix} \frac{2}{5} & \frac{-i}{8} \\ \frac{i}{8} & \frac{3}{5} \end{pmatrix}$$

(A) *Show that this matrix is Hermitian.*
(B) *Verify that the eigenvalues are $\lambda_{1,2} = 20 \pm \sqrt{41}/40$.*
(C) *Does this matrix represent a valid density matrix?*
(D) *Show that the probability of finding the system in the $|0\rangle$ state is 0.66.*
(E) *Compute the components of the Bloch vector, and show that this is a mixed state.*

6

QUANTUM MEASUREMENT THEORY

When it comes to the measurement of physical observables, quantum mechanics can tell us what measurement results are possible and what the probability is of obtaining each measurement result. It is also important to focus on what the state of the system is after a measurement is made. While measurement generally has no effect on a system in classical mechanics (i.e., macroscopic systems in general), measurement has a profound impact on a quantum mechanical system—altering its state in an irreversible way. Much of the material in this chapter is a review of concepts already introduced. Nevertheless, if this is your first exposure to quantum theory, it is important that you master these topics. Measurement plays a fundamental role in quantum computation because, at some point, we have to be able to get information out of the computational system. In this chapter we will learn the basics about different measurement models used in quantum theory.

DISTINGUISHING QUANTUM STATES AND MEASUREMENT

Measurement plays a central role in quantum mechanics. An act of measurement disturbs a quantum system in a fundamental way. Consider once again a general qubit

$$|\psi\rangle = \alpha|0\rangle + \beta|1\rangle \tag{6.1}$$

When a measurement is made, the qubit will be forced into the state $|\psi\rangle \mapsto |0\rangle$ or $|\psi\rangle \mapsto |1\rangle$. After measurement the original state (6.1) is lost. It isn't possible to make a measurement and determine what α and β are.

The measurement of a quantum system involves some type of interaction or coupling of that system with a measuring device. That device can be thought of as part of the larger *environment* which the quantum system is a part of. Frequently the measuring apparatus or larger environment is known as the *ancilla.* A system coupled to an environment is known as an *open system.*

In Chapters 3 and 5 we discussed the time evolution of a quantum system. The systems considered in that case were *closed* quantum systems—that is, systems that were isolated from the larger environment. The time evolution of a closed quantum system can be is governed by the Schrödinger equation (3.90), and we say that closed quantum systems evolve over time via unitary evolution, (3.91) and (3.92).

Let's summarize the axioms of quantum mechanics briefly here, focusing on what we need to begin a discussion of measurement. First we know that a quantum system is described by a vector (the "state vector") in a Hilbert space. The state of the system at time t is denoted by $|\psi(t)\rangle$.

The dynamical behavior of a quantum system is determined by the Hamiltonian operator H, which describes the total energy of the system. The time evolution of an isolated or closed system is described by the Schrödinger equation

$$i\hbar\frac{\partial}{\partial t}|\psi\rangle = H\psi \tag{6.2}$$

The actual form of the Hamiltonian operator depends on the specific nature of the system being studied. However, the general solution of (6.2) gives us the state of the system at time t. If we let the state of the system at the initial time $t = 0$ be $|\psi(0)\rangle$, then the solution of (6.2) tells us how quantum states evolve with time:

$$|\psi(t)\rangle = e^{-iHt/\hbar}|\psi(0)\rangle \tag{6.3}$$

Looking at (6.3) and recalling that the Hamiltonian is Hermitian, we see that $e^{-iHt/\hbar}$ is a unitary operator. This operator is called *unitary evolution operator*:

$$U = e^{-iHt/\hbar} \tag{6.4}$$

In the last chapter we described the evolution of a quantum system in terms of density operators. If the system is initially described by some density operator ρ_0, then the state of the system at time t will be

$$\rho_t = U\rho_0 U^\dagger \tag{6.5}$$

The *dynamics* of a quantum system is *trace-preserving*. This means that if the system is initially described by some density operator ρ_0 with $Tr(\rho_0) = 1$, then after the system has evolved to a final state described by ρ_t, then $Tr(\rho_t) = 1$.

While time evolution is trace-preserving, *measurement* is described by *trace-decreasing* quantum operations. A quantum operation involving measurement, described by a measurement operator that we will denote by M_m, transforms a density operator ρ according to $\rho' = M_m \rho M_m{}^\dagger$. In this case, $Tr(\rho') \leq 1$.

We begin our detailed discussion of measurement by considering projective or Von Neumann measurements.

PROJECTIVE MEASUREMENTS

The first measurement model we will explore involves *projective measurements.* We begin with this type of measurement because it's the easiest to understand, it's how introductory quantum mechanics is usually taught, and historically it's the oldest type of measurement model. Projective measurements are also known as *Von Neumann* measurements, after the mathematician who first described this type of measurement. We have already introduced some basic notions of projective measurements in previous chapters, so some of this material will be review. It plays a fundamental role in quantum information theory, however, and a review won't hurt.

The idea of making a projective measurement is based on the following notion: Given a set of mutually exclusive possible states, what state is the system is in? For example, an atom could have two mutually exclusive states—a lower energy "ground state" denoted by $|g\rangle$ and an "excited state" denoted by $|e\rangle$. We can use a projective measurement to determine if the atom is in the state $|g\rangle$ or in the state $|e\rangle$. As another example, we may be interested in the position of a particle. Is it located at position x_1 or at position x_2? For a qubit, we could ask: Is the qubit $|0\rangle$, or is it $|1\rangle$?

Such mutually exclusive possibilities are described by projection operators in quantum measurement theory. A projection operator P is Hermitian

$$P = P^\dagger \tag{6.6}$$

and equal to its own square

$$P^2 = P \tag{6.7}$$

We say that two projection operators P_1 and P_2 are *orthogonal* if their product is zero. That is, for every state $|\psi\rangle$, P_1 and P_2 are orthogonal if

$$P_1 P_2 |\psi\rangle = 0 \tag{6.8}$$

A set of mutually exclusive measurement results corresponds to a set of orthogonal projection operators that act on the state space of the system. A complete set of orthogonal projection operators is one for which

$$\sum_i P_i = I \tag{6.9}$$

Every complete set of orthogonal projectors specifies a measurement that can be realized. If a set of orthogonal projection operators is complete, (6.9) implies that at least one of the possible measurement results must be true. This is also another expression of the fact that probabilities must sum to one. The number of projection operators is determined by the dimension of the Hilbert space that describes the system. If the dimension of the Hilbert space is d and there are m projection operators, it must be true that

$$m \leq d \tag{6.10}$$

For example, if we are talking about a qubit where $|\psi\rangle = \alpha|0\rangle + \beta|1\rangle$, the dimension of the space is 2, and the projection operators corresponding to the mutually exclusive measurement results $|0\rangle$ and $|1\rangle$ are

$$P_0 = |0\rangle\langle 0|, \quad P_1 = |1\rangle\langle 1| \tag{6.11}$$

The projection operators corresponding to the ground and excited states of an atom would be written as

$$P_g = |g\rangle\langle g|, \quad P_e = |e\rangle\langle e| \tag{6.12}$$

If two projection operators commute, then their product $P_1 P_2$ is also a projection operator. However, the sum of two or more projection operators is, in general, not a projector. The necessary and sufficient condition needed for the sum of a set of projection operators to be a projection operator is that they be mutually orthogonal. If we have a set of projection operators $\{P_1, P_2, P_3, \ldots\}$, we can indicate that they are mutually orthogonal by writing

$$P_i P_j = \delta_{ij} P_i \tag{6.13}$$

Now let the dimension of the system be n, and consider a set of mutually orthogonal projection operators $\{P_1, P_2, P_3, \ldots, P_n\}$. Let the system be prepared in a state $|\psi\rangle$. The probability of finding the ith outcome when a measurement is made is

$$\Pr(i) = |P_i|\psi\rangle|^2 = (P_i|\psi\rangle)^\dagger (P_i|\psi\rangle) = \langle\psi|P_i^2|\psi\rangle = \langle\psi|P_i|\psi\rangle \tag{6.14}$$

Next consider some observable set of projection operators, which we denote by A, and let the eigenvectors of A be denoted by $|u_i\rangle$ each with eigenvalue a_i. The spectral decomposition of A allows us to write the operator as

$$A = \sum_{i=1}^{n} a_i |u_i\rangle\langle u_i| = \sum_{i=1}^{n} a_i P_i \tag{6.15}$$

where the projection operator corresponding to measurement outcome a_i is given by $P_i = |u_i\rangle\langle u_i|$. We can expand the state of the system $|\psi\rangle$ in terms of the eigenvectors of A as

$$|\psi\rangle = \sum_{i=1}^{n} (\langle u_i|\psi\rangle)|u_i\rangle = \sum_{i=1}^{n} c_i|u_i\rangle \tag{6.16}$$

Here $c_i = \langle u_i|\psi\rangle$ is the probability amplitude for obtaining measurement result a_i when the system is in the state $|\psi\rangle$. The actual probability for the given measurement result is found by computing the modulus squared of this quantity, that is,

$$\Pr(i) = |\langle u_i|\psi\rangle|^2 \tag{6.17}$$

(assuming the state is normalized, if not then we need to divide by $\langle\psi|\psi\rangle$). Result (6.17) is often called the *Born rule*. If the eigenvalue is degenerate, then the probability is found by summing over all eigenvectors that correspond to that eigenvalue

$$\Pr(i) = \sum_j |\langle u_j|\psi\rangle|^2 \tag{6.18}$$

Looking at (6.14) and recalling that the trace turns outer products into inner products (i.e., $Tr(A|\psi\rangle\langle\phi|) = \langle\phi|A|\psi\rangle$), notice that the probability of obtaining measurement result a_i can be written as

$$\Pr(i) = \langle\psi|P_i|\psi\rangle = Tr(P_i|\psi\rangle\langle\psi|) \tag{6.19}$$

Now consider the state of the system after measurement. When discussing projective measurements, one frequently hears about the mysterious *collapse of the wave function*. What this means is that while the state of the system prior to measurement could be a superposition of basis states as written in (6.16), after measurement the system collapses to the basis state that corresponds to the measurement result that was obtained. Formally, we write the state of the system after measurement $|\psi'\rangle$ as

$$|\psi'\rangle = \frac{P_i|\psi\rangle}{\sqrt{\langle\psi|P_i|\psi\rangle}} \tag{6.20}$$

The presence of the factor $\langle\psi|P_i|\psi\rangle$ in the denominator is to ensure that $|\psi'\rangle$ is normalized. The expectation value or average of an observable A with respect to a state $|\psi\rangle$ is given by

$$\langle A\rangle = \sum_i a_i\langle\psi|P_i|\psi\rangle \tag{6.21}$$

Example 6.1

A system is in the state

$$|\psi\rangle = \frac{2}{\sqrt{19}}|u_1\rangle + \frac{2}{\sqrt{19}}|u_2\rangle + \frac{1}{\sqrt{19}}|u_3\rangle + \frac{2}{\sqrt{19}}|u_4\rangle + \sqrt{\frac{6}{19}}|u_5\rangle$$

where $\{|u_1\rangle, |u_2\rangle, |u_3\rangle, |u_4\rangle, |u_5\rangle\}$ are a complete and orthonormal set of vectors. Each $|u_i\rangle$ is an eigenstate of the system's Hamiltonian corresponding to the possible measurement result $H|u_n\rangle = n\varepsilon|u_n\rangle$, where $n = 1, 2, 3, 4, 5$.

(a) Describe the set of projection operators corresponding to the possible measurement results.
(b) Determine the probability of obtaining each measurement result. What is the state of the system after measurement if we measure the energy to be 3ε?
(c) What is the average energy of the system?

Solution

(a) The possible measurement results are $\varepsilon, 2\varepsilon, 3\varepsilon, 4\varepsilon$, and 5ε. These measurement results correspond to the basis states $|u_1\rangle, |u_2\rangle, |u_3\rangle, |u_4\rangle$, and $|u_5\rangle$, respectively. Hence the projection operators corresponding to each measurement result are

$$P_1 = |u_1\rangle\langle u_1|$$

$$P_2 = |u_2\rangle\langle u_2|$$

$$P_3 = |u_3\rangle\langle u_3|$$

$$P_4 = |u_4\rangle\langle u_4|$$

$$P_5 = |u_5\rangle\langle u_5|$$

Since the $|u_i\rangle$'s are a set of orthnormal basis vectors, the completeness relation is satisfied and

$$\sum_i P_i = I$$

(b) We can calculate the probability of obtaining each measurement result using (6.14) or (6.17). Let's apply (6.17) to calculate the probability of finding ε or 2ε. First we need to check and see if the state is normalized. This is done by calculating

$$\sum_{i=1}^{5} |c_i|^2$$

and seeing if the result is 1. We have

$$\begin{aligned}\sum_{i=1}^{5} |c_i|^2 &= \left|\frac{2}{\sqrt{19}}\right|^2 + \left|\frac{2}{\sqrt{19}}\right|^2 + \left|\frac{1}{\sqrt{19}}\right|^2 + \left|\frac{2}{\sqrt{19}}\right|^2 + \left|\sqrt{\frac{6}{19}}\right|^2 \\ &= \frac{4}{19} + \frac{4}{19} + \frac{1}{19} + \frac{4}{19} + \frac{6}{19} \\ &= \frac{19}{19} = 1\end{aligned}$$

The state is normalized, so we can proceed. Before doing so, recall that the fact that the basis states are orthonormal means that

$$\langle u_i | u_j \rangle = \delta_{ij}$$

So, in the first case, applying the Born rule we have

$$\begin{aligned}
\Pr(\varepsilon) = |\langle u_2|\psi\rangle|^2 &= \left| \langle u_1| \left(\frac{2}{\sqrt{19}}|u_1\rangle + \frac{2}{\sqrt{19}}|u_2\rangle + \frac{1}{\sqrt{19}}|u_3\rangle + \frac{2}{\sqrt{19}}|u_4\rangle + \sqrt{\frac{6}{19}}|u_5\rangle \right) \right|^2 \\
&= \left| \frac{2}{\sqrt{19}}\langle u_1|u_1\rangle + \frac{2}{\sqrt{19}}\langle u_1|u_2\rangle + \frac{1}{\sqrt{19}}\langle u_1|u_3\rangle + \frac{2}{\sqrt{19}}\langle u_1|u_4\rangle + \sqrt{\frac{6}{19}}\langle u_1|u_5\rangle \right|^2 \\
&= \left| \frac{2}{\sqrt{19}}(1) + \frac{2}{\sqrt{19}}(0) + \frac{1}{\sqrt{19}}(0) + \frac{2}{\sqrt{19}}(0) + \sqrt{\frac{6}{19}}(0) \right|^2 \\
&= \left| \frac{2}{\sqrt{19}} \right|^2 \\
&= \frac{4}{19}
\end{aligned}$$

The probability of obtaining the second measurement result is

$$\begin{aligned}
\Pr(\varepsilon) = |\langle u_2|\psi\rangle|^2 &= \left| \langle u_2| \left(\frac{2}{\sqrt{19}}|u_1\rangle + \frac{2}{\sqrt{19}}|u_2\rangle + \frac{1}{\sqrt{19}}|u_3\rangle + \frac{2}{\sqrt{19}}|u_4\rangle + \sqrt{\frac{6}{19}}|u_5\rangle \right) \right|^2 \\
&= \left| \frac{2}{\sqrt{19}}\langle u_2|u_2\rangle \right|^2 \\
&= \left| \frac{2}{\sqrt{19}} \right|^2 \\
&= \frac{4}{19}
\end{aligned}$$

To calculate the remaining probabilities, let's use the projection operators and apply (6.14). We find that

$$\begin{aligned}
P_3|\psi\rangle &= (|u_3\rangle\langle u_3|) \left(|\psi\rangle = \frac{2}{\sqrt{19}}|u_1\rangle + \frac{2}{\sqrt{19}}|u_2\rangle + \frac{1}{\sqrt{19}}|u_3\rangle + \frac{2}{\sqrt{19}}|u_4\rangle + \sqrt{\frac{6}{19}}|u_5\rangle \right) \\
&= |u_3\rangle \left(\frac{1}{\sqrt{19}}\langle u_3|u_3\rangle \right) = \frac{1}{\sqrt{19}}|u_3\rangle
\end{aligned}$$

Therefore

$$\Pr(3\varepsilon) = \langle \psi | P_3 | \psi \rangle$$

$$= \left(\frac{2}{\sqrt{19}} \langle u_1| + \frac{2}{\sqrt{19}} \langle u_2| + \frac{1}{\sqrt{19}} \langle u_3| + \frac{2}{\sqrt{19}} \langle u_4| + \sqrt{\frac{6}{19}} \langle u_5| \right) \left(\frac{1}{\sqrt{19}} |u_3\rangle \right)$$

$$= \left(\frac{2}{\sqrt{19}} \right) \left(\frac{1}{\sqrt{19}} \right) \langle u_1|u_3\rangle + \left(\frac{2}{\sqrt{19}} \right) \left(\frac{1}{\sqrt{19}} \right) \langle u_2|u_3\rangle + \frac{1}{19} \langle u_3|u_3\rangle$$

$$+ \left(\frac{2}{\sqrt{19}} \right) \left(\frac{1}{\sqrt{19}} \right) \langle u_4|u_3\rangle + \left(\sqrt{\frac{6}{19}} \right) \left(\frac{1}{\sqrt{19}} \right) \langle u_5|u_3\rangle$$

$$= \frac{1}{19}$$

Similarly we find that

$$P_4|\psi\rangle = (|u_4\rangle\langle u_4|)|\psi\rangle = \frac{2}{\sqrt{19}} |u_4\rangle$$

$$P_5|\psi\rangle = (|u_5\rangle\langle u_5|)|\psi\rangle = \sqrt{\frac{6}{19}} |u_4\rangle$$

So we write

$$\Pr(4\varepsilon) = \langle\psi|P_4|\psi\rangle = \frac{4}{19}$$

$$\Pr(5\varepsilon) = \langle\psi|P_5|\psi\rangle = \frac{6}{19}$$

If a measurement is made and we find the energy to be 3ε, we apply (6.20). The state of the system after measurement is

$$|\psi'\rangle = \frac{P_3|\psi\rangle}{\sqrt{\langle\psi|P_3|\psi\rangle}} = \frac{1/\sqrt{19}[|\mu_3\rangle}{\sqrt{1/19}} = |\mu_3\rangle$$

(c) The average energy of the system is found using (6.21). We find that

$$\langle H\rangle = \sum_{i=1}^{5} E_i \langle\psi|P_i|\psi\rangle = \varepsilon\langle\psi|P_1|\psi\rangle + 2\varepsilon\langle\psi|P_2|\psi\rangle + 3\varepsilon\langle\psi|P_3|\psi\rangle$$

$$+ 4\varepsilon\langle\psi|P_4|\psi\rangle + 5\varepsilon\langle\psi|P_5|\psi\rangle$$

$$= \varepsilon\frac{4}{19} + 2\varepsilon\frac{4}{19} + 3\varepsilon\frac{1}{19} + 4\varepsilon\frac{4}{19} + 5\varepsilon\frac{6}{19}$$

$$= \frac{61}{19}\varepsilon$$

Example 6.2

A qubit is in the state

$$|\psi\rangle = \frac{\sqrt{3}}{2}|0\rangle - \frac{1}{2}|1\rangle$$

A measurement with respect to Y is made. Given that the eigenvalues of the Y matrix are ± 1, determine the probability that the measurement result is $+1$ and the probability that the measurement result is -1.

Solution

First we verify that the state is normalized

$$\begin{aligned}\langle\psi|\psi\rangle &= \left(\frac{\sqrt{3}}{2}\langle 0| - \frac{1}{2}\langle 1|\right)\left(\frac{\sqrt{3}}{2}|0\rangle - \frac{1}{2}|1\rangle\right) \\ &= \frac{3}{4}\langle 0|0\rangle - \frac{\sqrt{3}}{4}\langle 1|0\rangle - \frac{\sqrt{3}}{4}\langle 0|1\rangle + \frac{1}{4}\langle 1|1\rangle \\ &= \frac{3}{4} + \frac{1}{4} = 1\end{aligned}$$

Since $\langle\psi|\psi\rangle = 1$ the state is normalized. Recall that $Y = \begin{pmatrix} 0 & i \\ -i & 0 \end{pmatrix}$. You need to show that the eigenvectors of the Y matrix are

$$|u_1\rangle = \frac{1}{\sqrt{2}}\begin{pmatrix} 1 \\ i \end{pmatrix}, \quad |u_2\rangle = \frac{1}{\sqrt{2}}\begin{pmatrix} 1 \\ -i \end{pmatrix}$$

corresponding to the eigenvalues ± 1, respectively. The dual vectors in each case, found by computing the transpose of each vector can taking the complex conjugate of each element, are

$$\langle u_1| = (|u_1\rangle)^\dagger = \frac{1}{\sqrt{2}}(1 \quad -i), \quad \langle u_2| = (|u_2\rangle)^\dagger = \frac{1}{\sqrt{2}}(1 \quad i)$$

The projection operators corresponding to each possible measurement result are

$$\begin{aligned}P_{+1} &= |u_1\rangle\langle u_1| = \frac{1}{2}\begin{pmatrix} 1 \\ i \end{pmatrix}(1 \quad -i) = \frac{1}{2}\begin{pmatrix} 1 & -i \\ i & 1 \end{pmatrix} \\ P_{-1} &= |u_2\rangle\langle u_2| = \frac{1}{2}\begin{pmatrix} 1 \\ -i \end{pmatrix}(1 \quad i) = \frac{1}{2}\begin{pmatrix} 1 & i \\ -i & 1 \end{pmatrix}\end{aligned}$$

Writing the state $|\psi\rangle$ as a column vector, we have

$$|\psi\rangle = \frac{\sqrt{3}}{2}|0\rangle - \frac{1}{2}|1\rangle = \frac{\sqrt{3}}{2}\begin{pmatrix} 1 \\ 0 \end{pmatrix} - \frac{1}{2}\begin{pmatrix} 0 \\ 1 \end{pmatrix} = \frac{1}{2}\begin{pmatrix} \sqrt{3} \\ -1 \end{pmatrix}$$

Hence

$$\begin{aligned}P_{+1}|\psi\rangle &= \frac{1}{2}\begin{pmatrix} 1 & -i \\ i & 1 \end{pmatrix}\frac{1}{2}\begin{pmatrix} \sqrt{3} \\ -1 \end{pmatrix} = \frac{1}{4}\begin{pmatrix} 1 & -i \\ i & 1 \end{pmatrix}\begin{pmatrix} \sqrt{3} \\ -1 \end{pmatrix} = \frac{1}{4}\begin{pmatrix} \sqrt{3} + i \\ -1 + i\sqrt{3} \end{pmatrix} \\ P_{-1}|\psi\rangle &= \frac{1}{2}\begin{pmatrix} 1 & i \\ -i & 1 \end{pmatrix}\frac{1}{2}\begin{pmatrix} \sqrt{3} \\ -1 \end{pmatrix} = \frac{1}{4}\begin{pmatrix} 1 & i \\ -i & 1 \end{pmatrix}\begin{pmatrix} \sqrt{3} \\ -1 \end{pmatrix} = \frac{1}{4}\begin{pmatrix} \sqrt{3} - i \\ -1 - i\sqrt{3} \end{pmatrix}\end{aligned}$$

Now, if a measurement is made of the Y observable, the probability of finding $+1$ is

$$\Pr(+1) = \langle\psi|P_{+1}|\psi\rangle = \frac{1}{2}\begin{pmatrix}\sqrt{3} & -1\end{pmatrix}\frac{1}{4}\begin{pmatrix}\sqrt{3}+i \\ -1+i\sqrt{3}\end{pmatrix}$$
$$= \frac{1}{8}(3+i\sqrt{3}+1-i\sqrt{3}) = \frac{1}{8}(3+1) = \frac{1}{2}$$

Similarly find

$$\Pr(-1) = \langle\psi|P_{-1}|\psi\rangle = \frac{1}{2}(\sqrt{3} \quad -1)\frac{1}{4}\begin{pmatrix}\sqrt{3}-i \\ -1-i\sqrt{3}\end{pmatrix}$$
$$= \frac{1}{8}(3-i\sqrt{3}+1+i\sqrt{3}) = \frac{1}{8}(3+1) = \frac{1}{2}$$

You Try It

Show that the eigenvectors of $Y = \begin{pmatrix}0 & i \\ -i & 0\end{pmatrix}$ are

$$|u_1\rangle = \frac{1}{\sqrt{2}}\begin{pmatrix}1 \\ i\end{pmatrix}, \quad |u_2\rangle = \frac{1}{\sqrt{2}}\begin{pmatrix}1 \\ -i\end{pmatrix}$$

Example 6.3

A system is in the state

$$|\psi\rangle = \frac{1}{\sqrt{6}}|0\rangle + \sqrt{\frac{5}{6}}|1\rangle$$

A measurement is made with respect to the observable X. What is the expectation or average value?

Solution

The eigenvectors of $X = \begin{pmatrix}0 & 1 \\ 1 & 0\end{pmatrix}$ are

$$|+_x\rangle = \frac{|0\rangle + |1\rangle}{\sqrt{2}}, \quad |-_x\rangle = \frac{|0\rangle - |1\rangle}{\sqrt{2}} \tag{6.22}$$

The projection operator corresponding to a measurement of $+1$ is

$$P_+ = |+_x\rangle\langle +_x| = \left(\frac{|0\rangle + |1\rangle}{\sqrt{2}}\right)\left(\frac{\langle 0| + \langle 1|}{\sqrt{2}}\right)$$
$$= \frac{1}{2}\left(|0\rangle\langle 0| + |0\rangle\langle 1| + |1\rangle\langle 0| + |1\rangle\langle 1|\right) \tag{6.23}$$

The projection operator corresponding to a measurement of -1 is

$$
\begin{aligned}
P_- = |-_x\rangle\langle -_x| &= \left(\frac{|0\rangle - |1\rangle}{\sqrt{2}}\right)\left(\frac{\langle 0| - \langle 1|}{\sqrt{2}}\right) \\
&= \frac{1}{2}(|0\rangle\langle 0| - |0\rangle\langle 1| - |1\rangle\langle 0| + |1\rangle\langle 1|)
\end{aligned}
\tag{6.24}
$$

The probability of finding each measurement result is

$$
\begin{aligned}
\Pr(+1) &= \langle\psi|P_+|\psi\rangle \\
&= \left(\frac{1}{\sqrt{6}}\langle 0| + \sqrt{\frac{5}{6}}\langle 1|\right)\left(\frac{1}{2}(|0\rangle\langle 0| + |0\rangle\langle 1| + |1\rangle\langle 0| + |1\rangle\langle 1|)\right)\left(\frac{1}{\sqrt{6}}|0\rangle + \sqrt{\frac{5}{6}}|1\rangle\right) \\
&= \left(\frac{1}{\sqrt{6}}\langle 0| + \sqrt{\frac{5}{6}}\langle 1|\right)\left(\frac{1+\sqrt{5}}{2\sqrt{6}}|0\rangle + \frac{1+\sqrt{5}}{2\sqrt{6}}|1\rangle\right) \\
&= \frac{6+2\sqrt{5}}{12} \\
\Pr(-1) &= \langle\psi|P_-|\psi\rangle \\
&= \left(\frac{1}{\sqrt{6}}\langle 0| + \sqrt{\frac{5}{6}}\langle 1|\right)\left(\frac{1}{2}(|0\rangle\langle 0| - |0\rangle\langle 1| - |1\rangle\langle 0| + |1\rangle\langle 1|)\right)\left(\frac{1}{\sqrt{6}}|0\rangle + \sqrt{\frac{5}{6}}|1\rangle\right) \\
&= \left(\frac{1}{\sqrt{6}}\langle 0| + \sqrt{\frac{5}{6}}\langle 1|\right)\left(\frac{1-\sqrt{5}}{2\sqrt{6}}|0\rangle + \frac{-1+\sqrt{5}}{2\sqrt{6}}|1\rangle\right) \\
&= \frac{6-2\sqrt{5}}{12}
\end{aligned}
$$

Notice that the probabilities sum to 1:

$$
\langle\psi|P_+|\psi\rangle + \langle\psi|P_-|\psi\rangle = \frac{6+2\sqrt{5}}{12} + \frac{6-2\sqrt{5}}{12} = 1
$$

The average value is

$$
\begin{aligned}
\langle X\rangle &= (+1)\Pr(+1) + (-1)\Pr(-1) \\
&= \frac{6+2\sqrt{5}}{12} - \left(\frac{6-2\sqrt{5}}{12}\right) = \frac{\sqrt{5}}{3} \approx 0.75
\end{aligned}
$$

For a system with a single qubit, we can write the orthogonal projection operators as

$$
P_\pm = \frac{I \pm \vec{n}\cdot\vec{\sigma}}{2}
\tag{6.25}
$$

Here $\vec{n}$ is an axis on the Bloch sphere. Recall that the outer product representation of $X = \sigma_x = |1\rangle\langle 0| + |0\rangle\langle 1|$ and suppose that we choose $\vec{n}$ to be a unit vector in the x direction. Then

$$
P_\pm = \frac{I \pm \hat{x}\cdot\vec{\sigma}}{2} = \frac{1}{2}(|0\rangle\langle 0| + |1\rangle\langle 1| \pm |0\rangle\langle 1| \pm |1\rangle\langle 0|)
$$

Comparison with (6.23) and (6.24) shows that these are the correct projection operators onto the eignvectors of X, namely $|\pm\rangle$.

MEASUREMENTS ON COMPOSITE SYSTEMS

Composite systems were introduced in Chapter 4. Many applications in quantum computation involve composite systems. In this section we will go over some basic measurement operations on composite systems. The utility of some of the relations stated in the previous section will become clear when dealing with the example of composite systems provided below.

Example 6.4

Describe the action of the operators $P_0 \otimes I$ and $I \otimes P_1$ on the state

$$|\psi\rangle = \frac{|01\rangle - |10\rangle}{\sqrt{2}}.$$

Solution

The first operator, $P_0 \otimes I$, tells us to apply the projection operator, $P_0 = |0\rangle\langle 0|$, to the *first qubit* and to leave the second qubit alone. The result is

$$P_0 \otimes I|\psi\rangle = \frac{1}{\sqrt{2}}[(|0\rangle\langle 0|0\rangle) \otimes |1\rangle - (|0\rangle\langle 0|1\rangle \otimes |0\rangle)] = \frac{|01\rangle}{\sqrt{2}}$$

Interestingly, applying a projective measurement to the first qubit causes the second qubit to assume a definite state. As we will see in the next chapter, this is a property of entangled systems. Apparently it doesn't matter if the qubits are spatially separated for a collapse of the system to occur.

To find the properly normalized state of the system after measurement we use (6.20). We have

$$\langle\psi|P_0 \otimes I|\psi\rangle = \left(\frac{\langle 01| - \langle 10|}{\sqrt{2}}\right)\frac{|01\rangle}{\sqrt{2}} = \frac{\langle 0|0\rangle\langle 1|1\rangle - \langle 1|0\rangle\langle 0|1\rangle}{2} = \frac{1}{2}$$

The state after measurement is

$$|\psi'\rangle = \frac{P_0 \otimes I|\psi\rangle}{\sqrt{\langle\psi|P_0 \otimes I|\psi\rangle}} = \frac{|01\rangle/\sqrt{2}}{(1/\sqrt{2})} = |01\rangle$$

As can be seen, while applying (6.20) in the single qubit case can seem like overkill, in this case this allows us to quickly write down the properly normalized state after measurement.

The second operator, $I \otimes P_1$, tells us to leave the *first qubit alone* and to apply the projection operator $P_1 = |1\rangle\langle 1|$ to the second qubit. This gives

$$I \otimes P_1|\psi\rangle = \frac{1}{\sqrt{2}}[|0\rangle \otimes (|1\rangle\langle 1|1\rangle) - |1\rangle \otimes (|1\rangle\langle 1|0\rangle)] = \frac{|01\rangle}{\sqrt{2}}$$

We have therefore the same state, but this time doing the projective measurement represented by $P_1 = |1\rangle\langle 1|$ the second qubit has forced the first qubit into the state $|0\rangle$. Let's redo the calculation using matrices. The operator is

$$I \otimes P_1 = \begin{pmatrix} 1 \cdot P_1 & 0 \cdot P_1 \\ 0 \cdot P_1 & 1 \cdot P_1 \end{pmatrix} = \begin{pmatrix} 0 & 0 & 0 & 0 \\ 0 & 1 & 0 & 0 \\ 0 & 0 & 0 & 0 \\ 0 & 0 & 0 & 1 \end{pmatrix}$$

Then we have

$$|01\rangle = |0\rangle \otimes |1\rangle = \begin{pmatrix} 1 \\ 0 \end{pmatrix} \otimes \begin{pmatrix} 0 \\ 1 \end{pmatrix} = \begin{pmatrix} 0 \\ 1 \\ 0 \\ 0 \end{pmatrix}$$

$$|10\rangle = |1\rangle \otimes |0\rangle = \begin{pmatrix} 0 \\ 1 \end{pmatrix} \otimes \begin{pmatrix} 1 \\ 0 \end{pmatrix} = \begin{pmatrix} 0 \\ 0 \\ 1 \\ 0 \end{pmatrix}$$

So the state of the system prior to measurement is

$$|\psi\rangle = \frac{|01\rangle - |10\rangle}{\sqrt{2}} = \frac{1}{\sqrt{2}}\begin{pmatrix} 0 \\ 1 \\ -1 \\ 0 \end{pmatrix}$$

The action of the operator $I \otimes P_1$ is then computed as follows:

$$I \otimes P_1|\psi\rangle = \frac{1}{\sqrt{2}}\begin{pmatrix} 0 & 0 & 0 & 0 \\ 0 & 1 & 0 & 0 \\ 0 & 0 & 0 & 0 \\ 0 & 0 & 0 & 1 \end{pmatrix}\begin{pmatrix} 0 \\ 1 \\ -1 \\ 0 \end{pmatrix} = \frac{1}{\sqrt{2}}\begin{pmatrix} 0 \\ 1 \\ 0 \\ 0 \end{pmatrix} = \frac{1}{\sqrt{2}}|01\rangle$$

Example 6.5

A system is in the state

$$|\psi\rangle = \frac{1}{\sqrt{8}}|00\rangle + \sqrt{\frac{3}{8}}|01\rangle + \frac{1}{2}|10\rangle + \frac{1}{2}|11\rangle$$

(a) What is the probability that measurement finds the system in the state $|\phi\rangle = |01\rangle$?

(b) What is the probability that measurement finds the first qubit in the state $|0\rangle$? What is the state of the system after measurement?

Solution

(a) Given that the system is in the state $|\psi\rangle$, the probability of finding it in the state $|\phi\rangle = |01\rangle$ is calculated using the Born rule, which is $\Pr = |\langle\phi|\psi\rangle|^2$. Since $\langle 0|1\rangle = \langle 1|0\rangle = 0$, we have

$$
\begin{aligned}
\langle\phi|\psi\rangle &= \langle 01|\left(\frac{1}{\sqrt{8}}|00\rangle + \sqrt{\frac{3}{8}}|01\rangle + \frac{1}{2}|10\rangle + \frac{1}{2}|11\rangle\right) \\
&= \frac{1}{\sqrt{8}}\langle 0|0\rangle\langle 1|0\rangle + \sqrt{\frac{3}{8}}\langle 0|0\rangle\langle 1|1\rangle + \frac{1}{2}\langle 0|1\rangle\langle 1|0\rangle + \frac{1}{2}\langle 0|1\rangle\langle 1|1\rangle \\
&= \sqrt{\frac{3}{8}}
\end{aligned}
$$

Therefore the probability is

$$
\Pr = |\langle\phi|\psi\rangle|^2 = \frac{3}{8}
$$

(b) To find the probability that measurement finds the first qubit in the state $|0\rangle$, we can apply $P_0 \otimes I = |0\rangle\langle 0| \otimes I$ to the state. So the projection operator P_0 is applied to the first qubit and the identity operator to the second qubit, leaving the second qubit unchanged. This obtains

$$
\begin{aligned}
P_0 \otimes I|\psi\rangle &= (|0\rangle\langle 0| \otimes I)\left(\frac{1}{\sqrt{8}}|00\rangle + \sqrt{\frac{3}{8}}|01\rangle + \frac{1}{2}|10\rangle + \frac{1}{2}|11\rangle\right) \\
&= \frac{1}{\sqrt{8}}|0\rangle\langle 0|0\rangle \otimes |0\rangle + \sqrt{\frac{3}{8}}|0\rangle\langle 0|0\rangle \otimes |1\rangle + \frac{1}{2}|0\rangle\langle 0|1\rangle \otimes |0\rangle + \frac{1}{2}|0\rangle\langle 0|1\rangle \otimes |1\rangle \\
&= \frac{1}{\sqrt{8}}|00\rangle + \sqrt{\frac{3}{8}}|01\rangle
\end{aligned}
$$

The probability of obtaining this result is

$$
\begin{aligned}
\Pr &= \langle\psi|P_0 \otimes I|\psi\rangle \\
&= \left(\frac{1}{\sqrt{8}}\langle 00| + \sqrt{\frac{3}{8}}\langle 01| + \frac{1}{2}\langle 10| + \frac{1}{2}\langle 11|\right)\left(\frac{1}{\sqrt{8}}|00\rangle + \sqrt{\frac{3}{8}}|01\rangle\right) \\
&= \frac{1}{8} + \frac{3}{8} = \frac{1}{2}
\end{aligned}
$$

The state of the system after measurement using (6.20) is found to be

$$|\psi'\rangle = \frac{\frac{1}{\sqrt{8}}|00\rangle + \sqrt{\frac{3}{8}}|01\rangle}{\sqrt{\langle\psi|P_0 \otimes I|\psi\rangle}} = \sqrt{2}\left(\frac{1}{\sqrt{8}}|00\rangle + \sqrt{\frac{3}{8}}|01\rangle\right)$$

$$= \frac{1}{2}|00\rangle + \frac{\sqrt{3}}{2}|01\rangle$$

Example 6.6

A three-qubit system is in the state

$$|\psi\rangle = \left(\frac{\sqrt{2}+i}{\sqrt{20}}\right)|000\rangle + \frac{1}{\sqrt{2}}|001\rangle + \frac{1}{\sqrt{10}}|011\rangle + \frac{i}{2}|111\rangle$$

(a) Is the state normalized? What is the probability that the system is found in the state $|000\rangle$ if all 3 qubits are measured?

(b) What is the probability that a measurement on the first qubit only gives 0? What is the postmeasurement state of the system?

Solution

(a) To determine if the state is normalized, we compute the sum of the squares of the coefficients:

$$\sum_i |c_i|^2 = \left(\frac{\sqrt{2}+i}{\sqrt{20}}\right)\left(\frac{\sqrt{2}-i}{\sqrt{20}}\right) + \left(\frac{1}{\sqrt{2}}\right)\left(\frac{1}{\sqrt{2}}\right) + \left(\frac{1}{\sqrt{10}}\right)\left(\frac{1}{\sqrt{10}}\right) + \left(\frac{i}{2}\right)\left(-\frac{i}{2}\right)$$

$$= \frac{3}{20} + \frac{1}{2} + \frac{1}{10} + \frac{1}{4} = \frac{20}{20} = 1$$

So the state is normalized. The probability the system is found in state $|000\rangle$ if all three qubits are measured is

$$\Pr(000) = \left(\frac{\sqrt{2}+i}{\sqrt{20}}\right)\left(\frac{\sqrt{2}-i}{\sqrt{20}}\right) = \frac{3}{20} = 0.15$$

(b) The probability that a measurement on the first qubit is zero can be found by acting on the state with the operator $P_0 \otimes I \otimes I$ and computing $\langle\psi|P_0 \otimes I \otimes I|\psi\rangle$. This will project onto the $|0\rangle$ state for the first qubit while leaving the second and third qubits alone. We find that

$$P_0 \otimes I \otimes I|\psi\rangle = \left(\frac{\sqrt{2}+i}{\sqrt{20}}\right)(|0\rangle\langle 0| \otimes I \otimes I)|000\rangle + \frac{1}{\sqrt{2}}(|0\rangle\langle 0| \otimes I \otimes I)|001\rangle$$

$$+ \frac{1}{\sqrt{10}}(|0\rangle\langle 0| \otimes I \otimes I)|011\rangle + \frac{i}{2}(|0\rangle\langle 0| \otimes I \otimes I)|111\rangle$$

$$= \left(\frac{\sqrt{2}+i}{\sqrt{20}}\right)|000\rangle + \frac{1}{\sqrt{2}}|001\rangle + \frac{1}{\sqrt{10}}|011\rangle$$

The last term vanishes, since $\langle 0|1\rangle = 0$ So

$$\frac{i}{2}(|0\rangle\langle 0| \otimes I \otimes I)|111\rangle = \frac{i}{2}(|0\rangle\langle 0|1\rangle) \otimes |1\rangle \otimes |1\rangle = 0$$

Hence the probability that measurement on the first qubit finds 0 is

$$\langle\psi|P_0 \otimes I \otimes I|\psi\rangle = \left|\frac{\sqrt{2}+i}{\sqrt{20}}\right|^2 + \left|\frac{1}{\sqrt{2}}\right|^2 + \left|\frac{1}{\sqrt{10}}\right|^2 = \frac{3}{20} + \frac{1}{2} + \frac{1}{10} = \frac{3}{4}$$

The postmeasurement state is

$$|\psi'\rangle = \frac{P_0 \otimes I \otimes I|\psi\rangle}{\sqrt{\langle\psi|P_0 \otimes I \otimes I|\psi\rangle}} = \sqrt{\frac{4}{3}}\left(\left(\frac{\sqrt{2}+i}{\sqrt{20}}\right)|000\rangle + \frac{1}{\sqrt{2}}|001\rangle + \frac{1}{\sqrt{10}}|011\rangle\right)$$

$$= \left(\frac{\sqrt{2}+i}{\sqrt{15}}\right)|000\rangle + \sqrt{\frac{2}{3}}|001\rangle + \sqrt{\frac{2}{15}}|011\rangle$$

You Try It

Verify that the postmeasurement state in Example 6.6 is normalized.

Example 6.7

A system is in the GHZ state where

$$|\psi\rangle = \frac{1}{\sqrt{2}}(|000\rangle + |111\rangle)$$

Suppose that an observable $A = \sigma_x \otimes \sigma_y \otimes \sigma_z$. If $|\pm\rangle = |0\rangle \pm |1\rangle/\sqrt{2}$, what is the probability that measurement finds the system in the state $|{+}{+}{+}\rangle$ and in the state $|{-}{-}{-}\rangle$ when the system is in the state $A|\psi\rangle$? What is the expectation value of A?

Solution

Recall the action of the Pauli operators:

$$\sigma_x|0\rangle = |1\rangle, \qquad \sigma_x|1\rangle = |0\rangle$$
$$\sigma_y|0\rangle = -i|1\rangle, \quad \sigma_y|1\rangle = i|0\rangle$$
$$\sigma_z|0\rangle = |0\rangle, \qquad \sigma_z|1\rangle = -|1\rangle$$

Write

$$A|\psi\rangle = (\sigma_x \otimes \sigma_y \otimes \sigma_z)\left(\frac{1}{\sqrt{2}}(|000\rangle + |111\rangle)\right)$$
$$= -\frac{i}{\sqrt{2}}(|110\rangle + |001\rangle)$$

Then rewrite this in the $|\pm\rangle = |0\rangle \pm |1\rangle/\sqrt{2}$ basis. For the first term,

$$|110\rangle = \left(\frac{|+\rangle - |-\rangle}{\sqrt{2}}\right) \otimes \left(\frac{|+\rangle - |-\rangle}{\sqrt{2}}\right) \otimes \left(\frac{|+\rangle + |-\rangle}{\sqrt{2}}\right)$$
$$= \left(\frac{|+\rangle - |-\rangle}{\sqrt{2}}\right) \otimes \left(\frac{|++\rangle + |+-\rangle - |-+\rangle - |--\rangle}{2}\right)$$
$$= \frac{1}{2\sqrt{2}}(|+++\rangle + |++-\rangle - |+-+\rangle - |+--\rangle - |-++\rangle$$
$$- |-+-\rangle + |--+\rangle + |---\rangle)$$

Similarly for the second term,

$$|001\rangle = \left(\frac{|+\rangle + |-\rangle}{\sqrt{2}}\right) \otimes \left(\frac{|+\rangle + |-\rangle}{\sqrt{2}}\right) \otimes \left(\frac{|+\rangle - |-\rangle}{\sqrt{2}}\right)$$
$$= \frac{1}{2\sqrt{2}}(|+++\rangle - |++-\rangle + |+-+\rangle - |+--\rangle + |-++\rangle$$
$$- |-+-\rangle + |--+\rangle - |---\rangle)$$

Therefore in the $|\pm\rangle$ basis the state is

$$A|\psi\rangle = -\frac{i}{2}(|+++\rangle - |+--\rangle - |-+-\rangle + |--+\rangle)$$

The probability that the system is found in the $|---\rangle$ state is zero, while the probability it is found in the $|+++\rangle$ state is

$$|\langle + + + |A|\psi\rangle|^2 = \left|-\frac{i}{2}\right|^2 = \frac{1}{4}$$

In the $|\pm\rangle$ basis the initial state is

$$|\psi\rangle = \frac{1}{2\sqrt{2}}(|{+}{+}{+}\rangle + |{+}{-}{-}\rangle + |{-}{+}{-}\rangle + |{-}{-}{+}\rangle)$$

The expectation value is

$$\begin{aligned}\langle\psi|A|\psi\rangle &= \frac{1}{2\sqrt{2}}(\langle{+}{+}{+}| + \langle{+}{-}{-}| + \langle{-}{+}{-}| + \langle{-}{-}{+}|)\\ &\quad -\frac{i}{2}(|{+}{+}{+}\rangle - |{+}{-}{-}\rangle - |{-}{+}{-}\rangle + |{-}{-}{+}\rangle)\\ &= -\frac{i}{4\sqrt{2}}(1-1-1+1) = 0\end{aligned}$$

You Try It

Write the GHZ state in the $|\pm\rangle$ basis.

Example 6.8

A two qubit system is in the state

$$|\phi\rangle = \frac{\sqrt{3}}{2}|00\rangle + \frac{1}{2}|11\rangle$$

A Y gate is applied to the first qubit. After this is done, what are the possible measurement results if both qubits are measured, and what are the respective probabilities of each measurement result?

Solution

The action of the Y gate on the computational basis states is

$$Y|0\rangle = i|1\rangle, \quad Y|1\rangle = -i|0\rangle$$

Hence

$$Y \otimes I|\phi\rangle = \frac{\sqrt{3}}{2}(Y \otimes I)|00\rangle + \frac{1}{2}(Y \otimes I)|11\rangle = i\frac{\sqrt{3}}{2}|10\rangle - \frac{i}{2}|01\rangle$$

If both qubits are measured, the possible measurement results are 10 and 01. The probability of finding 10 is

$$\left|i\frac{\sqrt{3}}{2}\right|^2 = \left(i\frac{\sqrt{3}}{2}\right)\left(-i\frac{\sqrt{3}}{2}\right) = \frac{3}{4}$$

The probability of finding 01 is

$$\left|\frac{i}{2}\right|^2 = \left(\frac{i}{2}\right)\left(-\frac{i}{2}\right) = \frac{1}{4}$$

These probabilities sum to one, as they should.

GENERALIZED MEASUREMENTS

Measurements can be described in a more general way. Following common notation, we denote a measurement operator by M_m, where m is an index that denotes a possible measurement result. Given a state $|\psi\rangle$, the probability that we find measurement result m is

$$\Pr(m) = \langle\psi|M_m^\dagger M_m|\psi\rangle \tag{6.26}$$

We can see how projection operators fit into this formalism by comparing (6.14) and (6.26) and remembering that projection operators are Hermitian and equal to their own square.

After a measurement the state of the system is

$$|\psi'\rangle = \frac{M_m|\psi\rangle}{\sqrt{\langle\psi|M_m^\dagger M_m|\psi\rangle}} \tag{6.27}$$

As we saw with projection operators, measurement operators satisfy a completeness relation that follows from the fact that probabilities sum to one. For general measurement operators we write this as

$$\sum_m M_m^\dagger M_m = I \tag{6.28}$$

Now let's look at measurements when a system is described by a density operator. If a quantum system is described by a density operator ρ, the probability of finding measurement result m is

$$\Pr(m) = Tr(M_m^\dagger M_m \rho) \tag{6.29}$$

If the measurement in question is described by a set of orthogonal projection operators $P_i = |u_i\rangle\langle u_i|$ corresponding to measurement result i then the probability of finding that measurement result is

$$\Pr(i) = Tr(P_i^\dagger P_i \rho) = Tr(|u_i\rangle\langle u_i|\rho) = \langle u_i|\rho|u_i\rangle \tag{6.30}$$

The state of a system described by a density operator after obtaining measurement result m is given by

$$\rho' = \frac{M_m \rho M_m^\dagger}{Tr(M_m^\dagger M_m \rho)} \tag{6.31}$$

In the case where the measurements are described by a set of orthogonal projection operators and measurement result i is obtained, the state of the system after measurement is

$$\rho' = \frac{|u_i\rangle\langle u_i|\rho|u_i\rangle\langle u_i|}{\langle u_i|\rho|u_i\rangle} \tag{6.32}$$

Example 6.9

A quantum system has a density matrix given by

$$\rho = \frac{5}{6}|0\rangle\langle 0| + \frac{1}{6}|1\rangle\langle 1|$$

What is the probability that the system is in the state $|0\rangle$?

Solution

Of course, one can read off that the probability the system is in the state $|0\rangle$ is 5/6. But let's see how we can apply (6.30). It's easy enough:

$$\begin{aligned}
\Pr(0) &= Tr(|0\rangle\langle 0|\rho) = \langle 0|\rho|0\rangle \\
&= \langle 0|\left(\frac{5}{6}|0\rangle\langle 0| + \frac{1}{6}|1\rangle\langle 1|\right)|0\rangle \\
&= \frac{5}{6}\langle 0|0\rangle\langle 0|0\rangle + \frac{1}{6}\langle 0|1\rangle\langle 1|0\rangle = \frac{5}{6}
\end{aligned}$$

Example 6.10

A system has the density operator

$$\rho = \frac{1}{3}|u_1\rangle\langle u_1| - i\frac{\sqrt{2}}{3}|u_1\rangle\langle u_2| + i\frac{\sqrt{2}}{3}|u_2\rangle\langle u_1| + \frac{2}{3}|u_2\rangle\langle u_2|$$

where the $|u_k\rangle$ constitute an orthonormal basis. What is the probability that a measurement finds the system in the state $|u_2\rangle$?

Solution

The projection operator corresponding to this measurement result is

$$P_2 = |u_2\rangle\langle u_2|$$

The probability is

$$\Pr(|u_2\rangle) = Tr(|u_2\rangle\langle u_2|\rho) = \langle u_2|\rho|u_2\rangle = \frac{2}{3}$$

POSITIVE OPERATOR-VALUED MEASURES

A type of measurement that is more general than the projective measurements we have focused on so far is known as a *positive operator-valued measure* or *POVM*. A POVM consists of a set of positive operators commonly denoted by E_m. The probability of obtaining measurement result m in this case is given by

$$\Pr(m) = \langle\psi|E_m|\psi\rangle \tag{6.33}$$

When the system is a mixed state described by a density operator ρ, the probability of obtaining measurement result m is given by $Tr(E_m\rho)$. In addition to being positive operators, the set of E_m satisfy

$$\sum_m E_m = I \tag{6.34}$$

The measurement operators in a POVM can be constructed from an arbitrary measurement operator by taking

$$E_m = M_m^\dagger M_m \tag{6.35}$$

It is possible to form a POVM to describe a projective measurement, but the operators E_m *do not have to be* projection operators. The POVM allows us to construct a more general type of measurement operator to describe measurements where projective measurements do not apply in the real world. For example, if a system is in a state $|\psi\rangle = \sum_i c_i|u_i\rangle$, a projective measurement $|u_k\rangle\langle u_k|$ collapses the wavefunction to the state $|u_k\rangle$. We are free to repeat another measurement immediately on the system—and that measurement will find the system in the state $|u_k\rangle$ with certainty.

In the laboratory not all measurements are repeatable. The quintessential example is the detection of a photon—after it has been detected, the photon is destroyed. Hence repeated measurements on the system are not possible. A POVM is applicable in this case because it allows us to describe measurements on the system without regard to the postmeasurement state.

Example 6.11

A system is in the state

$$|\psi\rangle = \frac{2}{\sqrt{5}}|0\rangle + \frac{1}{\sqrt{5}}|1\rangle$$

Describe the probabilities of measuring 0 and 1 for this state in the POVM formalism.

Solution

In the simple case of a single qubit, we actually have a POVM using the projection operators. In this example we denote

$$E_0 = |0\rangle\langle 0|, \quad E_1 = |1\rangle\langle 1|$$

Notice that $\sum_m E_m = E_0 + E_1 = |0\rangle\langle 0| + |1\rangle\langle 1| = I$. The matrix representations of these operators in the computational basis are

$$E_0 = \begin{pmatrix} 1 & 0 \\ 0 & 0 \end{pmatrix}, \quad E_1 = \begin{pmatrix} 0 & 0 \\ 0 & 1 \end{pmatrix}$$

Each matrix has two eigenvalues, namely {1, 0} indicating that these operators are positive semidefinite.

By (6.33), the respective probabilities are

$$\begin{aligned}
\Pr(0) = \langle\psi|E_0|\psi\rangle &= \left(\frac{2}{\sqrt{5}}\langle 0| + \frac{1}{\sqrt{5}}\langle 1|\right)(|0\rangle\langle 0|)\left(\frac{2}{\sqrt{5}}|0\rangle + \frac{1}{\sqrt{5}}|1\rangle\right) \\
&= \left(\frac{2}{\sqrt{5}}\langle 0| + \frac{1}{\sqrt{5}}\langle 1|\right)\left(\frac{2}{\sqrt{5}}|0\rangle\langle 0|0\rangle + \frac{1}{\sqrt{5}}|1\rangle\langle 0|1\rangle\right) \\
&= \left(\frac{2}{\sqrt{5}}\langle 0| + \frac{1}{\sqrt{5}}\langle 1|\right)\frac{2}{\sqrt{5}}|0\rangle \\
&= \frac{4}{5}\langle 0|0\rangle + \frac{2}{5}\langle 1|0\rangle = \frac{4}{5} \\
\Pr(1) = \langle\psi|E_1|\psi\rangle &= \left(\frac{2}{\sqrt{5}}\langle 0| + \frac{1}{\sqrt{5}}\langle 1|\right)(|1\rangle\langle 1|)\left(\frac{2}{\sqrt{5}}|0\rangle + \frac{1}{\sqrt{5}}|1\rangle\right) \\
&= \left(\frac{2}{\sqrt{5}}\langle 0| + \frac{1}{\sqrt{5}}\langle 1|\right)\left(\frac{2}{\sqrt{5}}|0\rangle\langle 1|0\rangle + \frac{1}{\sqrt{5}}|1\rangle\langle 1|1\rangle\right) \\
&= \left(\frac{2}{\sqrt{5}}\langle 0| + \frac{1}{\sqrt{5}}\langle 1|\right)\frac{1}{\sqrt{5}}|1\rangle \\
&= \frac{2}{5}\langle 0|1\rangle + \frac{1}{5}\langle 1|1\rangle = \frac{1}{5}
\end{aligned}$$

POVM's can be useful when projective measurements are not. For example, POVM's provide the ability to distinguish between nonorthogonal states.

Example 6.12

A system can be in one of two states $|\psi\rangle$ or $|\phi\rangle$. The states are not orthogonal; in fact $|\langle\psi|\phi\rangle| = \cos\theta$. Describe a POVM that can distinguish between the two states. Assume that the states are normalized.

Solution

Consider the POVM consisting of the following measurement operators:

$$E_1 = \frac{I - |\phi\rangle\langle\phi|}{1+\cos\theta}, \quad E_2 = \frac{I - |\psi\rangle\langle\psi|}{1+\cos\theta}, \quad E_3 = I - E_1 - E_2$$

Each of these operators corresponds to a different measurement outcome. These operators satisfy the completeness relation, since $\sum_m E_m = E_1 + E_2 + E_3 = I$. Now consider the first measurement outcome associated with E_1. The probabilities associated with each state are $\langle\psi|E_1|\psi\rangle$ and $\langle\phi|E_1|\phi\rangle$. In the first case,

$$\begin{aligned}\langle\psi|E_1|\psi\rangle &= \langle\psi|\frac{I - |\phi\rangle\langle\phi|}{1+\cos\theta}|\psi\rangle \\ &= \frac{\langle\psi\psi\rangle - \langle\psi\phi\rangle\langle\phi\psi\rangle}{1+\cos\theta} = \frac{1 - |\langle\psi\phi\rangle|^2}{1+\cos\theta} \\ &= \frac{1-\cos\theta^2}{1+\cos\theta} = \frac{(1-\cos\theta)(1+\cos\theta)}{1+\cos\theta} = 1 - \cos\theta\end{aligned}$$

Meanwhile, in the second case,

$$\begin{aligned}\langle\phi|E_1|\phi\rangle &= \langle\phi|\frac{I - |\phi\rangle\langle\phi|}{1+\cos\theta}|\phi\rangle \\ &= \frac{\langle\phi\phi\rangle - \langle\phi\phi\rangle\langle\phi\phi\rangle}{1+\cos\theta} = \frac{1-1}{1+\cos\theta} = 0\end{aligned}$$

Hence the operator E_1 allows us to identify the state $|\psi\rangle$ with probability $1 - \cos\theta$. If the system is in the state $|\phi\rangle$, the probability is zero—this measurement never identifies the state $|\phi\rangle$. A similar exercise shows that the operator

$$E_2 = \frac{1 - |\psi\rangle\langle\psi|}{1+\cos\theta}$$

never identifies the state $|\psi\rangle$, but it identifies the state $|\phi\rangle$ with probability $1 - \cos\theta$. These operators have provided a means of imperfectly distinguishing between two nonorthogonal quantum states.

If the measurement outcome E_3 is obtained, no information about the state is available.

Example 6.13

A POVM can be used to obtain information about a state with a *weak measurement*. This is a measurement that provides some information about the state while only mildly disturbing it, without forcing a "collapse of the wave function." Describe a POVM that will do this for a single qubit.

Solution

The system in this case is a single qubit

$$|\psi\rangle = a|0\rangle + b|1\rangle$$

where $|a|^2 + |b|^2 = 1$. Suppose that we have a small positive parameter $\varepsilon \ll 1$ and two measurement operators

$$A_0 = A|0\rangle\langle 0| + \sqrt{1-\varepsilon}|1\rangle\langle 1|$$
$$A_1 = \sqrt{\varepsilon}|1\rangle\langle 1|$$

We construct a POVM with the operators

$$E_0 = A_0^2 = |0\rangle\langle 0| + (1-\varepsilon)|1\rangle\langle 1|$$
$$E_1 = A_1^2 = \varepsilon|1\rangle\langle 1|$$

Notice that

$$E_0 + E_1 = |0\rangle\langle 0| + (1-\varepsilon)|1\rangle\langle 1| + \varepsilon|1\rangle\langle 1| = |0\rangle\langle 0| + |1\rangle\langle 1| = I$$

Hence the completeness relation is satisfied. The operators are positive semidefinite; for example, the eigenvalues of E_0 are $\{1, 1-\varepsilon\}$ while the eigenvalues of E_1 are $\{0, \varepsilon\}$. The probability of obtaining measurement result E_0 is

$$\langle\psi|E_0|\psi\rangle = (a^*\langle 0| + b^*\langle 1|)(a|0\rangle + b(1-\varepsilon)|1\rangle) = |a|^2 + |b|^2(1-\varepsilon)$$

The postmeasurement state is

$$\frac{E_0|\psi\rangle}{\sqrt{\langle\psi|E_0|\psi\rangle}} = \frac{a|0\rangle + b(1-\varepsilon)|1\rangle}{\sqrt{|a|^2 + |b|^2(1-\varepsilon)}} = \frac{a}{\sqrt{|a|^2 + |b|^2(1-\varepsilon)}}|0\rangle + \frac{b(1-\varepsilon)}{\sqrt{|a|^2 + |b|^2(1-\varepsilon)}}|1\rangle$$

Therefore we see that a measurement of E_0 has left the state in a superposition. Although the state has been disturbed from the initial state of the wave function, a "collapse" of the wave function to $|0\rangle$ or $|1\rangle$ has not occurred.

The probability of obtaining measurement result E_1 is

$$\langle\psi|E_1|\psi\rangle = (a^*\langle 0| + b^*\langle 1|)(b(\varepsilon)|1\rangle) = |b|^2(\varepsilon)$$

Since $\varepsilon \ll 1$, the probability of obtaining this measurement result is very small. Moreover the wave function has collapsed in this case—the postmeasurement state of the system is $|1\rangle$.

As we have seen, POVMs are a more general type of measurement that allow us to do things in quantum mechanics that are not possible using ordinary projective measurements. Three examples we have mentioned are correctly describing a system where we don't need or can't know the postmeasurement state, the possibility of imperfectly distinguishing between nonorthogonal states, and the possibility of making weak measurements on a system that give us some information about the state without causing the collapse of the wavefunction. In later chapters we will see more applications of POVMs in quantum computation and information.

EXERCISES

6.1. *Let P_1 and P_2 be two projection operators. Show that if their commutator $[P_1, P_2]=0$, then their product P_1P_2 is also a projection operator.*

6.2. *A system is in the state*

$$|\psi\rangle = \frac{1}{2}|u_1\rangle - \frac{\sqrt{2}}{2}|u_2\rangle + \frac{1}{2}|u_3\rangle$$

where the orthonormal basis states $|u_1\rangle$, $|u_2\rangle$, $|u_3\rangle$ correspond to possible measurement results $\hbar\omega, 2\hbar\omega,$ and$3\hbar\omega$, respectively. Write down the projection operators corresponding to each possible measurement result, and determine the probability of finding the system in each of the states $|u_1\rangle$, $|u_2\rangle$, $|u_3\rangle$. What is the average energy of the system?

6.3. *A qubit is in the state $|\psi\rangle = |1\rangle$. A measurement of X is made. What are the matrix representations of the projection operators corresponding to measurement results ± 1? What is the probability of finding measurement results ± 1?*

6.4. *A system is in the state*

$$|\psi\rangle = \frac{1}{\sqrt{3}}|00\rangle + \frac{1}{\sqrt{6}}|01\rangle + \frac{1}{\sqrt{2}}|11\rangle$$

(A) What is the probability that measurement finds the system in the state $|\phi\rangle = |01\rangle$?
(B) What is the probability that measurement finds the second qubit in the state $|1\rangle$, and what is the state of the system after measurement?

6.5. *A system is in the state*

$$|\psi\rangle = \frac{1}{\sqrt{6}}|0\rangle + \sqrt{\frac{5}{6}}|1\rangle$$

A measurement is made with respect to the observable Y. What is the expectation or average value?

6.6. *A three-qubit system is in the state*

$$|\psi\rangle = \left(\frac{\sqrt{2}+i}{\sqrt{20}}\right)|000\rangle + \frac{1}{\sqrt{2}}|001\rangle + \frac{1}{\sqrt{10}}|011\rangle + \frac{i}{2}|111\rangle$$

(A) What is the probability that the system is found in the state $|011\rangle$ if all three qubits are measured?

(B) What is the probability that a measurement on the second qubit only gives 1? What is the postmeasurement state of the system? Show that the postmeasurement state is normalized.

6.7. *A two-qubit system is in the state*

$$|\phi\rangle = \frac{1}{\sqrt{6}}|01\rangle + \sqrt{\frac{5}{6}}|10\rangle$$

Is the state normalized? An X gate is applied to the second qubit. After this is done, what are the possible measurement results if both qubits are measured, and what are the respective probabilities of each measurement result?

6.8. *Suppose $|\psi\rangle = |1\rangle$ and $|\phi\rangle = |0\rangle + |1\rangle/\sqrt{2}$. Write down a POVM that allows for imperfect distinguishability between the two states.*

6.9. *Verify that the POVM used in Problem 6.8 satisfies the completeness relation.*

6.10. *Why don't the operators*

$$A_0 = A|0\rangle\langle 0| + \sqrt{1-\varepsilon}|1\rangle\langle 1|$$
$$A_1 = \sqrt{\varepsilon}|1\rangle\langle 1|$$

used in Example 6.13 constitute a POVM?

7 ENTANGLEMENT

One of the most unusual and fascinating aspects of quantum mechanics is the fact that particles or systems can become *entangled*. For the simplest two quantum systems case we denote the systems A and B. If these systems are entangled, this means that the values of certain properties of system A are correlated with the values that those properties will assume for system B. The properties can become correlated even when the two systems are spatially separated—leading to the phrase *spooky action* at a distance.

The roots of this idea go back a long way—all the way back to the year 1935 when Einstein and two colleages, Podolsky and Rosen (now commonly known as EPR), published a paper titled "Can the quantum-mechanical description of reality be considered complete?" This paper—written by the quantum skeptic Einstein—was actually designed to show that quantum theory is incomplete and to make absurd predictions.

A core value held by EPR and other "realists" was that the properties of physical systems have definite values (an *objective reality*) whether you observe the system or not. Another way to say this is that a given property of a system has a sharply defined value before a measurement is made.

Quantum mechanics, however, tells a different story. Suppose that we have a qubit in the state $|\psi\rangle = |0\rangle$. Quantum mechanics tells us that prior to measurement a property of the system does not have a definite or sharply defined value. For our qubit, say that we want to measure X. We know that the state is in a superposition of the eigenstates of the X operator as

$$|\psi\rangle = |0\rangle = \frac{|+\rangle + |-\rangle}{\sqrt{2}}$$

Hence measurement of X will find the system in $|+\rangle$ 50% of the time and $|-\rangle$ 50% of the time. *After* measurement the system does assume a definite state—either $|+\rangle$or $|-\rangle$—but before measurement this is not the case. This is in direct contradiction to the values held by EPR.

EPR thought predictions such as this one for single systems by quantum mechanics were absurd enough. But things only get worse when you consider composite systems. By making a very clever observation about what quantum theory tells us, EPR demonstrated that quantum mechanics predicts that if two particles interact and then separate, measurement of one of the particles will determine the values that the properties of the other particle must assume. This is the case even though the particles are spatially separated and noninteracting at the time of measurement.

In their brilliant 1935 paper, EPR focused on measurements of position and momentum. The first thing to note about this emphasis is that position x and momentum p do not commute:

$$[x, p] = i\hbar \tag{7.1}$$

As (7.1) tells us, x and p do not have simultaneous eigenstates. A consequence of this fact is that if a definite value is assumed for one of the variables—momentum, say—then the other variable (position in our example) cannot be in a definite eigenstate.

In discussions of this type it is common to denote the two quantum systems as being in the possession of Alice and Bob. We will denote the position and momentum for the particle in possession of Alice by x_A and p_A, respectively, and we will denote the position and momentum of the particle in Bob's possession by x_B and p_B. The basic scenario proposed by EPR is the following:

- The particles interact, then the particles spatially separate.
- There are no more interactions between the particles. We can even assume that they are so far away from each other that no signal—not even a light ray—can connect them over the time span when measurements are made.

The EPR system has definite values for the following properties:

- The relative position of the two particles, which is given by $x_A - x_B$.
- The total momentum of the particles, which is $p_A + p_B$.

However, prior to measurement, the values of each parameter, x_A, x_B or p_A, p_B, are not determined according to quantum mechanics. If we allow Alice to make measurements, she can choose to measure whatever she wants. If she measures momentum, she obtains a definite value p_A. Then, since $p_A + p_B$, Alice knows the exact value of p_B even though she has not made any measurements or disturbed Bob's particle in any way.

Alice could instead choose to measure the position of her particle. When she measures position and finds a definite value of x_A, Alice then knows the value of x_B. Again, this is true even though Alice has not disturbed, looked at, or measured Bob's particle in any way. In principle, this is true even if Bob is on the other side of the galaxy.

EPR believed that because Alice can determine or know the values of position and momentum of Bob's particle, these properties have definite values regardless of whether or not we measure them (position and momentum are "elements of reality"—i.e., objective properties). In contrast, quantum theory tells us that the wavefunctions of each particle exist in superpositions and each property does not have a definite value until we measure it. Since quantum theory cannot tell one the definite values of each property prior to measurement, EPR said the theory was *incomplete*. There could be some other physical variables we don't yet know about that would allow us to describe the definite properties of each particle using the theory. Because we don't know what those variables are, they are called *hidden variables*.

We can summarize the conventional or classical view as *local realism* and the theories that are based on this philosophy as *local realistic theories*. Let's formally define each term before moving on:

- Locality. Measurement of particle A in no way disturbs the state of spatially separated particle B.
- Realism. The values of measurable properties of each particle are objectively real. They have definite values prior to measurement and regardless of whether or not they are observed.

For technical reasons the scenario depicted by EPR is difficult to test experimentally. A simpler version of this thought experiment was put forward by David Bohm in 1952 and involved particles with correlated spins. This thought experiment considered a spin-0 particle that decays into two spin-1/2 particles. The decay products have to travel in opposite directions to conserve momentum, and their total spin has to remain zero in order to conserve angular momentum. A total spin-0 state, which consists of a system of two spin-1/2 particles, can be described in terms of the computational basis with the *singlet* state

$$|\psi\rangle = \frac{|0\rangle|1\rangle - |1\rangle|0\rangle}{\sqrt{2}} \tag{7.2}$$

We can make a measurement of Z on the first particle, while leaving the second particle alone using the operator $Z \otimes I$. From (7.2) it's also clear that if we measure 0 for the first particle, the state of the second particle *must be* $|1\rangle$. On the other hand, if we measure 1 for the first particle, the state of the second particle must be $|0\rangle$.

An interesting property of (7.2) is revealed if we rewrite this state in terms of the eigenvectors of the X operator, the $|\pm\rangle$ states. Recall that

$$|0\rangle = \frac{|+\rangle + |-\rangle}{\sqrt{2}}, \quad |1\rangle = \frac{|+\rangle - |-\rangle}{\sqrt{2}} \tag{7.3}$$

The first term in (7.2) can be written as

$$|0\rangle|1\rangle = \left(\frac{|+\rangle + |-\rangle}{\sqrt{2}}\right)\left(\frac{|+\rangle - |-\rangle}{\sqrt{2}}\right) = \frac{1}{2}(|++\rangle + |-+\rangle - |+-\rangle - |--\rangle)$$

and the second term as

$$|1\rangle|0\rangle = \left(\frac{|+\rangle - |-\rangle}{\sqrt{2}}\right)\left(\frac{|+\rangle + |-\rangle}{\sqrt{2}}\right) = \frac{1}{2}(|++\rangle - |-+\rangle + |+-\rangle - |--\rangle)$$

Hence

$$\begin{aligned}
|\psi\rangle &= \frac{|0\rangle|1\rangle - |1\rangle|0\rangle}{\sqrt{2}} \\
&= \frac{1}{\sqrt{2}}\frac{1}{2}(|++\rangle + |-+\rangle - |+-\rangle - |--\rangle - |++\rangle + |-+\rangle - |+-\rangle + |--\rangle) \\
&= \frac{1}{\sqrt{2}}\frac{1}{2}(2|-+\rangle - 2|+-\rangle) \\
&= -\frac{(|+-\rangle - |-+\rangle)}{\sqrt{2}}
\end{aligned}$$

Notice that if we look in the x direction—so we have the same state. If we measure the X operator for the first particle and obtain the result $+$, the second particle must be in the state $|-\rangle$, and vice versa. Finding a "$-$" for the first particle means that the second particle must be in the $|+\rangle$state.

In this state there seem to be some strange correlations between the particles. At first, if we only consider the z direction, the correlations probably do not seem all that mysterious. We could assume that the particles had definite states prior to measurement. But things start looking fishy when we find that the measurement results are correlated in different directions. The fact that the states are also correlated in the x direction indicates that the particles cannot be in definite states prior to measurement and that the states are part of a larger composite system. Remember the commutation relations satisfied by the Pauli operators:

$$[X, Y] = 2iZ, \quad [Y, Z] = 2iX, \quad [Z, X] = 2iY \tag{7.4}$$

The fact that $[Z, X] = 2iY \neq 0$ reminds us that the state of the second particle cannot be an eigenstate of both the Z operator and the X operator. However, it is possible to assume that the particles had definite spins for both the x and z directions prior to measurement and that measurement of one particle does not disturb the spin of the distant, spatially separated particle (local realism) and reproduce the results of quantum mechanics—assuming only that conservation of angular momentum hold. If we only assume that conservation of angular momentum holds and say nothing about quantum mechanics, then the measurement results with respect to the Pauli operators Z, X will be found as listed in Table 7.1.

As Table 7.1 shows the results agree with quantum mechanics. To illuminate the difference between quantum mechanics and local realistic theories, we are going to have to consider a more complicated situation. We will do so in the next section for measurements of spin along three nonorthogonal directions, and arrive at Bell's theorem.

When a system is entangled, this means that the individual component systems are really linked together as a *single* entity. Any measurement that measures a part of the system—in our case the first particle—is really a measurement on the entire system. The wavefunction for the system then collapses, and both particles assume definite states. In summary, if two systems are entangled, the description of each system has to be made with reference to the state of the other system, even if the component systems are spatially separated and noninteracting.

BELL'S THEOREM

In 1964 these ideas were taken much further by the brilliant theoretical physicist John S. Bell. By considering spin measurements along three nonorthogonal directions, which are conventionally labeled **a, b,** and **c,** and derived an inequality that is satisfied by local realistic theories but violated by quantum mechanics. This inequality is experimentally testable, and to date all experimental evidence has come down in favor of quantum mechanics.

The following simple derivation of Bell's inequality has been stated in many quantum mechanics textbooks such as Sakuri (1985) and Townsend (2000). We begin our thought experiment by imagining that an ensemble or large number of systems have been prepared so that Alice and Bob can measure spin along the three directions **a, b,** and **c**. With three directions to consider and 2 possible states ($\pm$) along each direction, there will be $2^3 = 8$ different populations in total. Here we consider the local realist position, based only on conservation of angular momentum. We only make the simple assumption that if Alice measures + along a given

TABLE 7.1 Conservation of momentum by particles Alice and Bob

Alice z Direction	Alice x Direction	Bob z Direction	Bob x Direction
+1	+1	−1	−1
+1	−1	−1	+1
−1	+1	+1	−1
−1	−1	+1	+1

TABLE 7.2 Measurements for particles Alice and Bob

	Alice			Bob		
Population	a	b	c	a	b	c
N_1	+	+	+	−	−	−
N_2	+	+	−	−	−	+
N_3	+	−	+	−	+	−
N_4	+	−	−	−	+	+
N_5	−	+	+	+	−	−
N_6	−	−	+	+	+	−
N_7	−	+	−	+	−	+
N_8	−	−	−	+	+	+

direction, then Bob will measure – along that same direction, should he choose to make that given measurement. The total number of particles is denoted by N, where there are N_i particles found in state i as described in Table 7.2.

$$N = N_1 + N_2 + N_3 + N_4 + N_5 + N_6 + N_7 + N_8$$

The direction along which Alice and Bob choose to measure is completely random. But, when deriving Bell's inequality, we assume that Alice and Bob measure along different directions. Let's consider some individual populations starting with N_1. From the table we see that if Alice chooses to measure along the **a** direction, she finds $+1$. Again, assuming that Bob measures along a different direction, he will measure along **b** or along **c**. In this case no matter what, Bob obtains the result -1. Look at population N_2, where Alice again measures along the **a** direction, again giving $+1$. This time Bob obtains -1 as he measures along **b,** and he will obtain $+1$ as he measures along **c**. This gives you an idea of how the measurements work.

Now let's see how many cases occur when Alice measures $+ 1$ along **a** and Bob measures -1 along **b**. The table shows that this occurs in populations N_1 and N_2. If Bob chooses to measure along **c,** then he obtains a -1 for populations N_1 and N_3.

In each population there are six possible measurement results if we require Alice and Bob to measure along different axes. If we look at population N_4, where Alice measures along **a,** she obtains $+1$. If Bob measures along **b** or **c,** he also obtains $+1$. Now, if Alice instead measures along **b,** Bob can measure along **a** (in which case he gets $- 1$) or along **c** (in which case he obtains $+1$). All the measurement possibilities are listed in Table 7.3.

From the table we see that Alice and Bob get opposite measurement results two out of six times or 1/3 of the time. It turns out that Alice and Bob will get opposite measurement results at least 1/3 of the time. If all populations occur with equal frequency then half of the measurements will be such that Alice and Bob obtain opposite measurement results.

By using some basic math regarding real numbers, we can derive some inequalities for the data in Table 7.2. We make no assumptions as to the size of each

TABLE 7.3 All possible measurements for particles Alice and Bob

Direction			
Alice	Bob	Alice	Bob
a	**b**	+1	+1
a	**c**	+1	+1
b	**a**	−1	−1
b	**c**	−1	+1
c	**a**	−1	−1
c	**b**	−1	+1

population in the ensemble, but it must be true that $N_i \geq 0$ and $N = \sum_{i=1}^{8} N_i$. Now it's a basic fact that if x, y, z are real numbers such that $x \geq 0$, $y \geq 0$, $z \geq 0$, then $x + y \leq x + y + z$. So it must be true that

$$N_3 + N_4 \leq (N_3 + N_4) + (N_2 + N_7) = (N_2 + N_4) + (N_3 + N_7) \tag{7.5}$$

We can divide by N, the total number in the ensemble,

$$\frac{N_3 + N_4}{N} \leq \frac{(N_2 + N_4)}{N} + \frac{(N_3 + N_7)}{N} \tag{7.6}$$

Referring back to Table 7.2, we see that N_3 and N_4 both have Alice measuring +1 along **a** and Bob measuring +1 along **b**. Hence

$$\frac{N_3 + N_4}{N} = \Pr(+\mathbf{a}; +\mathbf{b}) \tag{7.7}$$

That is, this is the probability that Alice measures +1 along **a** *and* Bob measures +1 along **b**. Now let's look at N_2 and N_4. Again, we see that Alice measures +1 along **a**. This time Bob gets +1 along **c** in both cases. Hence

$$\frac{N_2 + N_4}{N} = \Pr(+\mathbf{a}; +\mathbf{c}) \tag{7.8}$$

Finally, consider N_3 and N_7. Alice measures a + 1 when measuring along **c** in both cases, while Bob obtains a + 1 when measuring along **b** in both cases. So

$$\frac{N_3 + N_7}{N} = \Pr(+\mathbf{c}; +\mathbf{b}) \tag{7.9}$$

Putting the results (7.7), (7.8), and (7.9) into (7.6) gives us *Bell's inequality*

$$\Pr(+\mathbf{a}; +\mathbf{b}) \leq \Pr(+\mathbf{a}; +\mathbf{c}) + \Pr(+\mathbf{c}; +\mathbf{b}) \tag{7.10}$$

Local realistic theories *satisfy* or obey Bell's inequality. As we'll see in a moment, quantum mechanics *does not*.

To see what quantum mechanics says about the situation, we need to consider a qubit oriented in an arbitrary direction. Consider a unit vector $\vec{n} = \sin\theta\cos\phi\hat{x} + \sin\theta\sin\phi\hat{y} + \cos\theta\hat{z}$. The eigenvectors of $\vec{\sigma}\cdot\vec{n}$ are

$$
\begin{aligned}
|+_n\rangle &= \cos\frac{\theta}{2}|0\rangle + e^{i\phi}\sin\frac{\theta}{2}|1\rangle \\
|-_n\rangle &= \cos\frac{\theta}{2}|0\rangle - e^{i\phi}\sin\frac{\theta}{2}|1\rangle
\end{aligned}
\tag{7.11}
$$

Notice that (7.11) works by considering the x and y axes. For example, to get the eigenvectors of $\vec{\sigma}\cdot\hat{x}$, set $\theta = \pi/2$ and $\phi = 0$.

If the system is in the state $|+_n\rangle$, then we have

$$
\langle 0|+_n\rangle = \cos\frac{\theta}{2} \tag{7.12}
$$

Therefore the probability that measurement finds $|0\rangle$ given that the system is in the state $|+_n\rangle$ is

$$
|\langle 0|+_n\rangle|^2 = \cos^2\left(\frac{\theta}{2}\right) \tag{7.13}
$$

Similarly the probability to find $|1\rangle$ given that the system is in the state $|-_n\rangle$ is

$$
|\langle 1|+_n\rangle|^2 = \sin^2\left(\frac{\theta}{2}\right) \tag{7.14}
$$

Now that we see how to relate some axis defined by $\vec{n}$ to the z axis, we can write down relations for arbitrary axes **a, b,** and **c**. We define the angles between each of these axes as θ_{ab}, θ_{cb}, and θ_{ac}, respectively. We prepare the system in the singlet state. The form of the state is invariant under rotations, so we consider it along the **a** axis:

$$
|\psi\rangle = \frac{|+_a\rangle|-_a\rangle - |-_a\rangle|+_a\rangle}{\sqrt{2}} \tag{7.15}
$$

Then we can calculate $\Pr(+\mathbf{a}; +\mathbf{c})$ by looking at the inner product

$$
\begin{aligned}
\langle +_a +_c|\psi\rangle &= \frac{\langle +_a|+_a\rangle\langle +_c|-_a\rangle - \langle +_a|-_a\rangle\langle +_c|+_a\rangle}{\sqrt{2}} \\
&= \frac{\langle +_c|-_a\rangle}{\sqrt{2}} = \frac{1}{\sqrt{2}}\sin\frac{\theta_{ac}}{2}
\end{aligned}
\tag{7.16}
$$

Hence the probability is

$$
\Pr(+\mathbf{a}; +\mathbf{c}) = \left|\frac{\langle +_c|-_a\rangle}{\sqrt{2}}\right|^2 = \frac{1}{2}\sin^2\left(\frac{\theta_{ac}}{2}\right) \tag{7.17}
$$

Similarly we find that

$$\Pr(+\mathbf{a}; +\mathbf{b}) = \frac{1}{2}\sin^2\left(\frac{\theta_{ab}}{2}\right), \quad \Pr(+\mathbf{c}; +\mathbf{b}) = \frac{1}{2}\sin^2\left(\frac{\theta_{cb}}{2}\right) \tag{7.18}$$

This means that we can write Bell's inequality (7.10) as

$$\sin^2\left(\frac{\theta_{ab}}{2}\right) \leq \sin^2\left(\frac{\theta_{ac}}{2}\right) + \sin^2\left(\frac{\theta_{cb}}{2}\right) \tag{7.19}$$

Following Sakuri, we take **a, b,** and **c** to lie in a plane with **c** bisecting the angle θ_{ab}. We then obtain the simplification $\theta_{ac} = \theta_{cb} = \theta$, $\theta_{ab} = 2\theta$, and Bell's inequality becomes $\sin^2(\theta) \leq 2\sin^2(\theta/2)$. Bell's inequality is violated when

$$0 < \theta < \frac{\pi}{2} \tag{7.20}$$

Suppose that Alice and Bob design their system such that $\theta = \pi/3$. Bell's inequality is then the nonsensical statement that $0.75 \leq 0.5$. Therefore quantum mechanics clearly predicts a violation of Bell's inequality, which was derived under the assumption of local realism. A clear distinction has been made between simple counting arguments based on local realism (Table 7.2) and quantum mechanics. Experiment agrees with the predictions of quantum mechanics, so theories of the type that Einstein favored based on local realism are *ruled out* as descriptions of nature. Bell's inequality does not rule out nonlocal theories, however.

BIPARTITE SYSTEMS AND THE BELL BASIS

Now that we have an idea of what it means for two systems to be entangled, let's see how to work with the basic description and mathematics of such systems. When a system consists of two subsystems we say it is a *bipartite system*. An example of this is when Alice and Bob each have one member of an entangled pair of particles. Let's review some of the results from Chapter 4 where we described composite systems in quantum mechanics. The Hilbert space of the composite system is the tensor product of the Hilbert space that describes Alice's system and the Hilbert space that describes Bob's system. If we denote these as H_A and H_B, respectively, then the Hilbert space of the composite system is

$$H = H_A \otimes H_B \tag{7.21}$$

If we denote the basis states for Alice as $|a_i\rangle$ and the basis states for Bob by $|b_j\rangle$, then the basis states for the composite system are found by taking the tensor product of the Alice and Bob basis states:

$$|\alpha_{ij}\rangle = |a_i\rangle \otimes |b_j\rangle = |a_i\rangle|b_j\rangle = |a_i b_j\rangle \tag{7.22}$$

The basis for Alice and the basis for Bob are both orthonormal, so the basis states for the composite system are

$$\langle\alpha_{ij}|\alpha_{kl}\rangle = \langle a_i b_j|a_k b_l\rangle = \langle a_i|a_k\rangle\langle b_j|b_l\rangle = \delta_{ik}\delta_{jl} \tag{7.23}$$

Now consider a quantum state $|\psi\rangle$ of the composite system. It can be expanded in terms of the basis states (7.22) as follows:

$$|\psi\rangle = \sum_{i,j} c_{ij}|\alpha_{ij}\rangle = \sum_{i,j} |a_i b_j\rangle\langle a_i b_j|\psi\rangle \tag{7.24}$$

The coefficients in the expansion (7.24) $\langle a_i b_j|\psi\rangle$ are the probability amplitudes to find the system in the state $|a_i b_j\rangle$. So the probability of finding the system in this state is given by

$$\Pr(a_i b_j) = |\langle a_i b_j|\psi\rangle|^2 \tag{7.25}$$

Finally, we can write down the representation of an operator A in the basis $|a_i b_j\rangle$ as

$$A = \sum_{i,j,k,l} |a_i b_j\rangle\langle a_i b_j|A|a_k b_l\rangle\langle a_k b_l| \tag{7.26}$$

The matrix element $A_{ijkl} = \langle a_i b_j|A|a_k b_l\rangle$.

For an example of a basis for a bipartite system, consider the *Bell Basis*. The members of the Bell basis, sometimes called the *Bell states* or the *EPR states*, are

$$|\beta_{00}\rangle = \frac{|00\rangle + |11\rangle}{\sqrt{2}} \tag{7.27}$$

$$|\beta_{01}\rangle = \frac{|01\rangle + |10\rangle}{\sqrt{2}} \tag{7.28}$$

$$|\beta_{10}\rangle = \frac{|00\rangle - |11\rangle}{\sqrt{2}} \tag{7.29}$$

$$|\beta_{11}\rangle = \frac{|01\rangle - |10\rangle}{\sqrt{2}} \tag{7.30}$$

The state $|\beta_{01}\rangle$is known as the *triplet state* (there are three triplet states—the other two are $|11\rangle$ and $|00\rangle$), while $|\beta_{11}\rangle$ is known as the *singlet* state. For those with a physics background, the triplet states are spin-1 states (and can have $m_s = \pm 1$, 0, while the singlet state is a spin-0 state.

We can write the Bell states compactly as

$$|\beta_{xy}\rangle = \frac{|0y\rangle + (-1)^x|1\bar{y}\rangle}{\sqrt{2}} \tag{7.31}$$

where $\overline{y}$ denotes "not" y (if y is 0, then $\overline{y}$ is 1, and vice versa). In (7.31), x is called the *phase bit* and y is called the *parity bit*.

Example 7.1

Show that the operator $Z \otimes Z$ acts on (7.31) via the parity bit as $Z \otimes Z|\beta_{xy}\rangle = (-1)^y|\beta_{xy}\rangle$.

Solution

Recall the action of the Z operator

$$Z|0\rangle = |0\rangle, \quad Z|1\rangle = -|1\rangle$$

This can be written more abstractly as $Z|a\rangle = (-1)^a|a\rangle$. Look at the first term in (7.31). The operator on the first qubit does nothing because it's $|0\rangle$. So we obtain

$$Z \otimes Z|0y\rangle = (-1)^y|0y\rangle$$

In the second case, we have

$$(Z \otimes Z)(-1)^x|1\overline{y}\rangle = (-1)^x(Z|1\rangle) \otimes (Z|\overline{y}\rangle) = (-1)^x(-1)(-1)^{\overline{y}}|1\overline{y}\rangle$$

Now, if $\overline{y} = 0$, then $(-1)(-1)^0 = (-1)(+1) = -1$. But, if $\overline{y} = 0$, then obviously $y = 1$, and this is the same as $(-1)^y$. If $y = 1$, then $(-1)(-1)^1 = (-1)(-1) = +1 = (-1)^y$, since $y = 0$ in that case. So we've found that $(Z \otimes Z)(-1)^x|1\overline{y}\rangle = (-1)^x(-1)^y|1\overline{y}\rangle$. Putting everything together we have

$$Z \otimes Z|\beta_{xy}\rangle = \frac{(-1)^y|0y\rangle + (-1)^x(-1)^y|1\overline{y}\rangle}{\sqrt{2}} = (-1)^y|\beta_{xy}\rangle$$

WHEN IS A STATE ENTANGLED?

Not all states $|\psi\rangle \in H_A \otimes H_B$ are entangled. When two systems are entangled, the state of each composite system can only be described with reference to the other state. If two states are not entangled, we say that they are a *product state* or *separable*. If $|\psi\rangle \in H_A$ and $|\phi\rangle \in H_B$ and $|\chi\rangle = |\psi\rangle \otimes |\phi\rangle$, then $|\chi\rangle$ is a product state.

One simple test that can be applied to states in $\mathbb{C}^4$ is the following: Let

$$|\psi\rangle = \begin{pmatrix} a \\ b \\ c \\ d \end{pmatrix}$$

This state is separable if and only if

$$ad = bc \tag{7.32}$$

Example 7.2

Are the Bell states given in (7.27) through (7.30) entangled?

Solution

The Bell states are clearly entangled (in fact they could be said to be the quintessential entangled state), but let's apply criterion (7.32) to show they are not separable. Writing each state as a column vector, we have

$$|\beta_{00}\rangle = \frac{|00\rangle + |11\rangle}{\sqrt{2}} = \frac{1}{\sqrt{2}}\begin{pmatrix}1\\0\\0\\1\end{pmatrix}, \quad |\beta_{01}\rangle = \frac{|01\rangle + |10\rangle}{\sqrt{2}} = \frac{1}{\sqrt{2}}\begin{pmatrix}0\\1\\1\\0\end{pmatrix}$$

$$|\beta_{10}\rangle = \frac{|00\rangle - |11\rangle}{\sqrt{2}} = \frac{1}{\sqrt{2}}\begin{pmatrix}1\\0\\0\\-1\end{pmatrix}, \quad |\beta_{11}\rangle = \frac{|01\rangle - |10\rangle}{\sqrt{2}} = \frac{1}{\sqrt{2}}\begin{pmatrix}0\\1\\-1\\0\end{pmatrix}$$

For $|\beta_{00}\rangle$, we have $a = d = 1/\sqrt{2}, b = c = 0$, so $ad = 1/2 \neq bc$. So $|\beta_{00}\rangle$ is not a product state and must be entangled. For $|\beta_{01}\rangle$, $a = d = 0, b = c = 1/\sqrt{2}, \Rightarrow ad = 0 \neq bc = 1/2$. We conclude that $|\beta_{01}\rangle$ is also entangled. For $|\beta_{10}\rangle$, we find that $ad = -1/2 \neq bc = 0$, and for $|\beta_{11}\rangle$, we have $ad = 0 \neq bc = -1/2$, so these states are also entangled by (7.32).

Example 7.3

A system of two qubits is in the state $|00\rangle$. We operate on this state with $H \otimes H$, where H is the Hadamard matrix. Is the state $H \otimes H|00\rangle$ entangled?

Solution

First let's write down the matrix representation of $H \otimes H$ in the computational basis. The Hadamard matrix is given by

$$H = \frac{1}{\sqrt{2}}\begin{pmatrix}1 & 1\\1 & -1\end{pmatrix} \tag{7.33}$$

So we find that

$$H \otimes H = \frac{1}{\sqrt{2}}\begin{pmatrix}H & H\\H & -H\end{pmatrix} = \frac{1}{2}\begin{pmatrix}1 & 1 & 1 & 1\\1 & -1 & 1 & -1\\1 & 1 & -1 & -1\\1 & -1 & -1 & 1\end{pmatrix}$$

The state in question has a column vector representation

$$|00\rangle = \begin{pmatrix} 1 \\ 0 \\ 0 \\ 0 \end{pmatrix}$$

In this case $ad = (1)(0) = 0 = bc$ so $|00\rangle$ is clearly a product state according to (7.32).

Now let's calculate $H \otimes H\,|00\rangle$:

$$H \otimes H|00\rangle = \frac{1}{2}\begin{pmatrix} 1 & 1 & 1 & 1 \\ 1 & -1 & 1 & -1 \\ 1 & 1 & -1 & -1 \\ 1 & -1 & -1 & 1 \end{pmatrix}\begin{pmatrix} 1 \\ 0 \\ 0 \\ 0 \end{pmatrix} = \frac{1}{2}\begin{pmatrix} 1 \\ 1 \\ 1 \\ 1 \end{pmatrix}$$

Using (7.32), we see that this is also a product state, since

$$ad = \left(\frac{1}{2}\right)\left(\frac{1}{2}\right) = \frac{1}{4}$$
$$bc = \left(\frac{1}{2}\right)\left(\frac{1}{2}\right) = \frac{1}{4}$$
$$\Rightarrow ad = bc$$

In fact this state is the tensor product

$$\left(\frac{|0\rangle + |1\rangle}{\sqrt{2}}\right)\left(\frac{|0\rangle + |1\rangle}{\sqrt{2}}\right) = \frac{1}{2}(|00\rangle + |01\rangle + |10\rangle + |11\rangle)$$

Example 7.4

Alice and Bob each possess one member of a pair of interacting magnetic dipoles (spin-1/2 particles). The interaction Hamiltonian for two interacting magnetic dipoles separated by a distance r is given by

$$H_I = \frac{\mu^2}{r^3}(\vec{\sigma}_A \cdot \vec{\sigma}_B - 3Z_A Z_B)$$

where $\vec{\sigma}_A = X_A\,\hat{x} + Y_A\,\hat{y} + Z_A\,\hat{z}$ and similarly for Bob. Find the allowed energies of the system and show that the eigenvectors of the Hamiltonian include entangled states. Then rewrite the Hamiltonian using it's eigenvectors.

Solution

First we express the Hamiltonian in matrix form. We have

$$\vec{\sigma}_A \cdot \vec{\sigma}_B = X_A \otimes X_B + Y_A \otimes Y_B + Z_A \otimes Z_B$$

The first term is

$$X_A \otimes X_B = \begin{pmatrix} 0 \cdot X_B & 1 \cdot X_B \\ 1 \cdot X_B & 0 \cdot X_B \end{pmatrix} = \begin{pmatrix} 0 & 0 & 0 & 1 \\ 0 & 0 & 1 & 0 \\ 0 & 1 & 0 & 0 \\ 1 & 0 & 0 & 0 \end{pmatrix}$$

Similarly we find that

$$Y_A \otimes Y_B = \begin{pmatrix} 0 & 0 & 0 & -1 \\ 0 & 0 & 1 & 0 \\ 0 & 1 & 0 & 0 \\ -1 & 0 & 0 & 0 \end{pmatrix} \quad \text{and} \quad Z_A \otimes Z_B = \begin{pmatrix} 1 & 0 & 0 & 0 \\ 0 & -1 & 0 & 0 \\ 0 & 0 & -1 & 0 \\ 0 & 0 & 0 & 1 \end{pmatrix}$$

So

$$\begin{aligned} \vec{\sigma}_A \cdot \vec{\sigma}_B &= X_A \otimes X_B + Y_A \otimes Y_B + Z_A \otimes Z_B \\ &= \begin{pmatrix} 0 & 0 & 0 & 1 \\ 0 & 0 & 1 & 0 \\ 0 & 1 & 0 & 0 \\ 1 & 0 & 0 & 0 \end{pmatrix} + \begin{pmatrix} 0 & 0 & 0 & -1 \\ 0 & 0 & 1 & 0 \\ 0 & 1 & 0 & 0 \\ -1 & 0 & 0 & 0 \end{pmatrix} + \begin{pmatrix} 1 & 0 & 0 & 0 \\ 0 & -1 & 0 & 0 \\ 0 & 0 & -1 & 0 \\ 0 & 0 & 0 & 1 \end{pmatrix} \\ &= \begin{pmatrix} 1 & 0 & 0 & 0 \\ 0 & -1 & 2 & 0 \\ 0 & 2 & -1 & 0 \\ 0 & 0 & 0 & 1 \end{pmatrix} \end{aligned}$$

Therefore the matrix representation of the Hamiltonian is

$$\begin{aligned} H_I &= \frac{\mu^2}{r^3}(\vec{\sigma}_A \cdot \vec{\sigma}_B - 3Z_A Z_B) = \frac{\mu^2}{r^3}\left[\begin{pmatrix} 1 & 0 & 0 & 0 \\ 0 & -1 & 2 & 0 \\ 0 & 2 & -1 & 0 \\ 0 & 0 & 0 & 1 \end{pmatrix} - 3\begin{pmatrix} 1 & 0 & 0 & 0 \\ 0 & -1 & 0 & 0 \\ 0 & 0 & -1 & 0 \\ 0 & 0 & 0 & 1 \end{pmatrix}\right] \\ &= \frac{\mu^2}{r^3}\begin{pmatrix} -2 & 0 & 0 & 0 \\ 0 & 2 & 2 & 0 \\ 0 & 2 & 2 & 0 \\ 0 & 0 & 0 & -2 \end{pmatrix} \end{aligned}$$

Keep in mind that the matrix is written in the following way:

$$H_I = \begin{pmatrix} \langle 00|H_I|00\rangle & \langle 00|H_I|01\rangle & \langle 00|H_I|10\rangle & \langle 00|H_I|11\rangle \\ \langle 01|H_I|00\rangle & \langle 01|H_I|01\rangle & \langle 01|H_I|10\rangle & \langle 01|H_I|11\rangle \\ \langle 10|H_I|00\rangle & \langle 10|H_I|01\rangle & \langle 10|H_I|10\rangle & \langle 10|H_I|11\rangle \\ \langle 11|H_I|00\rangle & \langle 11|H_I|01\rangle & \langle 11|H_I|10\rangle & \langle 11|H_I|11\rangle \end{pmatrix}$$

And a state vector $|\psi\rangle = a|00\rangle + b|01\rangle + c|10\rangle + d|11\rangle$is given by

$$|\psi\rangle = \begin{pmatrix} a \\ b \\ c \\ d \end{pmatrix}$$

The eigenvalues of this matrix are the energies that the system can assume. With a 4×4matrix, it's easiest to find the eigenvalues and eigenvectors using a computer. Using Mathematica® we find that the eigenvalues of H_I are $\mu^2/r^3\{4, -2, -2, 0\}$. The eigenvectors corresponding to each of these eigenvalues are in turn

$$|\phi_1\rangle = \frac{1}{\sqrt{2}} \begin{pmatrix} 0 \\ 1 \\ 1 \\ 0 \end{pmatrix} = \frac{|01\rangle + |10\rangle}{\sqrt{2}}$$

$$|\phi_2\rangle = \begin{pmatrix} 0 \\ 0 \\ 0 \\ 1 \end{pmatrix} = |11\rangle, \quad |\phi_3\rangle = \begin{pmatrix} 1 \\ 0 \\ 0 \\ 0 \end{pmatrix} = |00\rangle$$

and

$$|\phi_4\rangle = \frac{1}{\sqrt{2}} \begin{pmatrix} 0 \\ 1 \\ -1 \\ 0 \end{pmatrix} = \frac{|01\rangle - |10\rangle}{\sqrt{2}}$$

Comparison with (7.28) and (7.30) shows that two of the eigenvectors are Bell states—and hence represent states in which the particles in possession of Alice and Bob are entangled. Specifically, $|\phi_1\rangle = |\beta_{01}\rangle$ and $|\phi_4\rangle = |\beta_{11}\rangle$.

In the basis of its eigenstates, H_I is diagonal with the entries along the diagonal given by its eigenvalues

$$H_I = \begin{pmatrix} \langle\phi_1|H_I|\phi_1\rangle & \langle\phi_1|H_I|\phi_2\rangle & \langle\phi_1|H_I|\phi_3\rangle & \langle\phi_1|H_I|\phi_4\rangle \\ \langle\phi_2|H_I|\phi_1\rangle & \langle\phi_2|H_I|\phi_2\rangle & \langle\phi_2|H_I|\phi_3\rangle & \langle\phi_2|H_I|\phi_4\rangle \\ \langle\phi_3|H_I|\phi_1\rangle & \langle\phi_3|H_I|\phi_2\rangle & \langle\phi_3|H_I|\phi_3\rangle & \langle\phi_3|H_I|\phi_4\rangle \\ \langle\phi_4|H_I|\phi_1\rangle & \langle\phi_4|H_I|\phi_2\rangle & \langle\phi_4|H_I|\phi_3\rangle & \langle\phi_4|H_I|\phi_4\rangle \end{pmatrix}$$

$$= \frac{\mu^2}{r^3} \begin{pmatrix} 4 & 0 & 0 & 0 \\ 0 & -2 & 0 & 0 \\ 0 & 0 & -2 & 0 \\ 0 & 0 & 0 & 0 \end{pmatrix}$$

You Try It

Show that

$$Y_A \otimes Y_B = \begin{pmatrix} 0 & 0 & 0 & -1 \\ 0 & 0 & 1 & 0 \\ 0 & 1 & 0 & 0 \\ -1 & 0 & 0 & 0 \end{pmatrix} \quad \text{and} \quad Z_A \otimes Z_B = \begin{pmatrix} 1 & 0 & 0 & 0 \\ 0 & -1 & 0 & 0 \\ 0 & 0 & -1 & 0 \\ 0 & 0 & 0 & 1 \end{pmatrix}$$

You Try It

Verify that for the Hamiltonian in Example 7.4,

$$H_I|\beta_{01}\rangle = 4\frac{\mu^2}{r^3}|\beta_{01}\rangle \quad \text{and} \quad H_I|\beta_{00}\rangle = 0$$

THE PAULI REPRESENTATION

The *Pauli representation* provides a means to write down a density operator of a single qubit or two-qubit systems in terms of Pauli matrices. For a single qubit the *Pauli representation* is given by

$$\rho = \frac{1}{2}\sum_{i=0}^{3} c_i \sigma_i \tag{7.34}$$

The coefficients in this expansion are given by

$$c_i = Tr(\rho\sigma_i) = \langle \sigma_i \rangle$$

Example 7.5

A certain density matrix is given by

$$\rho = \frac{3}{4}|0\rangle\langle 0| + \frac{1}{4}|1\rangle\langle 1|$$

Find its Pauli representation.

Solution

Proceeding by brute force, and recalling that $\sigma_0 = I$ the identity operator, we have

$$c_0 = Tr(\rho\sigma_0) = Tr\begin{pmatrix} \frac{3}{4} & 0 \\ 0 & \frac{1}{4} \end{pmatrix} = \frac{3}{4} + \frac{1}{4} = 1$$

Next we find

$$c_1 = Tr(\rho\sigma_1) = Tr\begin{pmatrix} \frac{3}{4} & 0 \\ 0 & \frac{1}{4} \end{pmatrix}\begin{pmatrix} 0 & 1 \\ 1 & 0 \end{pmatrix} = Tr\begin{pmatrix} 0 & \frac{3}{4} \\ \frac{1}{4} & 0 \end{pmatrix} = 0$$

Similarly $c_2 = Tr(\rho\sigma_2) = 0$. Finally,

$$c_3 = Tr(\rho\sigma_3) = Tr\begin{pmatrix} \frac{3}{4} & 0 \\ 0 & \frac{1}{4} \end{pmatrix}\begin{pmatrix} 1 & 0 \\ 0 & -1 \end{pmatrix} = tr\begin{pmatrix} \frac{3}{4} & 0 \\ 0 & \frac{-1}{4} \end{pmatrix} = \frac{3}{4} - \frac{1}{4} = \frac{1}{2}$$

The Pauli representation is

$$\rho = \sigma_0 - \frac{1}{2}\sigma_3 = I - \frac{1}{2}Z$$

The Pauli representation for a system of two qubits is given by

$$\rho = \frac{1}{4}\sum_{i,j} c_{ij}\sigma_i \otimes \sigma_j \tag{7.35}$$

where $c_{ij} = \langle \sigma_i \otimes \sigma_j \rangle = Tr(\rho\, \sigma_i \otimes \sigma_j)$. If the density operator $\rho$is a separable state, then

$$|c_{11}| + |c_{22}| + |c_{33}| \leq 1 \tag{7.36}$$

Example 7.6

Show that $H \otimes H\,|00\rangle$ is a separable state using the criterion (7.36), and show that $|\beta_{00}\rangle$ is entangled.

Solution

First let's find the density operator for $H \otimes H\ |00\rangle$. This can be written down using

$$\rho = \frac{1}{2}(|00\rangle + |01\rangle + |10\rangle + |11\rangle)\frac{1}{2}(\langle 00| + \langle 01| + \langle 10| + \langle 11|)$$

The density matrix turns out to be

$$\rho = \frac{1}{4}\begin{pmatrix} 1 & 1 & 1 & 1 \\ 1 & 1 & 1 & 1 \\ 1 & 1 & 1 & 1 \\ 1 & 1 & 1 & 1 \end{pmatrix}$$

The first term is

$$c_{11} = \langle \sigma_1 \otimes \sigma_1 \rangle = \langle X \otimes X \rangle = Tr(\rho X \otimes X)$$

Now

$$X \otimes X = \begin{pmatrix} 0 & X \\ X & 0 \end{pmatrix} = \begin{pmatrix} 0 & 0 & 0 & 1 \\ 0 & 0 & 1 & 0 \\ 0 & 1 & 0 & 0 \\ 1 & 0 & 0 & 0 \end{pmatrix}$$

and

$$\rho X \otimes X = \frac{1}{4}\begin{pmatrix} 1 & 1 & 1 & 1 \\ 1 & 1 & 1 & 1 \\ 1 & 1 & 1 & 1 \\ 1 & 1 & 1 & 1 \end{pmatrix}\begin{pmatrix} 0 & 0 & 0 & 1 \\ 0 & 0 & 1 & 0 \\ 0 & 1 & 0 & 0 \\ 1 & 0 & 0 & 0 \end{pmatrix} = \frac{1}{4}\begin{pmatrix} 1 & 1 & 1 & 1 \\ 1 & 1 & 1 & 1 \\ 1 & 1 & 1 & 1 \\ 1 & 1 & 1 & 1 \end{pmatrix}$$

Hence

$$c_{11} = Tr(\rho X \otimes X) = \frac{1}{4} + \frac{1}{4} + \frac{1}{4} + \frac{1}{4} = 1$$

Using

$$Y \otimes Y = \begin{pmatrix} 0 & 0 & 0 & -1 \\ 0 & 0 & 1 & 0 \\ 0 & 1 & 0 & 0 \\ -1 & 0 & 0 & 0 \end{pmatrix}$$

we find that

$$\begin{aligned} c_{22} &= Tr(\rho \sigma_2 \otimes \sigma_2) = Tr(\rho Y \otimes Y) \\ &= Tr\frac{1}{4}\begin{pmatrix} 1 & 1 & 1 & 1 \\ 1 & 1 & 1 & 1 \\ 1 & 1 & 1 & 1 \\ 1 & 1 & 1 & 1 \end{pmatrix}\begin{pmatrix} 0 & 0 & 0 & -1 \\ 0 & 0 & 1 & 0 \\ 0 & 1 & 0 & 0 \\ -1 & 0 & 0 & 0 \end{pmatrix} \\ &= Tr\frac{1}{4}\begin{pmatrix} -1 & 1 & 1 & -1 \\ -1 & 1 & 1 & -1 \\ -1 & 1 & 1 & -1 \\ -1 & 1 & 1 & -1 \end{pmatrix} = \frac{1}{4}(-1+1+1-1) = 0 \end{aligned}$$

Finally, we have

$$Z \otimes Z = \begin{pmatrix} 1 & 0 & 0 & 0 \\ 0 & -1 & 0 & 0 \\ 0 & 0 & -1 & 0 \\ 0 & 0 & 0 & 1 \end{pmatrix}$$

and

$$\rho Z\otimes Z=\frac{1}{4}\begin{pmatrix}1&1&1&1\\1&1&1&1\\1&1&1&1\\1&1&1&1\end{pmatrix}\begin{pmatrix}1&0&0&0\\0&-1&0&0\\0&0&-1&0\\0&0&0&1\end{pmatrix}=\frac{1}{4}\begin{pmatrix}1&-1&-1&1\\1&-1&-1&1\\1&-1&-1&1\\1&-1&-1&1\end{pmatrix}$$

The trace of this matrix, which is the sum of the diagonal elements, also vanishes, so we have

$$|c_{11}|+|c_{22}|+|c_{33}|=1+0+0=1$$

Hence (7.36) is satisfied, and this is a separable state. Now let's check $|\beta_{00}\rangle$. The density operator for this state is

$$\rho=\frac{1}{2}(|00\rangle+|11\rangle)(\langle 00|+\langle 11|)=\frac{1}{2}(|00\rangle\langle 00|+|00\rangle\langle 11|+|11\rangle\langle 00|+|11\rangle\langle 11|)$$

The density matrix is thus

$$\rho=\begin{pmatrix}1&0&0&1\\0&0&0&0\\0&0&0&0\\1&0&0&1\end{pmatrix}$$

We find that

$$c_{11}=Tr(\rho X\otimes X)=Tr\frac{1}{2}\begin{pmatrix}1&0&0&1\\0&0&0&0\\0&0&0&0\\1&0&0&1\end{pmatrix}\begin{pmatrix}0&0&0&1\\0&0&1&0\\0&1&0&0\\1&0&0&0\end{pmatrix}$$

$$=Tr\frac{1}{2}\begin{pmatrix}1&0&0&1\\0&0&0&0\\0&0&0&0\\1&0&0&1\end{pmatrix}=\frac{1}{2}+0+0+\frac{1}{2}=1$$

Next we find that

$$c_{22}=Tr(\rho Y\otimes Y)=Tr\frac{1}{2}\begin{pmatrix}1&0&0&1\\0&0&0&0\\0&0&0&0\\1&0&0&1\end{pmatrix}\begin{pmatrix}0&0&0&-1\\0&0&1&0\\0&1&0&0\\-1&0&0&0\end{pmatrix}$$

$$=Tr\frac{1}{2}\begin{pmatrix}-1&0&0&-1\\0&0&0&0\\0&0&0&0\\-1&0&0&-1\end{pmatrix}=-\frac{1}{2}+0+0-\frac{1}{2}=-1$$

The last coefficient we need is

$$c_{33} = Tr(\rho Z \otimes Z) = Tr\frac{1}{2}\begin{pmatrix} 1 & 0 & 0 & 1 \\ 0 & 0 & 0 & 0 \\ 0 & 0 & 0 & 0 \\ 1 & 0 & 0 & 1 \end{pmatrix}\begin{pmatrix} 1 & 0 & 0 & 0 \\ 0 & -1 & 0 & 0 \\ 0 & 0 & -1 & 0 \\ 0 & 0 & 0 & 1 \end{pmatrix}$$

$$= Tr\frac{1}{2}\begin{pmatrix} 1 & 0 & 0 & 1 \\ 0 & 0 & 0 & 0 \\ 0 & 0 & 0 & 0 \\ 1 & 0 & 0 & 1 \end{pmatrix} = \frac{1}{2} + 0 + 0 + \frac{1}{2} = 1$$

So in this case

$$|c_{11}| + |c_{22}| + |c_{33}| = |1| + |-1| + |1| = 1 + 1 + 1 = 3$$

Hence by (7.36) the state is entangled.

ENTANGLEMENT FIDELITY

Consider a density operator for a single qubit that is diagonal with respect to the computational basis

$$\rho = f|0\rangle\langle 0| + (1 - f)|1\rangle\langle 1| \tag{7.37}$$

The parameter f is known as the *entanglement fidelity*. For example, if

$$\rho = \frac{3}{4}|0\rangle\langle 0| + \frac{1}{4}|1\rangle\langle 1|$$

the entanglement fidelity is $\frac{3}{4}$.

USING BELL STATES FOR DENSITY OPERATOR REPRESENTATION

A density operator for a two-qubit system that is diagonal with respect to the Bell basis can be represented in terms of the Bell states using the expansion

$$\begin{aligned} \rho &= \sum_{i,j} c_{ij}|\beta_{ij}\rangle\langle\beta_{ij}| \\ &= c_{00}|\beta_{00}\rangle\langle\beta_{00}| + c_{01}|\beta_{01}\rangle\langle\beta_{01}| + c_{10}|\beta_{10}\rangle\langle\beta_{10}| + c_{11}|\beta_{11}\rangle\langle\beta_{11}| \end{aligned} \tag{7.38}$$

This type of expansion is possible because we can write outer products of the Bell states in terms of the Pauli operators:

$$|\beta_{00}\rangle\langle\beta_{00}| = \frac{1}{4}(I \otimes I + X \otimes X - Y \otimes Y + Z \otimes Z) \tag{7.39}$$

$$|\beta_{01}\rangle\langle\beta_{01}| = \frac{1}{4}(I \otimes I + X \otimes X + Y \otimes Y - Z \otimes Z) \tag{7.40}$$

$$|\beta_{10}\rangle\langle\beta_{10}| = \frac{1}{4}(I \otimes I - X \otimes X + Y \otimes Y + Z \otimes Z) \tag{7.41}$$

$$|\beta_{11}\rangle\langle\beta_{11}| = \frac{1}{4}(I \otimes I - X \otimes X - Y \otimes Y - Z \otimes Z) \tag{7.42}$$

When written in the expansion (7.38) a density operator, $\rho$is separable if and only if

$$c_{00} \leq \frac{1}{2} \tag{7.43}$$

Example 7.7

A density matrix for a certain two-qubit system in the $\{|00\rangle, |01\rangle, |10\rangle, |11\rangle\}$ basis is

$$\rho = \begin{pmatrix} \frac{1}{8} & 0 & 0 & \frac{1}{8} \\ 0 & \frac{3}{8} & \frac{-3}{8} & 0 \\ 0 & \frac{-3}{8} & \frac{3}{8} & 0 \\ \frac{1}{8} & 0 & 0 & \frac{1}{8} \end{pmatrix}$$

Can this state be written in a diagonal form with respect to the Bell basis? Is this a separable state?

Solution

This matrix represents the density operator

$$\rho = \frac{1}{8}(|00\rangle\langle 00| + |00\rangle\langle 11| + |11\rangle\langle 00| + |11\rangle\langle 11|)$$
$$+ \frac{3}{8}(|01\rangle\langle 01| - |01\rangle\langle 10| - |10\rangle\langle 01| + |10\rangle\langle 10|)$$

In terms of the Bell basis, this operater is rewritten as

$$\rho = \frac{1}{4}|\beta_{00}\rangle\langle\beta_{00}| + \frac{3}{4}|\beta_{11}\rangle\langle\beta_{11}|$$

Since $c_{00} = 1/4 < 1/2$, by (7.43) this is a separable state.

SCHMIDT DECOMPOSITION

Consider a composite system $H_A \otimes H_B$, and let $|\psi\rangle \in H_A \otimes H_B$ be a pure state. Then there exists an expansion of $|\psi\rangle$ of the form

$$|\psi\rangle = \sum_i \lambda_i |a_i\rangle |b_i\rangle \tag{7.44}$$

where $|a_i\rangle$ are orthnormal states belonging to system A and $|b_i\rangle$ are orthnormal states belonging to system B (these states are called the *Schmidt bases* for system A and system B, respectively). The expansion coefficients λ_i are such that $\lambda_i \geq 0$and $\sum_i \lambda_i^2 = 1$. We call the λ_i *Schmidt coefficients* and the expansion (7.44) is known as the *Schmidt decomposition*.

The Schmidt coefficients are calculated from the matrix

$$Tr_B(|\psi\rangle\langle\psi|) \tag{7.45}$$

This matrix has eigenvalues λ_i^2. The *Schmidt number* is the number of nonzero eigenvalues λ_i. The Schmidt number is used in the following way:

- If a state is separable, then the Schmidt number is 1.
- If a state is entangled, then the Schmidt number is > 1.

We will denote the Schmidt number by *Sch*.

Example 7.8

Consider the state $|\psi\rangle = \frac{1}{2}(|00\rangle - |01\rangle - |10\rangle + |11\rangle)$. Is this state separable? What is the Schmidt number?

Solution

The density operator for this state is

$$\begin{aligned}\rho = |\psi\rangle\langle\psi| = \frac{1}{4}(&|00\rangle\langle 00| - |00\rangle\langle 01| - |00\rangle\langle 10| + |00\rangle\langle 11| - |01\rangle\langle 00| + |01\rangle\langle 01| \\ &+ |01\rangle\langle 10| - |01\rangle\langle 11| - |10\rangle\langle 00| + |10\rangle\langle 01| + |10\rangle\langle 10| - |10\rangle\langle 11| + |11\rangle\langle 00| \\ &- |11\rangle\langle 01| - |11\rangle\langle 10| + |11\rangle\langle 11|)\end{aligned}$$

We can trace out system B immediately giving

$$\begin{aligned}\rho_A = Tr_B(|\psi\rangle\langle\psi|) &= \langle 0|\psi\rangle\langle\psi|0\rangle + \langle 1|\psi\rangle\langle\psi|1\rangle \\ &= \frac{1}{4}(|0\rangle\langle 0| - |0\rangle\langle 1| - |1\rangle\langle 0| + |1\rangle\langle 1|) + \frac{1}{4}(|0\rangle\langle 0| - |0\rangle\langle 1| - |1\rangle\langle 0| + |1\rangle\langle 1|) \\ &= \frac{1}{2}(|0\rangle\langle 0| - |0\rangle\langle 1| - |1\rangle\langle 0| + |1\rangle\langle 1|)\end{aligned}$$

The matrix representation is

$$\rho_A = \frac{1}{2}\begin{pmatrix} 1 & -1 \\ -1 & 1 \end{pmatrix}$$

The eigenvalues of this matrix are $\lambda_1 = 1, \lambda_2 = 0$. The Schmidt number is the number of nonzero eigenvalues, and since Sch = 1, this is a separable state. In fact this is the product state given by

$$|\psi\rangle = \left(\frac{|0\rangle - |1\rangle}{\sqrt{2}}\right) \otimes \left(\frac{|0\rangle - |1\rangle}{\sqrt{2}}\right)$$

Example 7.9

Show that the singlet state

$$|S\rangle = \frac{|01\rangle - |10\rangle}{\sqrt{2}}$$

is entangled by computing its Schmidt number.

Solution

The density operator in this case is

$$\rho = |S\rangle\langle S| = \left(\frac{|01\rangle - |10\rangle}{\sqrt{2}}\right)\left(\frac{\langle 01| - \langle 10|}{\sqrt{2}}\right)$$

$$= \frac{1}{2}(|01\rangle\langle 01| - |01\rangle\langle 10| - |10\rangle\langle 01| + |10\rangle\langle 10|)$$

Tracing out system B, we obtain

$$\rho_A = Tr_B\left(\frac{1}{2}(|01\rangle\langle 01| - |01\rangle\langle 10| - |10\rangle\langle 01| + |10\rangle\langle 10|)\right)$$

$$= \langle 0|\frac{1}{2}(|01\rangle\langle 01| - |01\rangle\langle 10| - |10\rangle\langle 01| + |10\rangle\langle 10|)|0\rangle$$

$$+ \langle 1|\frac{1}{2}(|01\rangle\langle 01| - |01\rangle\langle 10| - |10\rangle\langle 01| + |10\rangle\langle 10|)|1\rangle$$

$$= \frac{1}{2}(|1\rangle\langle 1| + |0\rangle\langle 0|) = \frac{1}{2}I$$

This matrix has two nonzero eigenvalues, namely $\lambda_1 = \lambda_2 = 1/2$. Since the Schmidt number is $\text{Sch} = 2 > 1$, this is an entangled state.

PURIFICATION

Purification is the process by which we create a reference system B such that, given the system A, the state $|\phi_A\phi_B\rangle$is a pure state. The starting point for this procedure is the density matrix of the mixed state ρ_A. The state $|\phi_B\rangle$is a purification of ρ_A if

$$\rho_A = Tr_B(|\phi_B\rangle\langle\phi_B|) \tag{7.46}$$

If ρ_A is a mixed state, we can use purification to analyze the system as a pure state by expanding the Hilbert space to the larger space defined by $|\phi_A\phi_B\rangle$. Suppose

$$\rho_A = \sum_i p_i |a_i\rangle\langle a_i|$$

Let $|b_i\rangle$ be an orthonormal basis for system B. The purification is then

$$|\phi_B\rangle = \sum_i \sqrt{p_i}|a_i\rangle \otimes |b_i\rangle \tag{7.47}$$

We will see how to calculate a purification explicitly and with some applications in Chapter 10.

EXERCISES

7.1. *Derive the result (7.11).*

7.2. *The eigenstates of the Y operator are*

$$|\pm_y\rangle = \frac{|0\rangle \pm i|1\rangle}{\sqrt{2}}$$

Rewrite the singlet state (7.2) in terms of the Y eigenstates. Does it have a similar form?

7.3. *Verify that* $Z\otimes Z|\beta_{xy}\rangle = (-1)^y|\beta_{xy}\rangle$ *for*

$$|\beta_{00}\rangle = \frac{|00\rangle + |11\rangle}{\sqrt{2}} \quad \text{and} \quad |\beta_{01}\rangle = \frac{|01\rangle + |10\rangle}{\sqrt{2}}$$

7.4. *Show that* $X\otimes X|\beta_{xy}\rangle = (-1)^x|\beta_{xy}\rangle$.

7.5. *Show that* $Y\otimes Y|\beta_{xy}\rangle = (-1)^{x+y}|\beta_{xy}\rangle$.

7.6. *Show that* $X\otimes X$ *commutes with* $Z\otimes Z$.

7.7. *Consider the eigenvectors in Example 7.4. Show that* $[H_I, \vec{\sigma}_A\cdot\vec{\sigma}_B] = 0$, *and hence show that the eigenvectors of the Hamiltonian are eigenvectors of the* $\vec{\sigma}_A\cdot\vec{\sigma}_B$ *operator. In particular, show that* $\vec{\sigma}_A\cdot\vec{\sigma}_B|\phi_i\rangle = |\phi_i\rangle$ *for* $i = 1, 2, 3$ *and* $\vec{\sigma}_A\cdot\vec{\sigma}_B|\phi_4\rangle = -3|\phi_4\rangle$.

7.8. *Is the state* $X\otimes Z|\beta_{00}\rangle$*entangled?*

7.9. *Find the Pauli representation of*

$$\rho = \begin{pmatrix} \sin^2\theta & e^{-i\phi}\sin\theta\cos\theta \\ e^{i\phi}\sin\theta\cos\theta & \cos^2\theta \end{pmatrix}$$

7.10. *Use (7.36) to show that $|\beta_{10}\rangle$ is entangled. Apply the same criterion to test $X \otimes Z|\beta_{00}\rangle$.*

7.11. *Derive (7.39).*

7.12. *Can the following state be written in diagonal form in terms of the Bell basis?*

$$\rho = \begin{pmatrix} \frac{1}{2} & 0 & 0 & \frac{-1}{8} \\ 0 & 0 & 0 & 0 \\ 0 & 0 & 0 & 0 \\ \frac{-1}{8} & 0 & 0 & \frac{1}{2} \end{pmatrix}$$

Using (7.43), determine if this is a separable state.

7.13. *Verify that*

$$|\psi\rangle = \left(\frac{|0\rangle - |1\rangle}{\sqrt{2}}\right) \otimes \left(\frac{|0\rangle - |1\rangle}{\sqrt{2}}\right)$$

is a product state using (7.36).

7.14. *Verify that the state*

$$|\psi\rangle = \frac{1}{\sqrt{2}}|\beta_{00}\rangle - \frac{1}{\sqrt{2}}|\beta_{01}\rangle$$

is entangled by calculating the Schmidt number.

8

QUANTUM GATES AND CIRCUITS

In a classical computer, at the most fundamental level there are two basic tasks that we can use when manipulating information. We can move it from one place to another, or we can do some type of basic processing on the information using a *logic gate*. Sets of logic gates can be connected together to construct digital circuits. In this chapter we will be introduced to the equivalent notions of logic gates and circuits in a quantum computer. We begin with a brief overview of classical logic gates.

CLASSICAL LOGIC GATES

The basic purpose of a logic gate is to manipulate or process information at the bit level in some way. A simple example is the *NOT* gate. The NOT gate is a single input gate. As its name implies, the NOT gate simply flips or inverts the value of the input bit. That is, if the input to the gate is a 0, the output is a 1, and if the input is a 1, the output is a 0. We can write down the action of the NOT gate schematically as

$$0 \mapsto 1$$

$$1 \mapsto 0$$

Quantum Computing Explained, by David McMahon

In more complicated situations we will need a more systematic way to write down the action of a gate. This is done using a *truth table*, which is a table that lists the inputs together with the corresponding outputs of the gate. For a NOT gate, this is very easy to do. We write the values of the single input bit on the left side of the table and the corresponding outputs on the right:

Input	NOT
0	1
1	0

Now we will consider more complicated operations that are applied to pairs of bits. There are several two input gates, including the OR gate, the AND gate, and the exclusive-OR (XOR) gates. Let's examine each one in turn.

The OR gate accepts two bits as input, which we label A and B. The output of the OR gate is a 1 if A OR B is 1, and is 0 otherwise. That is,

A	B	A OR B
0	0	0
0	1	1
1	0	1
1	1	1

The AND gate returns a 1 only if both inputs are 1:

A	B	A AND B
0	0	0
0	1	0
1	0	0
1	1	1

The XOR gate returns a 1 when A OR B is 1, but does not return a 1 if they are both 1. The XOR operation can be indicated using the $\oplus$ symbol:

A	B	$A \oplus B$
0	0	0
0	1	1
1	0	1
1	1	0

The next two input gate that is important in classical computing is the NOT-AND or NAND gate. This gate works by inverting the result of the AND gate. The truth table in this case is as follows:

A	B	A NAND B
0	0	1
0	1	1
1	0	1
1	1	0

This gate has the interesting property of being *universal*. That is, all computing operations can be completed using only NAND gates. In fact you can construct an entire computer using nothing but NAND gates, or the combination of NOT and AND gates. As a simple example of how other logic operations can be implemented using only NAND gates, suppose that happens if we supply the *same* bit to both inputs of the NAND gate. The truth table is reduced to the following:

A	A	A OR B
0	0	1
1	1	0

In this case the NAND gate has inverted the value of the input gate. In other words, we've constructed a NOT gate out of a NAND gate.

You Try It

Using only NAND gates, show how you could construct and OR gate.

The NAND gate is interesting because it is universal, but it is also *irreversible*. That is, looking at the output of a NAND gate, we cannot work backward to determine the values of the input bits once the gate has acted upon them. However, it turns out that there are reversible gates that can be used to construct a classical computer. The first of these is called the *Fredkin gate*. This gate has three input bits, the first of which is a *control bit*. We denote the control bit by C; its function is to determine whether or not a given operation will be applied to the other input bits. In the Fredkin gate, if $C = 0$, then nothing is done to the input bits—they simply pass through the circuit unchanged. However, if $C = 1$, then the values of the bits are interchanged or swapped. The truth table for the Fredkin gate is given as follows:

C	A	B	A′	B′
0	0	0	0	0
0	0	1	0	1
0	1	0	1	0
0	1	1	1	1
1	0	0	0	0
1	0	1	1	0
1	1	0	0	1
1	1	1	1	1

We've denoted the output bits by A′ and B′. Like the NAND gate, the Fredkin gate is universal.

The final classical gate we'll look at is called the *Toffoli gate*. This gate has two control bits, which we'll designate by C1 and C2. The gate operates by computing C1 AND C2, and then computes the XOR of the result with a target bit, which we denote by T. The truth table is as follows:

C1	C2	T	T′
0	0	0	0
0	0	1	1
0	1	0	0
0	1	1	1
1	0	0	0
1	0	1	1
1	1	0	1
1	1	1	0

SINGLE-QUBIT GATES

A gate can be thought of as an abstraction that represents information processing. Now that we have a basic idea of how bits can be processed using logic gates, let's move on to consider the analogous process in a quantum computer. In a quantum computer, information is also processed using gates, but in this case the "gates" are unitary operations. Since quantum gates are just unitary operators, we'll often go back and forth between the words gate and operator—so keep in mind they mean the same thing in this context. Recall that a unitary operator U is one where the adjoint is equal to the inverse, meaning $U^\dagger = U^{-1}$. The defining relation for a unitary operator is thus

$$UU^\dagger = U^\dagger U = I \tag{8.1}$$

In addition, if H is a Hermitian operator, then $U = e^{iHt}$ is unitary.

Recall that quantum operators can be represented by matrices. A quantum gate with n inputs and outputs can be represented by a matrix of degree 2^n. For a single qubit, we require a matrix of degree $2^1 = 2$. That is, a quantum gate acting on a single qubit will be a 2×2 unitary matrix. A two-qubit gate can be implemented with a matrix of degree $2^2 = 4$ or a 4×4 matrix.

Following the procedure used when thinking about classical logic gates, we begin by examining the simplest gate possible, the quantum NOT gate. It turns out that we have already come across the quantum NOT gate—in fact we've already seen many of the single qubit gates. The NOT operation can be implemented with the X Pauli matrix. It never hurts to review, so let's go back and write out a few basics. The standard or computational basis states are given by

$$|0\rangle = \begin{pmatrix} 1 \\ 0 \end{pmatrix}, \quad |1\rangle = \begin{pmatrix} 0 \\ 1 \end{pmatrix} \tag{8.2}$$

The Pauli X matrix, which we will often refer to as the NOT operator, is given in matrix form in the standard or computational basis as

$$X = U_{NOT} = \begin{pmatrix} 0 & 1 \\ 1 & 0 \end{pmatrix} \tag{8.3}$$

Hence we have

$$U_{NOT}|0\rangle = \begin{pmatrix} 0 & 1 \\ 1 & 0 \end{pmatrix}\begin{pmatrix} 1 \\ 0 \end{pmatrix} = \begin{pmatrix} 0 \\ 1 \end{pmatrix} = |1\rangle \tag{8.4}$$

$$U_{NOT}|1\rangle = \begin{pmatrix} 0 & 1 \\ 1 & 0 \end{pmatrix}\begin{pmatrix} 0 \\ 1 \end{pmatrix} = \begin{pmatrix} 1 \\ 0 \end{pmatrix} = |0\rangle \tag{8.5}$$

So with respect to the standards or computational basis, the X matrix acts as a NOT operator. The action of a NOT gate on an arbitrary state $|j\rangle$ can be written using the XOR operation as

$$X|j\rangle = |j \oplus 1\rangle \tag{8.6}$$

To see how this works, recall the exclusive OR returns 1 if one or the other of the inputs are 1, but 0 otherwise. Hence, if $j = 0$, then $X|0\rangle = |0 \oplus 1\rangle = |1\rangle$. Meanwhile, if $j = 1$, then $X|1\rangle = |1 \oplus 1\rangle = |0\rangle$.

Example 8.1

The NOT operator takes $|0\rangle \mapsto |1\rangle$ and $|1\rangle \mapsto |0\rangle$. Describe the unitary operator that will implement the NOT operation in outer product form, and find its matrix representation with respect to the basis

$$|+\rangle = \frac{1}{\sqrt{2}}\begin{pmatrix}1\\1\end{pmatrix}, \quad |-\rangle = \frac{1}{\sqrt{2}}\begin{pmatrix}1\\-1\end{pmatrix}$$

Solution

In the standard or computational basis, the matrix representation of the NOT operator is given by (8.3). We can also write this as

$$X = \begin{pmatrix}\langle 0|X|0\rangle & \langle 0|X|1\rangle\\ \langle 1|X|0\rangle & \langle 1|X|1\rangle\end{pmatrix}$$

You can see that this will work if

$$X = |0\rangle\langle 1| + |1\rangle\langle 0| \tag{8.7}$$

(check it). Then the action of the operator on the standard or computational basis states is

$$\begin{aligned}X|0\rangle &= (|0\rangle\langle 1| + |1\rangle\langle 0|)|0\rangle\\ &= |0\rangle\langle 1|0\rangle + |1\rangle\langle 0|0\rangle\\ &= |1\rangle\end{aligned}$$

and

$$\begin{aligned}X|1\rangle &= (|0\rangle\langle 1| + |1\rangle\langle 0|)|1\rangle\\ &= |0\rangle\langle 1|1\rangle + |1\rangle\langle 0|1\rangle\\ &= |0\rangle\end{aligned}$$

where we have used the orthonormality of the basis states. To find the representation of the NOT operator in the $\{|+\rangle, |-\rangle\}$ basis, we need to find the unitary transformation connecting this basis with the standard or computational basis. That is, we need to find a matrix with components given by

$$U_{trans} = \begin{pmatrix}\langle +|0\rangle & \langle +|1\rangle\\ \langle -|0\rangle & \langle -|1\rangle\end{pmatrix}$$

We know already the representation of the $\{|+\rangle, |-\rangle\}$ states in the standard or computational basis:

$$|+\rangle = \frac{1}{\sqrt{2}}\begin{pmatrix}1\\1\end{pmatrix}, \quad |-\rangle = \frac{1}{\sqrt{2}}\begin{pmatrix}1\\-1\end{pmatrix}$$

So it's a simple matter to compute each matrix component:

$$\langle +|0\rangle = \frac{1}{\sqrt{2}}\begin{pmatrix}1 & 1\end{pmatrix}\begin{pmatrix}1\\0\end{pmatrix} = \frac{1}{\sqrt{2}}$$

$$\langle +|1\rangle = \frac{1}{\sqrt{2}}\begin{pmatrix}1 & 1\end{pmatrix}\begin{pmatrix}0\\1\end{pmatrix} = \frac{1}{\sqrt{2}}$$

$$\langle -|0\rangle = \frac{1}{\sqrt{2}}\begin{pmatrix}1 & -1\end{pmatrix}\begin{pmatrix}1\\0\end{pmatrix} = \frac{1}{\sqrt{2}}$$

$$\langle -|1\rangle = \frac{1}{\sqrt{2}}\begin{pmatrix}1 & -1\end{pmatrix}\begin{pmatrix}1\\0\end{pmatrix} = -\frac{1}{\sqrt{2}}$$

Hence the transformation matrix between the two bases is given by

$$U_{trans} = \frac{1}{\sqrt{2}}\begin{pmatrix}1 & 1\\1 & -1\end{pmatrix} = H \tag{8.8}$$

This matrix is nothing other than the Hadamard matrix H. For this reason the $\{|+\rangle, |-\rangle\}$ basis is sometimes called the *Hadamard basis*. Now we can apply this unitary transformation to the matrix representation of the NOT gate to find its representation with respect to the Hadamard basis. It is easy to verify that $H = H^{\dagger} = H^{-1}$, so the unitary transformation that takes NOT from the standard or computational basis to the Hadamard basis is just

$$U_{NOT}^{H} = HU_{NOT}H = \frac{1}{2}\begin{pmatrix}1 & 1\\1 & -1\end{pmatrix}\begin{pmatrix}0 & 1\\1 & 0\end{pmatrix}\begin{pmatrix}1 & 1\\1 & -1\end{pmatrix}$$

$$= \frac{1}{2}\begin{pmatrix}1 & 1\\1 & -1\end{pmatrix}\begin{pmatrix}1 & -1\\1 & 1\end{pmatrix}$$

$$= \begin{pmatrix}1 & 0\\0 & -1\end{pmatrix}$$

You Try It

Show that U_{NOT}^{H} acts as a NOT gate when applied to the Hadamard basis, taking $|+\rangle \mapsto |-\rangle$ and $|-\rangle \mapsto |+\rangle$.

In the Bloch sphere picture, the X or NOT gate reflects a state vector twice—first about the x-y plane, and then about the x-z plane.

Example 8.2

A rotation matrix by an angle γ is given by

$$R(\gamma) = \begin{pmatrix}\cos\gamma & -\sin\gamma\\ \sin\gamma & \cos\gamma\end{pmatrix}$$

Describe the action of this operator on a qubit $|\psi\rangle = \cos\theta|0\rangle + \sin\theta|1\rangle$.

Solution

The rotation matrix acts on the state as follows:

$$R(\gamma)|\psi\rangle = \begin{pmatrix} \cos\gamma & -\sin\gamma \\ \sin\gamma & \cos\gamma \end{pmatrix} \begin{pmatrix} \cos\theta \\ \sin\theta \end{pmatrix} = \begin{pmatrix} \cos\gamma\cos\theta - \sin\gamma\sin\theta \\ \sin\gamma\cos\theta + \cos\gamma\sin\theta \end{pmatrix}$$

Now recall some basic trig identities:

$$\cos(\alpha+\beta) = \cos\alpha\cos\beta - \sin\alpha\sin\beta$$
$$\sin(\alpha+\beta) = \sin\alpha\cos\beta + \cos\alpha\sin\beta$$

So the rotated state can be written as

$$|\psi'\rangle = \begin{pmatrix} \cos(\gamma+\theta) \\ \sin(\gamma+\theta) \end{pmatrix} = \cos(\gamma+\theta)|0\rangle + \sin(\gamma+\theta)|1\rangle$$

Consider the Bloch sphere picture. The rotation operator has rotated the state vector relative to the z axis by the angle γ. More specifically for those who like a more concrete interpretation, the rotation operator has altered the relative length of each probability amplitude. If the original qubit were measured, the probability that we find the system in the state $|0\rangle$ is given by $\cos^2\theta$, while the probability that we find the system in the state $|1\rangle$ is given by $\sin^2\theta$. If we rotate the state before measurement, then these probabilities are changed to $\cos^2(\gamma+\theta)$ and $\sin^2(\gamma+\theta)$, respectively.

MORE SINGLE-QUBIT GATES

The other Pauli matrices, being unitary 2×2 matrices, are also valid single-qubit gates. The Z operator is sometimes called the *phase flip* gate because it takes a qubit $|\psi\rangle = \alpha|0\rangle + \beta|1\rangle$ into a state $|\psi'\rangle = \alpha|0\rangle - \beta|1\rangle$. This is easy to see using the matrix representation:

$$Z|\psi\rangle = \begin{pmatrix} 1 & 0 \\ 0 & -1 \end{pmatrix} \begin{pmatrix} \alpha \\ \beta \end{pmatrix} = \begin{pmatrix} \alpha \\ -\beta \end{pmatrix} \quad (8.9)$$

We can also show the phase flip using the outer product representation, where we have $Z = |0\rangle\langle 0| - |1\rangle\langle 1|$:

$$\begin{aligned} Z|\psi\rangle &= (|0\rangle\langle 0| - |1\rangle\langle 1|)(\alpha|0\rangle + \beta|1\rangle) \\ &= \alpha|0\rangle\langle 0|0\rangle + \beta|0\rangle\langle 0|1\rangle - \alpha|1\rangle\langle 1|0\rangle - \beta|1\rangle\langle 1|1\rangle \\ &= \alpha|0\rangle - \beta|1\rangle \end{aligned}$$

It's fairly easy to see that generally we can represent the action of the Z gate by

$$Z|j\rangle = (-1)^j|j\rangle$$

You Try It

Describe the action of the Y Pauli operator on an arbitrary qubit. Use the matrix and outer product representations.

More generally, the *phase shift* gate is given by

$$P = \begin{pmatrix} 1 & 0 \\ 0 & e^{i\theta} \end{pmatrix} \tag{8.10}$$

This gate shifts or alters the relative phase of the amplitudes α and β of a qubit. The Z gate is just the special case where $\theta = \pi$, and hence $e^{i\pi} = \cos \pi + i \sin \pi = -1$. In general, the action of the phase shift gate on a qubit is

$$P|\psi\rangle = \begin{pmatrix} 1 & 0 \\ 0 & e^{i\theta} \end{pmatrix} \begin{pmatrix} \alpha \\ \beta \end{pmatrix} = \begin{pmatrix} \alpha \\ e^{i\theta}\beta \end{pmatrix} \tag{8.11}$$

Example 8.3

Describe the action of the phase shift gate when considering the Bloch sphere representation of a qubit.

Solution

We write the qubit as

$$|\psi\rangle = \cos\theta|0\rangle + e^{i\phi}\sin\theta|1\rangle$$

The phase shift operator (using an angle γ) can be written in outer product notation as follows:

$$P = |0\rangle\langle 0| + e^{i\gamma}|1\rangle\langle 1| \tag{8.12}$$

Hence

$$\begin{aligned} P|\psi\rangle &= (|0\rangle\langle 0| + e^{i\gamma}|1\rangle\langle 1|)(\cos\theta|0\rangle + e^{i\phi}\sin\theta|1\rangle) \\ &= \cos\theta|0\rangle + e^{i(\gamma+\phi)}\sin\theta|1\rangle \end{aligned}$$

Therefore we see that the phase shift operator takes the azimuthal angle $\phi \to \phi + \gamma$.

We have seen that the Z gate is a special case of the phase shift operator where we take the angle to be π. There are other special cases of interest. The first of these is when we take $\theta = \pi/2$. By Euler's identity, $e^{i\pi/2} = \cos(\pi/2) + i\sin(\pi/2) = i$. The resulting gate is called the S gate, which has the matrix representation in the standard or computational basis given by

$$S = \begin{pmatrix} 1 & 0 \\ 0 & i \end{pmatrix} \tag{8.13}$$

If we let $\theta = \pi/4$, then we have the $\pi/8$ or T gate:

$$T = \begin{pmatrix} 1 & 0 \\ 0 & e^{i\pi/4} \end{pmatrix} = e^{i\pi/8} \begin{pmatrix} e^{-i\pi/8} & 0 \\ 0 & e^{i\pi/8} \end{pmatrix} \tag{8.14}$$

Of course, we have already seen the Hadamard matrix:

$$H = \frac{1}{\sqrt{2}} \begin{pmatrix} 1 & 1 \\ 1 & -1 \end{pmatrix} \tag{8.15}$$

Example 8.4

Write the Hadamard matrix in outer product form (using the standard or computational basis) and describe its action on the basis states $\{|0\rangle, |1\rangle\}$.

Solution

The matrix representation given in (8.15) can be rewritten as

$$H \doteq \begin{pmatrix} \langle 0|H|0\rangle & \langle 0|H|1\rangle \\ \langle 1|H|0\rangle & \langle 1|H|1\rangle \end{pmatrix}$$

Comparing this to (8.15), we see that the outer product representation of the Hadamard operator must be

$$H = \frac{1}{\sqrt{2}} (|0\rangle\langle 0| + |0\rangle\langle 1| + |1\rangle\langle 0| - |1\rangle\langle 1|) \tag{8.16}$$

Now let's see how the Hadamard operator acts on $|0\rangle$:

$$\begin{aligned} H|0\rangle &= \frac{1}{\sqrt{2}} (|0\rangle\langle 0| + |0\rangle\langle 1| + |1\rangle\langle 0| - |1\rangle\langle 1|)|0\rangle \\ &= \frac{1}{\sqrt{2}} (|0\rangle\langle 0|0\rangle + |0\rangle\langle 1|0\rangle + |1\rangle\langle 0|0\rangle - |1\rangle\langle 1|0\rangle) \\ &= \frac{|0\rangle + |1\rangle}{\sqrt{2}} \end{aligned}$$

Similarly we find that

$$H|1\rangle = \frac{|0\rangle - |1\rangle}{\sqrt{2}}$$

Therefore the action of the Hadamard gate on the standard or computational basis states is to map the $\{|0\rangle, |1\rangle\}$ states into the superposition states

$$\left\{ \frac{|0\rangle + |1\rangle}{\sqrt{2}}, \frac{|0\rangle - |1\rangle}{\sqrt{2}} \right\}$$

In general, the Hadamard gate takes the state $|\psi\rangle = \alpha|0\rangle + \beta|1\rangle$ into the state

$$H|\psi\rangle = \left(\frac{\alpha+\beta}{\sqrt{2}}\right)|0\rangle + \left(\frac{\alpha-\beta}{\sqrt{2}}\right)|1\rangle \tag{8.17}$$

This means that the probability of finding the qubit in the state $|0\rangle$ is changed from

$$|\alpha|^2 \text{to} \left|\frac{\alpha+\beta}{\sqrt{2}}\right|^2 = \left(\frac{\alpha^*+\beta^*}{\sqrt{2}}\right)\left(\frac{\alpha+\beta}{\sqrt{2}}\right) = \frac{1}{2}(|\alpha|^2 + |\beta|^2 + \text{Re}(\alpha\beta^*))$$

and similarly for the probability of finding the system in the $|1\rangle$ state. We can regroup the terms in (8.17) to give another interpretation of the output of a Hadamard gate:

$$H|\psi\rangle = \alpha\frac{|0\rangle+|1\rangle}{\sqrt{2}} + \beta\frac{|0\rangle-|1\rangle}{\sqrt{2}} = \alpha|+\rangle + \beta|-\rangle \tag{8.18}$$

That is, the Hadamard gate has turned a state that, with respect to the standard or computational basis, had the probability $|\alpha|^2$ of finding the system in the state $|0\rangle$ and the probability $|\beta|^2$ of finding the system in the state $|1\rangle$ into a state that has the probability $|\alpha|^2$ of finding the system in the state $|+\rangle$ and the probability $|\beta|^2$ of finding the system in the state $|-\rangle$.

You Try It

A Hadamard gate is applied to the qubit $|\psi\rangle = \cos\theta|0\rangle + e^{i\phi}\sin\theta|1\rangle$, and subsequently a measurement is made. What is the probability that this measurement finds the system in the state $|1\rangle$?

EXPONENTIATION

Now let's consider the construction of more single-qubit gates using exponentiation. If a matrix U is unitary and Hermitian, then

$$\exp(-i\theta U) = \cos\theta I - i\sin\theta U \tag{8.19}$$

This is very easy to prove, which we do in the next example.

Example 8.5

Prove that if an operator U is unitary and Hermitian, then $\exp(-i\theta U) = \cos\theta I - i\sin\theta U$.

Solution

If a matrix or operator U is unitary, then

$$UU^\dagger = U^\dagger U = I$$

If the operator is also Hermitian, then

$$U = U^\dagger$$

Combining these two relations yields

$$U^2 = UU = UU^\dagger = I$$

So we have

$$\exp(-i\theta U) = I - i\theta U + (-i)^2\frac{\theta^2}{2!}U^2 + (-i)^3\frac{\theta^3}{3!}U^3 + (-i)^4\frac{\theta^4}{4!}U^4 + (-i)^5\frac{\theta^5}{5!}U^5 + \cdots$$

Since $U^2 = I$ and $i^2 = -1$, this relation becomes

$$\begin{aligned}\exp(-i\theta U) &= \left(I - \frac{\theta^2}{2!}I + \frac{\theta^4}{4!}I - \cdots\right) - i\theta U + i\frac{\theta^3}{3!}U - i\frac{\theta^5}{5!}U + \cdots \\ &= \left(1 - \frac{\theta^2}{2!} + \frac{\theta^4}{4!} - \cdots\right)I - i\left(\theta - \frac{\theta^3}{3!} + \frac{\theta^5}{5!} + \cdots\right)U \\ &= \cos\theta I - \sin\theta U\end{aligned}$$

By exponentiating a given matrix, we can come up with more gates. In fact we can create rotation operators to represent rotation about the x, y, and z axes on the Bloch sphere by exponentiating the Pauli matrices. These are given by

$$R_x(\gamma) = e^{-i\gamma X/2} = \begin{pmatrix} \cos\left(\frac{\gamma}{2}\right) & -i\sin\left(\frac{\gamma}{2}\right) \\ -i\sin\left(\frac{\gamma}{2}\right) & \cos\left(\frac{\gamma}{2}\right) \end{pmatrix} \tag{8.20}$$

$$R_y(\gamma) = e^{-i\gamma Y/2} = \begin{pmatrix} \cos\left(\frac{\gamma}{2}\right) & -\sin\left(\frac{\gamma}{2}\right) \\ \sin\left(\frac{\gamma}{2}\right) & \cos\left(\frac{\gamma}{2}\right) \end{pmatrix} \tag{8.21}$$

$$R_z(\gamma) = e^{-i\gamma Z/2} = \begin{pmatrix} e^{-i\gamma/2} & 0 \\ 0 & e^{i\gamma/2} \end{pmatrix} \tag{8.22}$$

You Try It

Using (8.19), derive the matrix representation of $R_z(\gamma)$.

The rotation matrices given in (8.20), (8.21), and (8.22) allow us to generate different representations of other single-qubit gates related to the Bloch sphere picture. For example for the T gate of (8.14),

$$T = e^{i\pi/8}\begin{pmatrix} e^{-i\pi/8} & 0 \\ 0 & e^{i\pi/8} \end{pmatrix}$$

From (8.22) it is easy to see that

$$T = e^{i\pi/8}\begin{pmatrix} e^{-i\pi/8} & 0 \\ 0 & e^{i\pi/8} \end{pmatrix} = e^{i\pi/8}e^{-i\pi Z/8} = e^{i\pi/8}R_z\left(\frac{\pi}{4}\right) \tag{8.23}$$

This shows us that the T gate is equivalent to a 45 degree rotation around the z axis.

THE Z–Y DECOMPOSITION

Given a single-qubit operator U, we can find real numbers a, b, c, d such that

$$U = e^{ia}R_z(b)R_y(c)R_z(d) \tag{8.24}$$

BASIC QUANTUM CIRCUIT DIAGRAMS

We can represent the action of a quantum gate by drawing a circuit diagram. Each unitary operator or gate is represented by a block with lines (or "wires") used to represent input and output. For example, the representations of the Pauli operators X, Y, and Z and their action on a single qubit are shown in Figure 8.1.

The circuit diagram representation of a Hadamard gate is shown in Figure 8.2. We represent measurement in a circuit with an encircled M. This is shown in Figure 8.3. A measurement in a circuit diagram should remind you of the

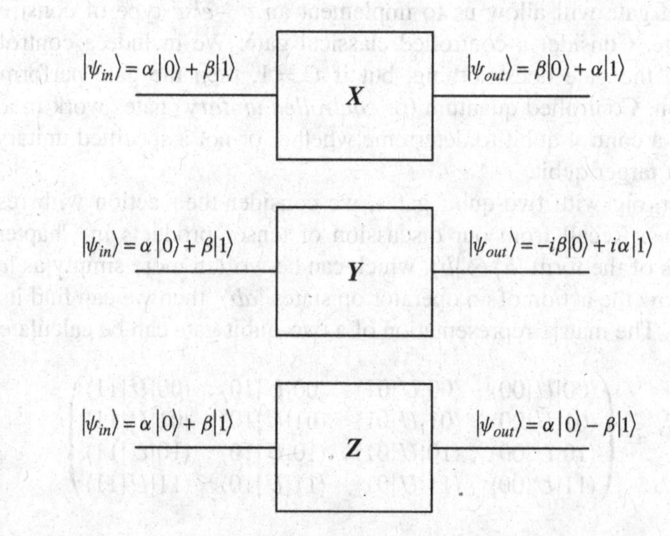

Figure 8.1 Circuit diagram representations of the Pauli operators and their actions on a single arbitrary qubit

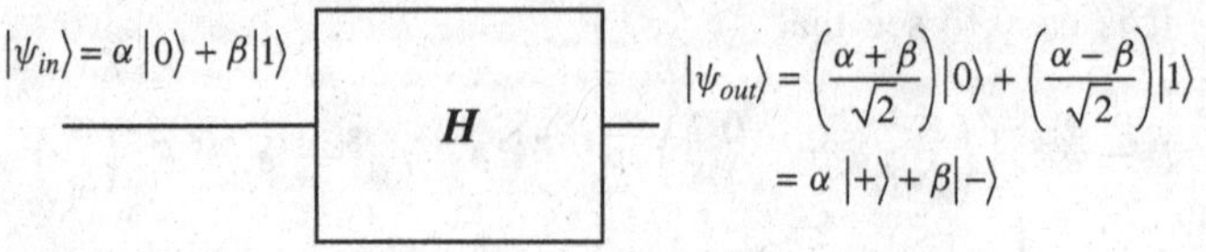

$$|\psi_{in}\rangle = \alpha|0\rangle + \beta|1\rangle$$

H

$$|\psi_{out}\rangle = \left(\frac{\alpha+\beta}{\sqrt{2}}\right)|0\rangle + \left(\frac{\alpha-\beta}{\sqrt{2}}\right)|1\rangle$$
$$= \alpha|+\rangle + \beta|-\rangle$$

Figure 8.2 The Hadamard gate

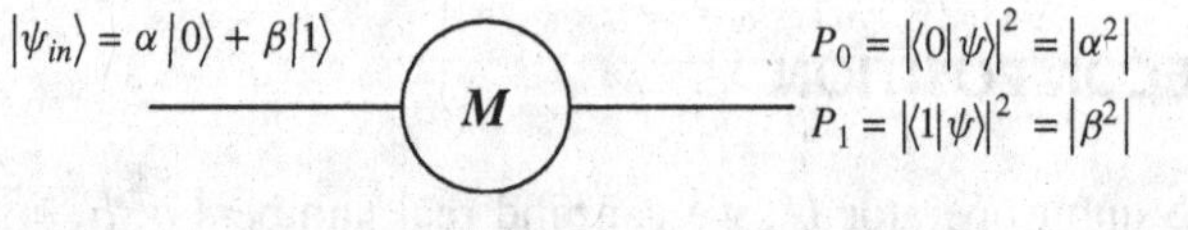

$$|\psi_{in}\rangle = \alpha|0\rangle + \beta|1\rangle$$

M

$$P_0 = |\langle 0|\psi\rangle|^2 = |\alpha^2|$$
$$P_1 = |\langle 1|\psi\rangle|^2 = |\beta^2|$$

Figure 8.3 Representation of measurement in a quantum circuit

Born rule—the probability of each measurement result is given by the corresponding squared amplitudes.

In other texts or papers the representation of measurement may be different, but it should be clear from the context. We will see how to incorporate measurement into quantum circuits in later chapters, for example, when we consider teleportation in Chapter 10.

CONTROLLED GATES

We are ready to move to the case of two-qubit gates. In this section the notion of a controlled gate will allow us to implement an *if −else* type of construct with a quantum gate. Consider a controlled classical gate. We include a control bit C. If $C = 0$, then the gate does nothing, but if $C = 1$, then the gate performs some specified action. Controlled quantum (or *controlled unitary*) gates work in a similar fashion, using a control qubit to determine whether or not a specified unitary action is applied to a target qubit.

When working with two-qubit gates, we consider their action with respect to two-qubit states. Recall from our discussion of tensor products in Chapter 4 that these are states of the form $|a\rangle \otimes |b\rangle$, which can be written more simply as $|a\rangle|b\rangle$ or $|ab\rangle$. If we know the action of an operator on states $|ab\rangle$, then we can find its matrix representation. The matrix representation of a two-qubit gate can be calculated using

$$U \doteq \begin{pmatrix} \langle 00|U|00\rangle & \langle 00|U|01\rangle & \langle 00|U|10\rangle & \langle 00|U|11\rangle \\ \langle 01|U|00\rangle & \langle 01|U|01\rangle & \langle 01|U|10\rangle & \langle 01|U|11\rangle \\ \langle 10|U|00\rangle & \langle 10|U|01\rangle & \langle 10|U|10\rangle & \langle 10|U|11\rangle \\ \langle 11|U|00\rangle & \langle 11|U|01\rangle & \langle 11|U|10\rangle & \langle 11|U|11\rangle \end{pmatrix} \tag{8.25}$$

The first two-qubit gate we will meet is the *controlled NOT* or *CNOT* gate. The first input to a controlled not gate acts as the control qubit. Using the notation

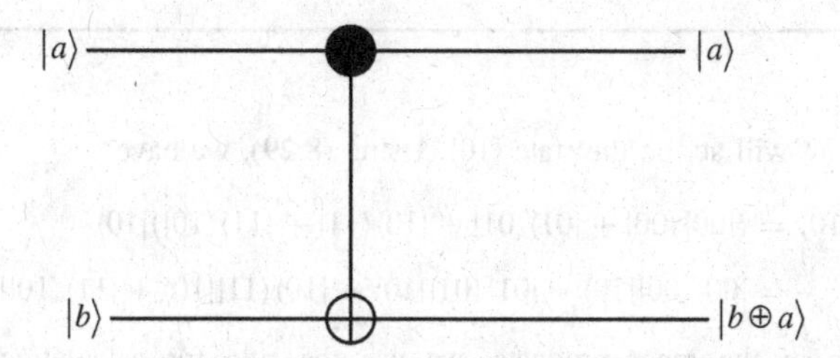

Figure 8.4 Circuit diagram representation of a controlled NOT gate

developed in Chapter 4 for tensor product states, the action of a CNOT gate can be described in terms of the XOR operation as follows:

$$|a, b\rangle \to |a, b \oplus a\rangle \tag{8.26}$$

If the control qubit is $|0\rangle$, then nothing happens to the target qubit. If the control qubit is $|1\rangle$, then the NOT or X matrix is applied to the target cubit. The possible input states to the CNOT gate are $|00\rangle$,$|01\rangle$,$|10\rangle$, and $|11\rangle$, and the action of the CNOT gate on these states is

$$\begin{aligned} |00\rangle &\mapsto |00\rangle \\ |01\rangle &\mapsto |01\rangle \\ |10\rangle &\mapsto |11\rangle \\ |11\rangle &\mapsto |10\rangle \end{aligned} \tag{8.27}$$

A common circuit representation of the controlled NOT gate is shown in Figure 8.4.

To write the matrix representation of the controlled NOT gate, we have to do it with respect to the states $|00\rangle$,$|01\rangle$,$|10\rangle$, and $|11\rangle$. The matrix will be a 4×4 matrix. The matrix representation of this gate is given by

$$CN = \begin{pmatrix} 1 & 0 & 0 & 0 \\ 0 & 1 & 0 & 0 \\ 0 & 0 & 0 & 1 \\ 0 & 0 & 1 & 0 \end{pmatrix} \tag{8.28}$$

If we use Dirac notation, the outer product representation of the controlled NOT is

$$CN = |00\rangle\langle 00| + |01\rangle\langle 01| + |10\rangle\langle 11| + |11\rangle\langle 10| \tag{8.29}$$

Example 8.6

Using Dirac notation, find the action of the controlled NOT gate when the control bit is $|1\rangle$ and the target qubit is given by $|0\rangle$,$|1\rangle$ and $\alpha|0\rangle + \beta|1\rangle$.

Solution

In the first case, CN will act on the state $|10\rangle$. Using (8.29), we have

$$CN|10\rangle = (|00\rangle\langle 00| + |01\rangle\langle 01| + |10\rangle\langle 11| + |11\rangle\langle 10|)|10\rangle$$
$$= |00\rangle\langle 00||10\rangle + |01\rangle\langle 01||10\rangle + |10\rangle\langle 11||10\rangle + |11\rangle\langle 10||10\rangle$$

To calculate each of the inner products, we use the rule for calculating inner products given in (4.8):

$$\langle ab|cd = \langle a|c\rangle\langle b|d\rangle$$

Hence

$$\begin{aligned}\langle 00|10\rangle &= \langle 0|1\rangle\langle 0|0\rangle = 0\\ \langle 01|10\rangle &= \langle 0|1\rangle\langle 1|0\rangle = 0\\ \langle 11|10\rangle &= \langle 1|1\rangle\langle 1|0\rangle = 0\\ \langle 10|10\rangle &= \langle 1|1\rangle\langle 0|0\rangle = 1\end{aligned}$$

We conclude that

$$CN|10\rangle = |11\rangle$$

When the target qubit is $|1\rangle$, we have

$$CN|11\rangle = (|00\rangle\langle 00| + |01\rangle\langle 01| + |10\rangle\langle 11| + |11\rangle\langle 10|)|11\rangle$$
$$= |00\rangle\langle 00||11\rangle + |01\rangle\langle 01||11\rangle + |10\rangle\langle 11||11\rangle + |11\rangle\langle 10||11\rangle$$
$$= |10\rangle$$

So we've confirmed that the controlled NOT gate flips the target qubit when the control bit is $|1\rangle$. Now we can use what we've learned to find the action on the target qubit when it's in the state $\alpha|0\rangle + \beta|1\rangle$. In this case

$$CN(\alpha|10\rangle + \beta|11\rangle) = \alpha CN|10\rangle + \beta CN|11\rangle = \alpha|11\rangle + \beta|10\rangle$$

Therefore the CN takes $\alpha|0\rangle + \beta|1\rangle$ to $\beta|0\rangle + \alpha|1\rangle$ when the control bit is $|1\rangle$.

Example 8.7

Describe a circuit that will generate the Bell states.

Solution

In Chapter 7 we learned that the Bell states are given by

$$|\beta_{00}\rangle = \frac{|00\rangle + |11\rangle}{\sqrt{2}} \tag{8.30}$$

$$|\beta_{01}\rangle = \frac{|01\rangle + |10\rangle}{\sqrt{2}} \tag{8.31}$$

$$|\beta_{10}\rangle = \frac{|00\rangle - |11\rangle}{\sqrt{2}} \tag{8.32}$$

$$|\beta_{11}\rangle = \frac{|01\rangle - |10\rangle}{\sqrt{2}} \tag{8.33}$$

To see how we can draw a circuit to generate these states, consider the action of the CNOT gate when the *control qubit* is in a superposition state. For example, let's take it to be $|c\rangle = |0\rangle + |1\rangle$, and let the target qubit be $|0\rangle$. Then the CNOT gate will act on the sum $|00\rangle + |10\rangle$. From (8.29), we see that the action of the CN gate on this state is

$$\begin{aligned}
CN(|00\rangle + |10\rangle) &= (|00\rangle\langle 00| + |01\rangle\langle 01| + |10\rangle\langle 11| + |11\rangle\langle 10|)(|00\rangle + |10\rangle) \\
&= |00\rangle\langle 00||00\rangle + |01\rangle\langle 01||00\rangle + |10\rangle\langle 11||00\rangle + |11\rangle\langle 10||00\rangle \\
&\quad + |00\rangle\langle 00||10\rangle + |01\rangle\langle 01||10\rangle + |10\rangle\langle 11||10\rangle + |11\rangle\langle 10||10\rangle \\
&= |00\rangle + |11\rangle
\end{aligned}$$

This is almost the Bell state given by (8.30). All we are missing is the normalization constant $1/\sqrt{2}$. Of course, if we look back at (8.16) and the action of the Hadamard gate on $|0\rangle$, we see that this gives us the factor we need. So we will start with a qubit given by $|0\rangle$ and act on it with a Hadamard gate. That should give us the state

$$\frac{|0\rangle + |1\rangle}{\sqrt{2}}$$

The resulting qubit output from the Hadamard gate is then passed as the control qubit for the CN gate. The result will be the state $|\beta_{00}\rangle$. We can use a similar thought process to see how to generate the other Bell states. The circuit required to do this, in general, is shown in Figure 8.5. One thing to notice about this circuit is that moving from left to right indicates the passage of *time*. So a wire in a quantum

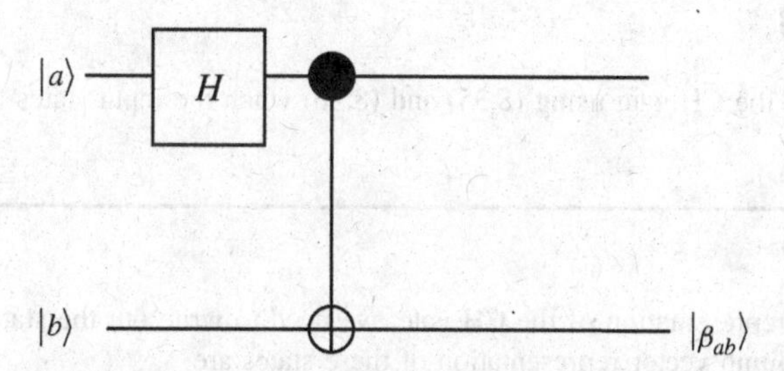

Figure 8.5 Diagram for a quantum circuit that creates Bell states. Time moves from left to right, with wires used to represent the passage of time where the state is left alone. First we apply a Hadamard gate to the qubit —a⟩ to generate a superposition state. This is then used as the control bit in a *CN* gate. The result is the bell state —$\beta_{ab}\rangle$

circuit is simply the time evolution of a quantum state. The operation of the circuit goes as follows, with the steps applied in order:

- Take a qubit $|a\rangle$ where $|a=0\rangle$ or $|a=1\rangle$, and act on it with a Hadamard gate.
- Use the resulting output as the control bit that is passed to a *CN* gate. If the target qubit is denoted by $|b\rangle$, where $b=\{0,1\}$, the target output of the *CN* gate will be the Bell state $|\beta_{ab}\rangle$

The Bell state $|\beta_{ab}\rangle$ can be written as

$$|\beta_{ab}\rangle = \frac{|0,b\rangle + (-1)^a|1,\bar{a}\rangle}{\sqrt{2}} \tag{8.34}$$

where $\bar{a}$ represents *NOT a*.

As we mentioned earlier, it is possible to generate any type of controlled U gate we wish. For example, we can have a *controlled-Hadamard* gate. Let's denote this by *CH*. The action of the controlled Hadamard gate is as follows: If the control qubit is $|0\rangle$, nothing happens to the target qubit. If the control qubit is $|1\rangle$, then we apply a Hadamard gate to the target qubit. The matrix representation of the controlled Hadamard gate is

$$CH = \begin{pmatrix} 1 & 0 & 0 & 0 \\ 0 & 1 & 0 & 0 \\ 0 & 0 & \frac{1}{\sqrt{2}} & \frac{1}{\sqrt{2}} \\ 0 & 0 & \frac{1}{\sqrt{2}} & \frac{-1}{\sqrt{2}} \end{pmatrix} \tag{8.35}$$

The Dirac notation representation of this operator is

$$CH = |00\rangle\langle 00| + |01\rangle\langle 01| + \frac{1}{\sqrt{2}}(|10\rangle\langle 10| + |10\rangle\langle 11| + |11\rangle\langle 10| - |11\rangle\langle 11|) \tag{8.36}$$

Example 8.8

Find the action of the CH gate using (8.35) and (8.36) when the input states are $|01\rangle$and$|11\rangle$.

Solution

Using the matrix representation of the CH gate, we need to write out the states $|01\rangle$ and $|11\rangle$. From (4.9) the column vector representation of these states are

$$|01\rangle = \begin{pmatrix} 1 \\ 0 \end{pmatrix} \otimes \begin{pmatrix} 0 \\ 1 \end{pmatrix} = \begin{pmatrix} 0 \\ 1 \\ 0 \\ 0 \end{pmatrix}$$

$$|11\rangle = \begin{pmatrix}0\\1\end{pmatrix} \otimes \begin{pmatrix}0\\1\end{pmatrix} = \begin{pmatrix}0\\0\\0\\1\end{pmatrix}$$

We should find that applied to the state $|01\rangle$, the CH gate does nothing, since the control qubit is set to 0. We have

$$CH|01\rangle = \begin{pmatrix}1 & 0 & 0 & 0\\ 0 & 1 & 0 & 0\\ 0 & 0 & \frac{1}{\sqrt{2}} & \frac{1}{\sqrt{2}}\\ 0 & 0 & \frac{1}{\sqrt{2}} & \frac{-1}{\sqrt{2}}\end{pmatrix}\begin{pmatrix}0\\1\\0\\0\end{pmatrix} = \begin{pmatrix}0\\1\\0\\0\end{pmatrix} = |01\rangle$$

In the second case, we have

$$CH|11\rangle = \begin{pmatrix}1 & 0 & 0 & 0\\ 0 & 1 & 0 & 0\\ 0 & 0 & \frac{1}{\sqrt{2}} & \frac{1}{\sqrt{2}}\\ 0 & 0 & \frac{1}{\sqrt{2}} & \frac{-1}{\sqrt{2}}\end{pmatrix}\begin{pmatrix}0\\0\\0\\1\end{pmatrix} = \begin{pmatrix}0\\0\\\frac{1}{\sqrt{2}}\\\frac{-1}{\sqrt{2}}\end{pmatrix}$$

Hence the target qubit has been taken into the state $|0\rangle - |1\rangle/\sqrt{2}$. Now let's redo the calculations using Dirac notation. For the first state, we have

$$\begin{aligned}CH|01\rangle &= \left[|00\rangle\langle 00| + |01\rangle\langle 01| + \frac{1}{\sqrt{2}}(|10\rangle\langle 10| + |10\rangle\langle 11| + |11\rangle\langle 10| - |11\rangle\langle 11|)\right]|01\rangle\\ &= |00\rangle\langle 00||01\rangle + |01\rangle\langle 01||01\rangle + \frac{1}{\sqrt{2}}(|10\rangle\langle 10||01\rangle + |10\rangle\langle 11||01\rangle + |11\rangle\langle 10||01\rangle\\ &\quad - |11\rangle\langle 11||01\rangle) = |01\rangle\end{aligned}$$

For the second state, we find that

$$\begin{aligned}CH|11\rangle &= \left[|00\rangle\langle 00| + |01\rangle\langle 01| + \frac{1}{\sqrt{2}}(|10\rangle\langle 10| + |10\rangle\langle 11| + |11\rangle\langle 10| - |11\rangle\langle 11|)\right]|01\rangle\\ &= |00\rangle\langle 00||11\rangle + |01\rangle\langle 01||11\rangle + \frac{1}{\sqrt{2}}(|10\rangle\langle 10||11\rangle + |10\rangle\langle 11||11\rangle + |11\rangle\langle 10||01\rangle\\ &\quad - |11\rangle\langle 11||11\rangle) = \frac{|10\rangle - |11\rangle}{\sqrt{2}}\end{aligned}$$

Example 8.9

We wish to investigate the use of the controlled NOT gate as a cloning machine. Specifically can it clone the state $a|0\rangle - b|1\rangle$? Begin by supposing that the superposition state given by $a|0\rangle - b|1\rangle$ is used as the control qubit and that the target qubit is given by $|1\rangle$. Then consider the case where the target qubit is $|0\rangle$.

Solution

First let's write down what the output state would be if the gate could clone the state. If it could, then it would make a copy of $a|0\rangle - b|1\rangle$, and the output state would be the product state given by

$$a|0\rangle - b|1\rangle \otimes a|0\rangle - b|1\rangle = a^2|00\rangle - ab|01\rangle - ba|10\rangle + b^2|11\rangle$$

The input state is

$$a|0\rangle - b|1\rangle \otimes |1\rangle = a|01\rangle - b|11\rangle$$

The controlled NOT gate acts on this state as follows:

$$CN(a|01\rangle - b|11\rangle) = aCN|01\rangle - bCN|11\rangle = a|01\rangle - b|10\rangle$$

You can see that $a|01\rangle - b|10\rangle \neq a|0\rangle - b|1\rangle \otimes a|0\rangle - b|1\rangle$. If the target qubit is $|0\rangle$, then we have

$$CN(a|00\rangle - b|10\rangle) = aCN|00\rangle - bCN|10\rangle = a|00\rangle - b|11\rangle$$

This state is also not equal to the product state $a|0\rangle - b|1\rangle \otimes a|0\rangle - b|1\rangle$, so we aren't any closer to cloning the state. This example shows that the CN gate can't clone, in general. Can you think of any specific states that the CN gate might clone?

GATE DECOMPOSITION

A large part of working with quantum circuits is decomposing an arbitrary controlled unitary operation U into a series of single-qubit operations and controlled NOT gates. This procedure is illustrated schematically in Figure 8.6.

The figure shows some arbitrary controlled-U operation. An equivalent circuit, consisting of two controlled NOT gates and the single-qubit gates A, B, and C, will result in the same output. To illustrate how circuits can be manipulated, we have the common example illustrated in Figure 8.7. We want to prove that a controlled NOT gate can be written in terms of two Hadamard gates and a controlled Z gate.

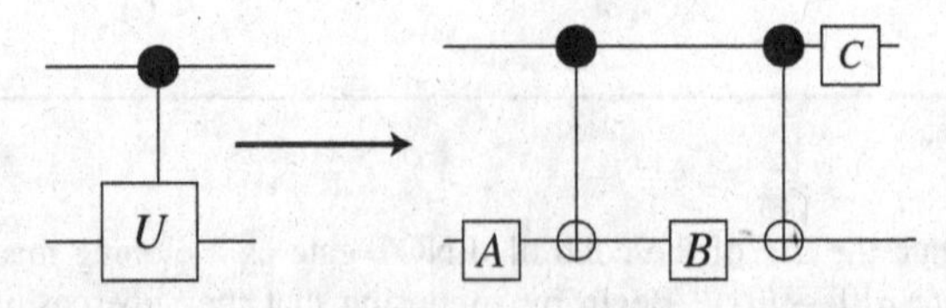

Figure 8.6 We replaced a controlled-U operation by an equivalent circuit consisting of controlled NOT gates and single-qubit gates

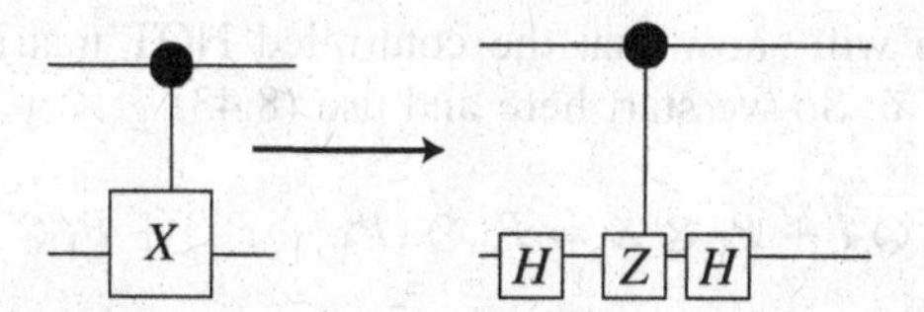

Figure 8.7 A controlled NOT gate is equivalent to a circuit consisting of two Hadamard gates and a controlled Z gate

To derive this result, we start by recalling the resolution of the identity operator in Dirac notation. We denote the projection operators onto the $|0\rangle$ and $|1\rangle$ states by

$$P_0 = |0\rangle\langle 0| \tag{8.37}$$

$$P_1 = |1\rangle\langle 1| \tag{8.38}$$

Then we write the identity operator as

$$I = |0\rangle\langle 0| + |1\rangle\langle 1| = P_0 + P_1 \tag{8.39}$$

Now consider the $+$X state:

$$|+\rangle = \frac{|0\rangle + |1\rangle}{\sqrt{2}} \tag{8.40}$$

The projection operator onto this state is

$$P_+ = |+\rangle\langle +| = \frac{1}{2}(|0\rangle\langle 0| + |0\rangle\langle 1| + |1\rangle\langle 0| + |1\rangle\langle 1|) \tag{8.41}$$

Similarly the projection operator onto the $-$X state is

$$P_- = |-\rangle\langle -| = \frac{1}{2}(|0\rangle\langle 0| - |0\rangle\langle 1| - |1\rangle\langle 0| + |1\rangle\langle 1|) \tag{8.42}$$

Notice that

$$P_+ + P_- = I \tag{8.43}$$

Now recall the Dirac notation representations for the X and Z operators and how they relate to (8.37), (8.38), (8.41), and (8.42):

$$X = |0\rangle\langle 1| + |1\rangle\langle 0| = P_+ - P_- \tag{8.44}$$

$$Z = |0\rangle\langle 0| - |1\rangle\langle 1| = P_0 - P_1 \tag{8.45}$$

In the exercises you will show that the controlled NOT matrix can be generated from $P_0 \otimes I + P_1 \otimes X$. So we start here and use (8.43):

$$P_0 \otimes I + P_1 \otimes X = P_0 \otimes (P_+ + P_-) + P_1 \otimes X$$

Next we use (8.44) to write X in terms of the same projection operators, giving

$$P_0 \otimes (P_+ + P_-) + P_1 \otimes (P_+ - P_-) = (P_0 + P_1) \otimes P_+ + (P_0 - P_1) \otimes P_-$$

From (8.39) and (8.45) we see that this is just

$$I \otimes P_+ + (P_0 - P_1) \otimes P_- = I \otimes P_+ + Z \otimes P_- \tag{8.46}$$

From (8.16) in Dirac notation the Hadamard operator is

$$H = \frac{1}{\sqrt{2}}(|0\rangle\langle 0| + |0\rangle\langle 1| + |1\rangle\langle 0| - |1\rangle\langle 1|)$$

Notice that

$$\begin{aligned} P_0 H &= (|0\rangle\langle 0|)\frac{1}{\sqrt{2}}(|0\rangle\langle 0| + |0\rangle\langle 1| + |1\rangle\langle 0| - |1\rangle\langle 1|) \\ &= \frac{1}{\sqrt{2}}(|0\rangle\langle 0| + |0\rangle\langle 1|) \end{aligned}$$

So we have

$$\begin{aligned} HP_0H &= \frac{1}{\sqrt{2}}(|0\rangle\langle 0| + |0\rangle\langle 1| + |1\rangle\langle 0| - |1\rangle\langle 1|)\frac{1}{\sqrt{2}}(|0\rangle\langle 0| + |0\rangle\langle 1|) \\ &= \frac{1}{2}(|0\rangle\langle 0| + |0\rangle\langle 1| + |1\rangle\langle 0| + |1\rangle\langle 1|) \\ &= P_+ \end{aligned}$$

This means that

$$I \otimes P_+ = I \otimes HP_0H = (I \otimes H)(I \otimes P_0H) = (I \otimes H)(I \otimes P_0)(I \otimes H)$$

It can also be shown that

$$Z \otimes P_- = (I \otimes H)(Z \otimes P_1)(I \otimes H)$$

Putting the two results together, we can rewrite (8.46) as

$$\begin{aligned} I \otimes P_+ + Z \otimes P_- &= (I \otimes H)(I \otimes P_0)(I \otimes H) + (I \otimes H)(Z \otimes P_1)(I \otimes H) \\ &= (I \otimes H)(I \otimes P_0 + Z \otimes P_1)(I \otimes H) \end{aligned}$$

This is the circuit shown on the right-hand side of Figure 8.7. What this tells us is to first apply the operation that leaves the first qubit alone and applies a Hadamard gate to the second qubit, then apply the controlled Z gate, and finally apply another Hadamard gate to the second qubit only.

EXERCISES

8.1. *Describe the action of the Y gate in terms of the Bloch sphere picture.*

8.2. *The Hubbard operators are given by*

$$X^{11} = |0\rangle\langle 0|, \quad X^{12} = |0\rangle\langle 1|$$
$$X^{21} = |1\rangle\langle 0|, \quad X^{22} = |1\rangle\langle 1|$$

(A) Write down the matrix representations of the Hubbard operators in the computational basis.
(B) Describe the action of the Hubbard operators on the Hadamard basis states.

8.3. *Find a way to write the Pauli operators X, Y, and Z in terms of the Hubbard operators.*

8.4. *Show that the controlled NOT gate is Hermitian and unitary.*

8.5. *Let* $|a\rangle = |1\rangle$*, and consider the circuit shown in Figure 8.5. Determine which Bell states are generated as output when* $|b\rangle = |0\rangle, |b\rangle = |1\rangle$.

8.6. *Write down the matrix representation for the controlled Z gate. Then write down its representation using Dirac notation.*

8.7. *Consider the single qubit operators X, Y, Z, S, and T. Find the square of each operator.*

8.8. *Show that the Hadamard gate is equivalent to a 180 degree rotation about the axis defined by* $(\vec{e}_x - \vec{e}_z)/\sqrt{2}$*, where* $\vec{e}_x$ *and* $\vec{e}_z$ *are unit vectors pointing along the x and z axes.*

8.9. *Show that* $X\vec{\sigma}X = X\,\vec{e}_x - Y\,\vec{e}_y - Z\,\vec{e}_z$.

8.10. *By using the tensor product methods developed in chapter 4, show that the controlled-NOT matrix can be generated from* $P_0 \otimes I + P_1 \otimes X$.

8.11. *Show that the following circuits are equivalent:*

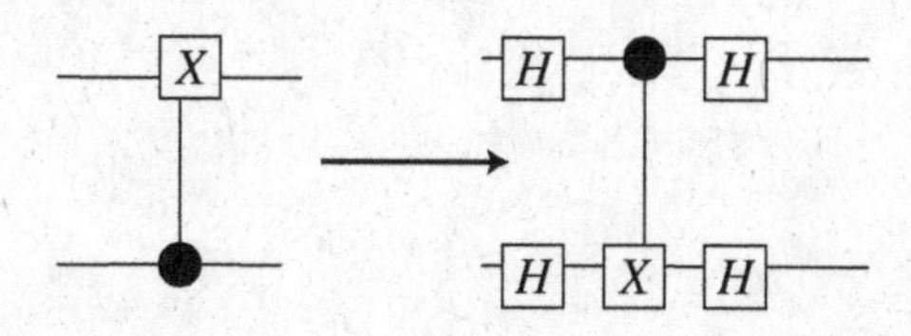

8.12. *The Toffoli gate can be implemented in a quantum circuit as*

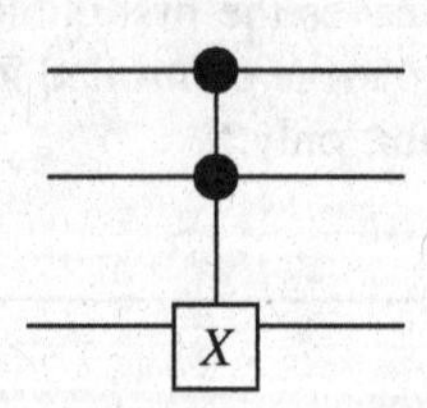

Show that if

$$V = (1 - i)\frac{(I + iX)}{2}$$

then this circuit can be replaced by

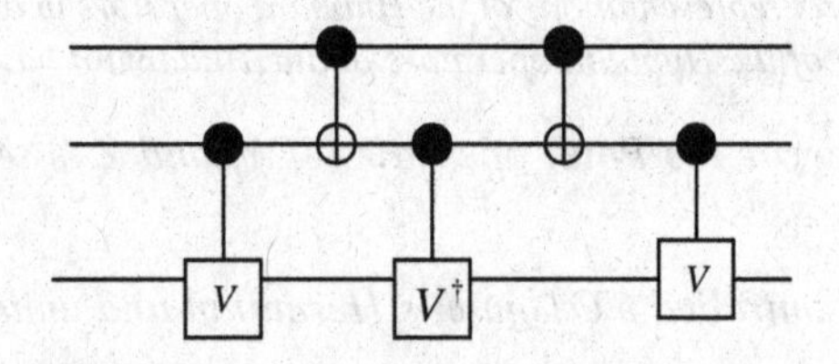

9

QUANTUM ALGORITHMS

An *algorithm* is a set of instructions used to perform some well-defined task on a computer. A large part of the desire to develop a quantum computer has come from the discovery that some algorithms work dramatically better on a quantum computer than they could ever work on a classical computer. This is because the nature of quantum systems—captured in superposition and interference of qubits—often allows a quantum system to compute in a parallel way that is not possible even, in principle, with a classical computer.

It was discovered that given a function $f(x)$, a quantum algorithm is capable of evaluating the function at multiple values of x simultaneously. As we will see below, a quantum algorithm highlights one of the central tugs-of-war that exist in quantum theory. A qubit can exist in a superposition of states, giving a quantum computer a hidden realm where exponential computations are possible. In other words, the fact that a quantum system can exist in a superposition or linear combination of states allows us to do simultaneous parallel computations that cannot be done even, in principle, on any classical computer. This feature allows a quantum computer to do parallel computations using a single circuit-providing a dramatic speedup in many cases. However, since measurement finds a qubit in one state *or* the other—frustratingly we find that if we give a quantum computer n inputs we only get n outputs.

Nature allows us to get around this to a certain extent and extract useful information using quantum interference. This is another feature of qubits that cannot be seen with classical bits—and it plays an important role in the development of useful algorithms. Because it's also true that a quantum computer can execute any classical algorithm, the hope is that as more algorithms are discovered and quantum computers become a reality, extremely efficient computing machines can be built. At the present time only a small set of truly quantum algorithms have been discovered, and this remains a very active area of research.

We begin by reviewing two quantum gates that are important in the development of quantum algorithms and explaining quantum interference. Then we explore our first truly quantum algorithm, called *Deutsch's algorithm* after its creator, the imaginative physicist David Deutsch.

HADAMARD GATES

An important step in quantum algorithms is to use Hadamard gates to create superposition states. Recall that the Hadamard gate H acts on the computational basis states in the following way:

$$H|0\rangle = \frac{|0\rangle + |1\rangle}{\sqrt{2}}, \quad H|1\rangle = \frac{|0\rangle - |1\rangle}{\sqrt{2}} \tag{9.1}$$

An interesting feature of the Hadamard gate is that two Hadamard gates in series act to reverse the operation and give back the original input. Let's consider a Hadamard gates applied to an arbitrary qubit $|\psi\rangle = \alpha|0\rangle + \beta|1\rangle$. We have

$$\begin{aligned} H|\psi\rangle &= \alpha H|0\rangle + \beta H|1\rangle = \alpha\left(\frac{|0\rangle + |1\rangle}{\sqrt{2}}\right) + \beta\left(\frac{|0\rangle - |1\rangle}{\sqrt{2}}\right) \\ &= \left(\frac{\alpha+\beta}{\sqrt{2}}\right)|0\rangle + \left(\frac{\alpha-\beta}{\sqrt{2}}\right)|1\rangle \end{aligned}$$

If we apply a Hadamard gate twice, we get the original state back:

$$\begin{aligned} H\left[\left(\frac{\alpha+\beta}{\sqrt{2}}\right)|0\rangle + \left(\frac{\alpha-\beta}{\sqrt{2}}\right)|1\rangle\right] &= \left(\frac{\alpha+\beta}{\sqrt{2}}\right)H|0\rangle + \left(\frac{\alpha-\beta}{\sqrt{2}}\right)H|1\rangle \\ &= \left(\frac{\alpha+\beta}{\sqrt{2}}\right)\left(\frac{|0\rangle + |1\rangle}{\sqrt{2}}\right) + \left(\frac{\alpha-\beta}{\sqrt{2}}\right)\left(\frac{|0\rangle - |1\rangle}{\sqrt{2}}\right) \\ &= \left(\frac{\alpha+\alpha+\beta-\beta}{2}\right)|0\rangle + \left(\frac{\alpha-\alpha+\beta+\beta}{2}\right)|1\rangle \\ &= \alpha|0\rangle + \beta|1\rangle = |\psi\rangle \end{aligned}$$

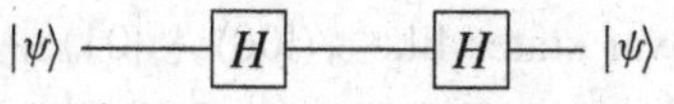

$|\psi\rangle$ — H — H — $|\psi\rangle$

Figure 9.1 Two Hadamard gates in series restore a qubit to its original state

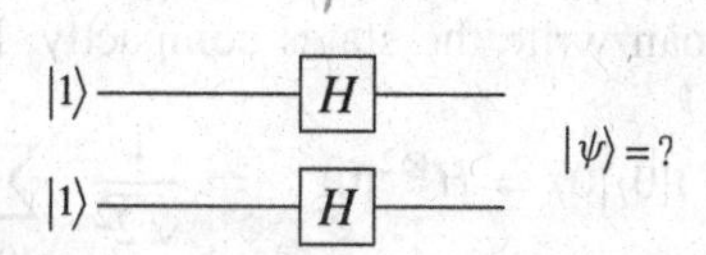

Figure 9.2 Two Hadamard gates used in parallel

So the action of two Hadamard gates in series, shown in Figure 9.1, restores a qubit to its original state.

Now consider what happens when we apply two Hadamard gates in parallel. Let's say that we do this for the state $|1\rangle|1\rangle$ (see Figure 9.2). The result of this action is the product state given by

$$\begin{aligned}(H \otimes H)|1\rangle|1\rangle &= (H|1\rangle)(H|1\rangle) = \left(\frac{|0\rangle - |1\rangle}{\sqrt{2}}\right)\left(\frac{|0\rangle - |1\rangle}{\sqrt{2}}\right) \\ &= \frac{1}{2}(|00\rangle - |01\rangle - |10\rangle + |11\rangle)\end{aligned} \tag{9.2}$$

As we will see, the operation of two or more Hadamard gates in parallel plays an important role in quantum algorithms. When n Hadamard gates act in parallel on n qubits, this is called a *Hadamard transform*. A shorthand notation that is sometimes used is to write $H^{\otimes n}$. So the operation shown in (9.2) can be denoted by $H^{\otimes 2}$. If we apply $H^{\otimes 2}$ to the product state $|0\rangle|0\rangle$, we obtain

$$\begin{aligned}(H \otimes H)|0\rangle|0\rangle &= (H|0\rangle)(H|0\rangle) = \left(\frac{|0\rangle + |1\rangle}{\sqrt{2}}\right)\left(\frac{|0\rangle + |1\rangle}{\sqrt{2}}\right) \\ &= \frac{1}{2}(|00\rangle + |01\rangle + |10\rangle + |11\rangle)\end{aligned} \tag{9.3}$$

In a similar way $H^{\otimes 3}|0\rangle|0\rangle|0\rangle$ gives us

$$\begin{aligned}(H \otimes H \otimes H)|0\rangle|0\rangle|0\rangle &= (H|0\rangle)(H|0\rangle)(H|0\rangle) \\ &= \left(\frac{|0\rangle + |1\rangle}{\sqrt{2}}\right)\left(\frac{|0\rangle + |1\rangle}{\sqrt{2}}\right)\left(\frac{|0\rangle + |1\rangle}{\sqrt{2}}\right) \\ &= \frac{1}{\sqrt{2^3}}(|000\rangle + |001\rangle + |010\rangle + |100\rangle + |101\rangle \\ &\quad + |110\rangle + |111\rangle)\end{aligned} \tag{9.4}$$

We can write a sum over states like $\frac{1}{2}(|00\rangle + |01\rangle + |10\rangle + |11\rangle)$ compactly in thefollowing way: We let $|x\rangle$ denote a general state where $x \in \{0, 1\}^2$; that is, $|x\rangle$ is one of $|00\rangle$, $|01\rangle$, $|10\rangle$, $|11\rangle$. If we write $x \in \{0, 1\}^3$, then we mean that $|x\rangle$ is one of the three qubit states $|000\rangle$, $|001\rangle$, $|010\rangle$, $|100\rangle$, $|101\rangle$, $|110\rangle$, $|111\rangle$. By summing over the variable $|x\rangle$, we can write the states compactly. For relation (9.3) can be written as

$$(H \otimes H)|0\rangle|0\rangle = H^{\otimes 2}|0\rangle^{\otimes 2} = \frac{1}{\sqrt{2^2}} \sum_{x \in \{0,1\}^2} |x\rangle \tag{9.5}$$

In general, the application of $H^{\otimes n}$ to a product state with n copies of $|0\rangle$ is

$$H^{\otimes n}(|0\rangle^{\otimes n}) = \frac{1}{\sqrt{2^n}} \sum_{x \in \{0,1\}^n} |x\rangle \tag{9.6}$$

Another useful operation is applying $H \otimes H$ to the product state $|0\rangle|1\rangle$. This gives

$$\begin{aligned}(H \otimes H)|0\rangle|1\rangle &= \left(\frac{|0\rangle + |1\rangle}{\sqrt{2}}\right)\left(\frac{|0\rangle - |1\rangle}{\sqrt{2}}\right) \\ &= \frac{1}{2}(|00\rangle - |01\rangle + |10\rangle - |11\rangle)\end{aligned} \tag{9.7}$$

If we think of x as running over the numbers 00, 01, 10, 11 (i.e, 0, 1, 2, 3), then we can write this compactly as

$$(H \otimes H)|0\rangle|1\rangle = \frac{1}{2} \sum_{x \in \{0,1\}} (-1)^x |x\rangle \tag{9.8}$$

Meanwhile

$$(H \otimes H)|1\rangle|1\rangle = \frac{1}{2}(|00\rangle - |01\rangle - |10\rangle + |11\rangle) = \frac{1}{2} \sum_{x \in \{0,1\}^2} (-1)^{x_0 \oplus x_1} |x\rangle \tag{9.9}$$

where $x_0 \oplus x_1$ is the exclusive-OR of the two qubits, and we are saying $|x\rangle$ is a two qubit state of the form $|x_0 x_1\rangle$, where x_0, x_1 are either 0 or 1.

Example 9.1

What is the result of apply the Hadamard transform to the state

$$|\psi\rangle = \frac{|0\rangle + (-1)^x |1\rangle}{\sqrt{2}}$$

where $x \in \{0, 1\}$.

Solution

Since

$$H|\psi\rangle = \alpha H|0\rangle + \beta H|1\rangle = \alpha\left(\frac{|0\rangle + |1\rangle}{\sqrt{2}}\right) + \beta\left(\frac{|0\rangle - |1\rangle}{\sqrt{2}}\right)$$

$$= \left(\frac{\alpha + \beta}{\sqrt{2}}\right)|0\rangle + \left(\frac{\alpha - \beta}{\sqrt{2}}\right)|1\rangle$$

The result is

$$H|\psi\rangle = \left(\frac{1 + (-1)^x}{2}\right)|0\rangle + \left(\frac{1 - (-1)^x}{2}\right)|1\rangle$$

If $x = 0$, this tells us that

$$H\left(\frac{|0\rangle + |1\rangle}{\sqrt{2}}\right) = |0\rangle \tag{9.10}$$

On the other hand, if $x = 1$, then

$$H\left(\frac{|0\rangle - |1\rangle}{\sqrt{2}}\right) = |1\rangle \tag{9.11}$$

THE PHASE GATE

Another useful gate that can be applied in the development of quantum algorithms is a variation of the phase gate we met in the last chapter, called the *discrete phase gate*. We denote the discrete phase gate by R_k, where

$$R_k = \begin{pmatrix} 1 & 0 \\ 0 & e^{(2\pi i/2^k)} \end{pmatrix} \tag{9.12}$$

MATRIX REPRESENTATION OF SERIAL AND PARALLEL OPERATIONS

When looking at quantum circuit diagrams or thinking about quantum algorithms, it is sometimes helpful to break down the problem in termsof matrices. There are two basic rules that can be followed. The first is that if a set of operations is performed in series, then this is represented by a *matrix product*. We represent a set of operations performed in series (i.e., in time) by starting with the first operation on the left followed by subsequent operations moving to the right. This is illustrated in Figure 9.3, where we first perform a phase gate with angle θ, followed by a Hadamard gate, which is then followed by the application of a Z gate.

The matrix representation of this sequence of operations is written down by multiplying the matrices in *reverse* order. Hence the operation shown in Figure 9.3 is written as

$$ZHP(\theta)$$

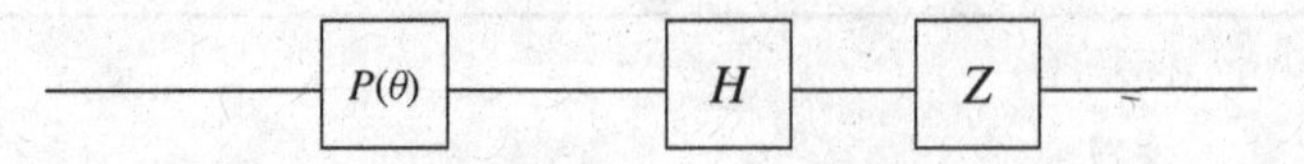

Figure 9.3 Circuit diagram for the application of a phase gate, followed by an application of a Hadamard gate, followed by the application of a Z gate

Explicitly we have

$$\begin{pmatrix} 1 & 0 \\ 0 & -1 \end{pmatrix} \frac{1}{\sqrt{2}} \begin{pmatrix} 1 & 1 \\ 1 & -1 \end{pmatrix} \begin{pmatrix} 1 & 0 \\ 0 & e^{i\theta} \end{pmatrix} = \frac{1}{\sqrt{2}} \begin{pmatrix} 1 & e^{i\theta} \\ -1 & e^{i\theta} \end{pmatrix}$$

When quantum operations are performed in parallel (i.e., at the *same time*), we compute the tensor product, which we already know how to do. So the matrix representation of $H \otimes H$ is

$$H \otimes H = \frac{1}{\sqrt{2}} \begin{pmatrix} H & H \\ H & -H \end{pmatrix} = \frac{1}{2} \begin{pmatrix} 1 & 1 & 1 & 1 \\ 1 & -1 & 1 & -1 \\ 1 & 1 & -1 & -1 \\ 1 & -1 & -1 & 1 \end{pmatrix}$$

As we will see when we start developing some quantum algorithms, interference and parallelism play a fundamental role. Before actually describing an algorithm we first describe quantum interference.

QUANTUM INTERFERENCE

The application of a Hadamard gate to an arbitrary qubit is an example of *quantum interference*. Let's recall what happens when we calculate $H|\psi\rangle$ for $|\psi\rangle = \alpha|0\rangle + \beta|1\rangle$. We get

$$H|\psi\rangle = \left(\frac{\alpha + \beta}{\sqrt{2}}\right)|0\rangle + \left(\frac{\alpha - \beta}{\sqrt{2}}\right)|1\rangle \tag{9.13}$$

Notice that the probability to obtain $|0\rangle$ upon measurement has been changed as

$$\alpha \to \frac{\alpha + \beta}{\sqrt{2}}$$

while the probability to find $|1\rangle$ has been changed as

$$\beta \to \frac{\alpha - \beta}{\sqrt{2}}$$

Specifically, looking at the state, we write

$$|\psi\rangle = \frac{|0\rangle + |1\rangle}{\sqrt{2}}$$

We saw in (9.10) that the Hadamard gate transforms $|\psi\rangle \rightarrow |0\rangle$. This is a manifestation of quantum interference—mathematically this means the addition of probability amplitudes. There are two types of interference, *positive interference* inwhich probability amplitudes add constructively to increase or *negative interference* in which probability amplitudes add destructively to decrease. In the case of (9.10), with the Hadamard transformation of the state we have the following:

- Positive interference with regard to the basis state $|0\rangle$. The two amplitudes add to increase the probability of finding 0 upon measurement. In fact in this case it goes to unity meaning we are certain to find 0.
- Negative interference where by the terms $|1\rangle$ and $-|1\rangle$ cancel. We go from a state where there was a 50% chance of finding 1 upon measurement to one where there is no chance of finding 1 upon measurement.

Quantum interference plays an important role in the development of quantum algorithms:

- Quantum interference allows us to gain information about a function $f(x)$ that depends on evaluating the function at many values of x. That is, interference allows us to deduce certain *global properties* of the function.

Now let's look at the final part of the basic quantum algorithm toolkit, quantum parallelism.

QUANTUM PARALLELISM AND FUNCTION EVALUATION

The first foray into the development of a quantum algorithm one can imagine is actually quite simple— but it demonstrates the overwhelming power of a quantum computer (i.e., should a practical one ever be constructed). The algorithm we are going to describe is called Deutsch's algorithm. Earlier we mentioned that quantum parallelism can be described as the ability to evaluate the function $f(x)$ at many values of x simultaneously.

Let's see how to do this by considering a very simple function, one that accepts a single bit as input and produces a single bit as output. That is, $x \in \{0, 1\}$. There are only a small number of functions that can act on the set $x \in \{0, 1\}$ and give a single bit as output. For example, we could have the identity function

$$f(x) = \begin{cases} 0 & \text{if } x = 0 \\ 1 & \text{if } x = 1 \end{cases}$$

Two more examples are the constant functions

$$f(x) = 0, \quad f(x) = 1$$

The final example is the bit flip function

$$f(x) = \begin{cases} 1 & \text{if } x = 0 \\ 0 & \text{if } x = 1 \end{cases}$$

The identity and bit flip functions are called *balanced* because the outputs are opposite for half the inputs. So a function on a single bit can be *constant* or *balanced*. Whether a function on a single bit is constant or balanced is a *global* property. What we're going to see in the following development is that Deutsch's algorithm will let us put together a state that has all of the output values of the function associated with each input value in a superposition state. Then we will use quantum interference to find out if the given function is constant or balanced.

The first step in developing this algorithm is to imagine a unitary operation denoted by U_f that acts on two qubits. It leaves the first qubit alone and produces the exclusive or (denoted by $\oplus$) of the second qubit with the function f evaluated with the first qubit as argument. That is,

$$U_f|x, y\rangle = |x, y \otimes f(x)\rangle \tag{9.14}$$

(note that $x, y \in \{0, 1\}$). Now, since $|x\rangle$ is a qubit, it can be in a superposition state. Let's specifically start with an initial state $|0\rangle$ and apply a Hadamard gate, as shown in Figure 9.4.

Using algebra, we write the action of the circuit shown in Figure 9.4 as

$$U_f\left(\frac{|0\rangle + |1\rangle}{\sqrt{2}}\right)|0\rangle = \frac{1}{\sqrt{2}}(U_f|00\rangle + U_f|10\rangle) = \frac{|0, 0 \oplus f(0)\rangle + |1, 0 \oplus f(1)\rangle}{\sqrt{2}}$$

This circuit has produced a superposition state that has information about every possible value of $f(x)$, in a *single* step. Suppose that $f(x)$ is the identity function. Then the final state is

$$U_f\left(\frac{|0\rangle + |1\rangle}{\sqrt{2}}\right)|0\rangle = \frac{|0, 0 \oplus f(0)\rangle + |1, 0 \oplus f(1)\rangle}{\sqrt{2}} = \frac{|00\rangle + |11\rangle}{\sqrt{2}}$$

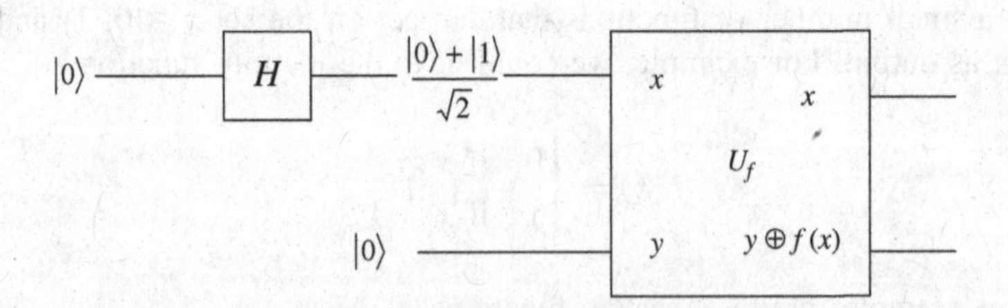

Figure 9.4 Circuit diagram for $U_f|x, y\rangle = |x, y \otimes f(x)\rangle$ where we take the first qubit to be in a superposition and the second qubit to be $|0\rangle$

Recall that $0 \oplus 0 = 1 \oplus 1 = 0$, $0 \oplus 1 = 1 \oplus 0 = 1$, so more generally the output of this circuit is

$$U_f\left(\frac{|0\rangle + |1\rangle}{\sqrt{2}}\right)|0\rangle = \frac{|0, f(0)\rangle + |1, f(1)\rangle}{\sqrt{2}}$$

This looks pretty nifty—we have a superposition state with all pairs of x, $f(x)$ represented. But remember how quantum measurement works. If we measure the state $\sum|x\rangle|f(x)\rangle$, we get one and only one value of x and $f(x)$. After measurement the system is in a state proportional to $|x\rangle|f(x)\rangle$ for that single and specific value of x. Moreover the value of x that we obtain is completely random. So what we have so far is the same as evaluating the function $f(x)$ at some randomly chosen x. That doesn't seem very useful. For a simple function on bits we can learn the value of $f(0)$ or $f(1)$, but not both simultaneously (even though they are simultaneously present in the premeasurement state). While it seems like we haven't gotten anything any more useful than what we can do with a classical computer, it's in fact worse: we can't choose which value to reveal, $f(0)$or $f(1)$, that occurs randomly.

Deutsch's algorithm takes what we have done so far to exploit the fact that the system is in a superposition state $\sum|x\rangle|f(x)\rangle$ to obtain information about a global property of the function—whether or not it is constant $f(0) = f(1)$ or balanced $f(0) \neq f(1)$. It does this by computing

$$|\psi_{out}\rangle = (H \otimes I)U_f(H \otimes H)|0\rangle|1\rangle \tag{9.15}$$

In words Deutsch's algorithm is implemented by the following steps:

1. Apply Hadamard gates to the input state $|0\rangle|1\rangle$ to produce a product state of two superpositions.
2. Apply U_f to that product state.
3. Apply a Hadamard gate to the first qubit leaving the second qubit alone.

We already know what the result of the first step is in (9.15). We calculated it in(9.7), which we restate here

$$\begin{aligned}(H \otimes H)|0\rangle|1\rangle &= \left(\frac{|0\rangle + |1\rangle}{\sqrt{2}}\right)\left(\frac{|0\rangle - |1\rangle}{\sqrt{2}}\right) \\ &= \frac{1}{2}(|00\rangle - |01\rangle + |10\rangle - |11\rangle)\end{aligned}$$

By doing this, we provide a superposition state as input that will contain all possible values of the combination $(x, f(x))$ once weapply U_f. Because it's a single state, we have these listed simultaneously. Now let's apply U_f to each term of $(H \otimes H)|0\rangle|1\rangle$. For the first term we have

$$U_f|00\rangle = |0, 0 \oplus f(0)\rangle = (1 - f(0))|100\rangle + f(0)|01\rangle \tag{9.16}$$

This result takes into account the possibilities $0 \oplus f(0) = 0$ and $0 \oplus f(0) = 1$. To see how this works, note that if $f(0) = 0$, then $0 \oplus f(0) = 0 \oplus 0 = 0$ and

$$|0, 0 \oplus f(0)\rangle = (1 - f(0))|00\rangle + f(0)|01\rangle = |00\rangle + (0)|01\rangle = |00\rangle$$

On the other hand, if $f(0) = 1$, then $0 \oplus f(0) = 0 \oplus 1 = 1$ and

$$|0, 0 \oplus f(0)\rangle = (1 - f(0))|00\rangle + f(0)|01\rangle = (0)|00\rangle + (1)|01\rangle = |01\rangle$$

Similar logic applied to the other terms gives

$$U_f|01\rangle = |0, 1 \oplus f(0)\rangle = f(0)|00\rangle + (1 - f(0))|01\rangle \tag{9.17}$$

$$U_f|10\rangle = |0, 1 \oplus f(0)\rangle = (1 - f(1))|00\rangle + f(1)|01\rangle \tag{9.18}$$

$$U_f|11\rangle = |0, 1 \oplus f(1)\rangle = f(1)|10\rangle + (1 - f(1))|11\rangle \tag{9.19}$$

Hence

$$\begin{aligned} |\psi'\rangle &= U_f(H \otimes H)|0\rangle|1\rangle \\ &= (1 - f(0))|00\rangle + f(0)|01\rangle + f(0)|00\rangle + (1 - f(0))|01\rangle \\ &\quad +(1 - f(1))|00\rangle + f(1)|01\rangle + f(1)|10\rangle + (1 - f(1))|11\rangle \end{aligned} \tag{9.20}$$

To get the final output state of Deutsch's algorithm, we apply $H \otimes I$ to $|\psi'\rangle$. The Hadamard gate is applied to the first qubit, and the second qubit is left alone. Let's look a the first couple of terms explicitly:

$$\begin{aligned} &(H \otimes I)[(1 - f(0))|00\rangle + f(0)|01\rangle] \\ &= (1 - f(0))(H|0\rangle) \otimes |0\rangle + f(0)(H|0\rangle) \otimes |1\rangle \\ &= (1 - f(0))\left(\frac{|0\rangle + |1\rangle}{\sqrt{2}}\right) \otimes |0\rangle + f(0)\left(\frac{|0\rangle + |1\rangle}{\sqrt{2}}\right) \otimes |1\rangle \\ &= (1 - f(0))\left(\frac{|00\rangle + |10\rangle}{\sqrt{2}}\right) + f(0)\left(\frac{|01\rangle + |11\rangle}{\sqrt{2}}\right) \end{aligned}$$

Applying $H \otimes I$ to all the terms in (9.20) and doing some algebra gives the final output state of Deutsch's algorithm as

$$|\psi_{out}\rangle = (1 - f(0) - f(1))|0\rangle\left(\frac{|0\rangle - |1\rangle}{\sqrt{2}}\right) + (f(1) - f(0))|1\rangle\left(\frac{|0\rangle - |1\rangle}{\sqrt{2}}\right) \tag{9.21}$$

Now suppose that the function is constant so that $f(0) = f(1)$. Then we obtain the final output state, using (9.21), as

$$|\psi_{out}\rangle = -|0\rangle\left(\frac{|0\rangle - |1\rangle}{\sqrt{2}}\right) \qquad (f(0) = f(1)) \tag{9.22}$$

On the other hand, if $f(0) \neq f(1)$, then $f(0)=0, f(1)=1$, or $f(0)=1, f(1)=0$, and the final output state is

$$|\psi_{out}\rangle = \pm|1\rangle\left(\frac{|0\rangle - |1\rangle}{\sqrt{2}}\right) \qquad (f(0) \neq f(1)) \tag{9.23}$$

In this algorithm, single-qubit interference is applied to the first qubit allowing us to distinguish between the two cases of the output of the function

DEUTSCH-JOZSA ALGORITHM

The *Deutsch-Jozsa algorithm* is a generalization of Deutsch's algorithm. Again, this algorithm allows us to determine whether a function $f(x)$ is constant or balanced, but this time the function has multiple input values. If $f(x)$ is constant, then the output is the same for all input values x. If the function is balanced, then $f(x)=0$ for half of the inputs and $f(x)=1$ for the other half of the inputs, and vice versa. We start with an initial state that includes n qubits in the state $|0\rangle$ and a singlequbit in the state $|1\rangle$. Hadamard gates are applied to all qubits. The circuit is illustrated in Figure 9.5.

We start off by calculating

$$|\psi'\rangle = (H^{\otimes n})(|0\rangle^{\otimes n}) \otimes (H|1\rangle) \tag{9.24}$$

From (9.6) we see that this is

$$|\psi'\rangle = \frac{1}{\sqrt{2^n}} \sum_{x\in\{0,1\}^n} |x\rangle\left(\frac{|0\rangle - |1\rangle}{\sqrt{2}}\right) \tag{9.25}$$

Next we apply (9.14), that is, $U_f|x, y\rangle = |x, y \otimes f(x)\rangle$, to evaluate the function. The first n qubits are the values of x and the last qubit plays the role of y as shown in the figure. The output state of the U_f gate is

$$|\psi''\rangle = \frac{1}{\sqrt{2^n}} \sum_{x} (-1)^{f(x)}|x\rangle\left(\frac{|0\rangle - |1\rangle}{\sqrt{2}}\right) \tag{9.26}$$

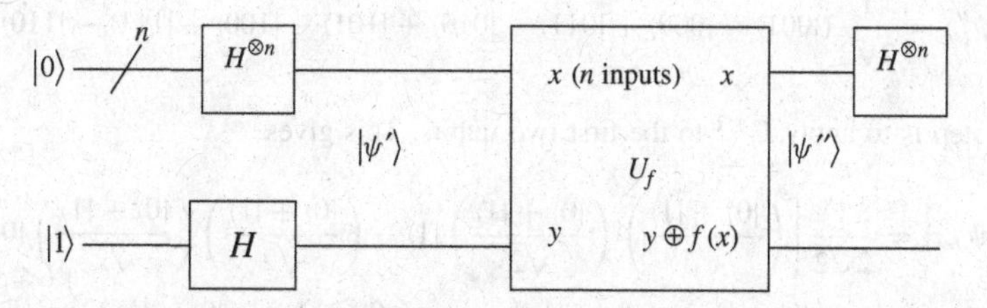

Figure 9.5 The Deutsch-Jozsa algorithm generalizes Deutsch's algorithm to handle a function with n input values and determine whether or not it is constant or balanced

Applying a Hadamard gate to an n qubit state $|x\rangle$ gives

$$H^{\otimes n}|x\rangle = \frac{1}{\sqrt{2^n}} \sum_y (-1)^{x \cdot y} |y\rangle \tag{9.27}$$

So the final output state is

$$|\psi_{out}\rangle = \frac{1}{2^n} \sum_y \sum_x (-1)^{x \cdot y + f(x)} |y\rangle \left(\frac{|0\rangle - |1\rangle}{\sqrt{2}} \right) \tag{9.28}$$

Now we measure the n inputs. It might not be immediately obvious looking at (9.28), but there are two possible measurement results on $|y\rangle$ (which is the state of n inputs at this stage) that are of interest. The possible results are as follows:

- Measurement of the first n input qubits in $|\psi_{out}\rangle$ returns all 0's. In this case $f(x)$ is constant.
- Otherwise, if at least one of the qubits in $|y\rangle$ is found to be a 1 on measurement, $f(x)$ is balanced.

Example 9.2

Consider a function with two inputs such that $f(x) = 1$. Explicitly show that the Deutsch-Jozsa algorithm works in this case by generating the vector $|y\rangle = |00\rangle$ as the final output.

Solution

The initial input state will be $|\psi_{in}\rangle = |0\rangle|0\rangle|1\rangle$. Applying Hadamard gates to this state gives

$$|\psi'\rangle = \frac{1}{2\sqrt{2}}(|000\rangle - |001\rangle + |010\rangle - |011\rangle + |100\rangle - |101\rangle + |110\rangle - |111\rangle)$$

Next we apply U_f to the system, obtaining

$$|\psi''\rangle = \frac{1}{2\sqrt{2}}(|001\rangle - |000\rangle + |011\rangle - |010\rangle + |101\rangle - |100\rangle + |111\rangle - |110\rangle)$$

The final step is to apply $H^{\otimes 2}$ to the first two qubits. This gives

$$|\psi_{out}\rangle = \frac{1}{2\sqrt{2}} \left[\left(\frac{|0\rangle + |1\rangle}{\sqrt{2}} \right) \left(\frac{|0\rangle + |1\rangle}{\sqrt{2}} \right) |1\rangle - \left(\frac{|0\rangle + |1\rangle}{\sqrt{2}} \right) \left(\frac{|0\rangle + |1\rangle}{\sqrt{2}} \right) |0\rangle \right.$$
$$+ \left(\frac{|0\rangle + |1\rangle}{\sqrt{2}} \right) \left(\frac{|0\rangle - |1\rangle}{\sqrt{2}} \right) |1\rangle - \left(\frac{|0\rangle + |1\rangle}{\sqrt{2}} \right) \left(\frac{|0\rangle - |1\rangle}{\sqrt{2}} \right) |0\rangle$$

$$+\left(\frac{|0\rangle-|1\rangle}{\sqrt{2}}\right)\left(\frac{|0\rangle+|1\rangle}{\sqrt{2}}\right)|1\rangle-\left(\frac{|0\rangle-|1\rangle}{\sqrt{2}}\right)\left(\frac{|0\rangle+|1\rangle}{\sqrt{2}}\right)|0\rangle$$
$$+\left(\frac{|0\rangle-|1\rangle}{\sqrt{2}}\right)\left(\frac{|0\rangle-|1\rangle}{\sqrt{2}}\right)|1\rangle-\left(\frac{|0\rangle-|1\rangle}{\sqrt{2}}\right)\left(\frac{|0\rangle-|1\rangle}{\sqrt{2}}\right)|0\rangle\Bigg]$$

Expanding out all the terms gives

$$\begin{aligned}|\psi_{out}\rangle = \frac{1}{4\sqrt{2}}[&(|00\rangle+|01\rangle+|10\rangle+|11\rangle)|1\rangle-(|00\rangle+|01\rangle+|10\rangle+|11\rangle)|0\rangle\\&+(|00\rangle-|01\rangle+|10\rangle-|11\rangle)|1\rangle-(|00\rangle-|01\rangle+|10\rangle-|11\rangle)|0\rangle\\&+(|00\rangle+|01\rangle-|10\rangle-|11\rangle)|1\rangle-(|00\rangle+|01\rangle-|10\rangle-|11\rangle)|0\rangle\\&+(|00\rangle-|01\rangle-|10\rangle+|11\rangle)|1\rangle-(|00\rangle-|01\rangle-|10\rangle+|11\rangle)|0\rangle]\end{aligned}$$

Now we want to factor these terms to put them in the form with the third qubit as $\left(|0\rangle-|1\rangle/\sqrt{2}\right)$. We get

$$\begin{aligned}|\psi_{out}\rangle = \frac{1}{4}\Bigg[&-(|00\rangle+|01\rangle+|10\rangle+|11\rangle)\left(\frac{|0\rangle-|1\rangle}{\sqrt{2}}\right)\\&-(|00\rangle-|01\rangle+|10\rangle-|11\rangle)\left(\frac{|0\rangle-|1\rangle}{\sqrt{2}}\right)\\&-(|00\rangle+|01\rangle-|10\rangle-|11\rangle)\left(\frac{|0\rangle-|1\rangle}{\sqrt{2}}\right)\\&-(|00\rangle-|01\rangle-|10\rangle+|11\rangle)\left(\frac{|0\rangle-|1\rangle}{\sqrt{2}}\right)\Bigg]\end{aligned}$$

Hence

$$|\psi_{out}\rangle = -|00\rangle\left(\frac{|0\rangle-|1\rangle}{\sqrt{2}}\right)$$

Measurement on the first two qubits gives 00, confirming that this is a constant function.

Example 9.3

Suppose that $f(00)=f(01)=0, f(10)=f(11)=1$. Apply the Deutsch-Josza algorithm and show that at least one of the first two qubits ends up as a 1.

Solution

The input state is again $|\psi_{in}\rangle=|0\rangle|0\rangle|1\rangle$, and the application of the Hadamard gates gives

$$|\psi'\rangle=\frac{1}{2\sqrt{2}}(|000\rangle-|001\rangle+|010\rangle-|011\rangle+|100\rangle-|101\rangle+|110\rangle-|111\rangle)$$

We apply U_f to the system, obtaining

$$|\psi''\rangle = \frac{1}{2\sqrt{2}}(|000\rangle - |001\rangle + |010\rangle + |011\rangle + |101\rangle - |100\rangle + |111\rangle - |110\rangle)$$

Applying Hadamard gates to the first two inputs gives

$$\begin{aligned}
|\psi_{out}\rangle = \frac{1}{4}\Bigg[&\left(\frac{|0\rangle + |1\rangle}{\sqrt{2}}\right)\left(\frac{|0\rangle + |1\rangle}{\sqrt{2}}\right)|0\rangle - \left(\frac{|0\rangle + |1\rangle}{\sqrt{2}}\right)\left(\frac{|0\rangle + |1\rangle}{\sqrt{2}}\right)|1\rangle \\
&+\left(\frac{|0\rangle + |1\rangle}{\sqrt{2}}\right)\left(\frac{|0\rangle - |1\rangle}{\sqrt{2}}\right)|0\rangle + \left(\frac{|0\rangle + |1\rangle}{\sqrt{2}}\right)\left(\frac{|0\rangle - |1\rangle}{\sqrt{2}}\right)|1\rangle \\
&+\left(\frac{|0\rangle - |1\rangle}{\sqrt{2}}\right)\left(\frac{|0\rangle + |1\rangle}{\sqrt{2}}\right)|1\rangle - \left(\frac{|0\rangle - |1\rangle}{\sqrt{2}}\right)\left(\frac{|0\rangle + |1\rangle}{\sqrt{2}}\right)|0\rangle \\
&+\left(\frac{|0\rangle - |1\rangle}{\sqrt{2}}\right)\left(\frac{|0\rangle - |1\rangle}{\sqrt{2}}\right)|1\rangle - \left(\frac{|0\rangle - |1\rangle}{\sqrt{2}}\right)\left(\frac{|0\rangle - |1\rangle}{\sqrt{2}}\right)|0\rangle\Bigg]
\end{aligned}$$

Again, we expand this result to obtain

$$\begin{aligned}
|\psi_{out}\rangle = \frac{1}{4\sqrt{2}}[&(|00\rangle + |01\rangle - |10\rangle - |11\rangle)|1\rangle + (|00\rangle + |01\rangle + |10\rangle + |11\rangle)|0\rangle \\
&- (|00\rangle + |01\rangle + |10\rangle + |11\rangle)|1\rangle + (|00\rangle - |01\rangle + |10\rangle - |11\rangle)|0\rangle \\
&+ (|00\rangle - |01\rangle + |10\rangle - |11\rangle)|1\rangle - (|00\rangle + |01\rangle - |10\rangle - |11\rangle)|0\rangle \\
&+ (|00\rangle - |01\rangle - |10\rangle + |11\rangle)|1\rangle - (|00\rangle - |01\rangle - |10\rangle + |11\rangle)|0\rangle]
\end{aligned}$$

We rearrange as follows:

$$\begin{aligned}
|\psi_{out}\rangle = \frac{1}{4}\Bigg[&(|00\rangle + |01\rangle + |10\rangle + |11\rangle)\left(\frac{|0\rangle - |1\rangle}{\sqrt{2}}\right) \\
&+ (|00\rangle - |01\rangle + |10\rangle - |11\rangle)\left(\frac{|0\rangle - |1\rangle}{\sqrt{2}}\right) \\
&- (|00\rangle + |01\rangle - |10\rangle - |11\rangle)\left(\frac{|0\rangle - |1\rangle}{\sqrt{2}}\right) \\
&- \left(|00\rangle - |01\rangle - |10\rangle + |11\rangle\right)\left(\frac{|0\rangle - |1\rangle}{\sqrt{2}}\right)\Bigg]
\end{aligned}$$

So we find that

$$|\psi_{out}\rangle = |10\rangle\left(\frac{|0\rangle - |1\rangle}{\sqrt{2}}\right)$$

Measurement on the first two qubits returns 10, since at least one qubit was 1. This shows that the function is balanced.

QUANTUM FOURIER TRANSFORM

A quantum computer can be used to compute a quantum analog of the Fourier transform. First let's get some preliminaries down. Consider two states of $d = n$ qubits

$$|x\rangle = |x_{n-1}x_{n-2}\ldots x_0\rangle$$
$$|y\rangle = |y_{n-1}y_{n-2}\ldots y_0\rangle$$

where each $x_i, y_i \in \{0,1\}$. Then

$$x \cdot y = x_0y_0 + x_1y_1 + \cdots + x_{n-1}y_{n-1} \tag{9.29}$$

The quantum Fourier transform maps the state $|\psi\rangle = |x_{n-1}x_{n-2}\cdots x_0\rangle$ into the state

$$\begin{aligned} &\frac{1}{2^{n/2}}(|0\rangle + e^{2\pi i[0.x_{n-1}]}|1\rangle) \otimes (|0\rangle + e^{2\pi i[0.x_{n-2}x_{n-1}]}|1\rangle) \\ &\otimes \cdots \otimes (|0\rangle + e^{2\pi i[0.x_0\ldots x_{n-2}x_{n-1}]}|1\rangle) \end{aligned} \tag{9.30}$$

where we have introduced the notation

$$\begin{aligned} \frac{x_0}{2} &\to 0.x_0 \\ \frac{x_0}{2^2} + \frac{x_1}{2} &\to 0.x_0x_1 \\ \frac{x_0}{2^3} + \frac{x_1}{2^2} + \frac{x_2}{2} &\to 0.x_0x_1x_2 \end{aligned} \tag{9.31}$$

and so on, to represent a binary fraction. Notice that we have two things going on here—superposition states and the introduction of phases. The first fact tells us that we will need to build a circuit with Hadamard gates to introduce superposition states. As for the phases, let's recall the discrete phase gate (9.12). The matrix representation of this gate in the computational basis is

$$R_k = \begin{pmatrix} 1 & 0 \\ 0 & e^{(2\pi i/2^k)} \end{pmatrix}$$

So we see that

$$R_k|0\rangle = |0\rangle,\ R_k|1\rangle = e^{(2\pi i/2^k)}|1\rangle \tag{9.32}$$

The quantum Fourier transform is implemented using Hadamard gates and controlled discrete phase gates. Let's denote the quantum Fourier transform operator by U_F. It acts on an arbitrary state $|x\rangle$ as

$$\begin{aligned} U_F|x\rangle &= \frac{1}{\sqrt{2^n}}\sum_{y=0}^{2^n-1} e^{2\pi i xy/2^n}|y\rangle \\ &= \frac{1}{\sqrt{2^n}}\sum_{y=0}^{2^n-1} e^{2\pi i(y_0(0.x_0x_1\ldots x_{n-1})+y_1(0.x_0x_1\ldots x_{n-2})+\cdots+y_{n-1}(0.x_0))}|y\rangle \end{aligned} \tag{9.33}$$

We use the property $e^{a+b} = e^a e^b$ and $|y\rangle = |y_0\rangle \otimes |y_1\rangle \otimes \cdots \otimes |y_{n-1}\rangle$ to write this as a tensor product state:

$$U_F|x\rangle = \frac{1}{\sqrt{2^n}} \sum_{y=0}^{2^n-1} e^{2\pi i y_0 (0.x_0 x_1 \dots x_{n-1})} |y_0\rangle \otimes \cdots \otimes e^{2\pi i y_{n-1}(0.x_0)} |y_{n-1}\rangle \tag{9.34}$$

Each $y_i \in \{0, 1\}$. If $y_i = 0$, then $e^{2\pi i y_i (.x_0 x_1 \dots x_j)} = e^0 = 1$. If $y_i = 1$, then the termsis left with $e^{2\pi i (.x_0 x_1 \dots x_j)}$. This is how we get the form (9.30).

The discussion so far has been pretty abstract. Next we will show how to compute the discrete quantum Fourier transform for a three-qubit state $|x\rangle = |x_2 x_1 x_0\rangle$. The circuit used to do this is shown in Figure 9.6.

The first step is to apply the Hadamard gate to $|x_2\rangle$. Note that $-1 = e^{i\pi}$, so the state on the top line in Figure 9.6 becomes

$$H|x_2\rangle = \frac{1}{\sqrt{2}} \sum_y (-1)^{x_2 y} |y\rangle = \frac{1}{\sqrt{2}} \sum_y e^{2\pi i x_2 y/2} |y\rangle = \frac{1}{\sqrt{2}} (|0\rangle + e^{2\pi i (0.x_2)} |1\rangle)$$

We proceed to act on this state with the controlled R_2 gate. The control bit for this gate is the state $|x_1\rangle$, which is $|0\rangle$ or $|1\rangle$. Since $1 = e^0 = e^{2\pi i 0}$, and $|x_1\rangle$ is either $|0\rangle$ or $|1\rangle$, we can write the action of the R_2 gate on the basis state $|1\rangle$ as $R_2|1\rangle = e^{i\pi x_1/2}|1\rangle = e^{2\pi i x_1/4}|1\rangle$. So the state of the entire system at this point is

$$I \otimes R_2 |x_1\rangle \otimes \frac{1}{\sqrt{2}} (|0\rangle + e^{2\pi i (0.x_2)} |1\rangle) = |x_1\rangle \otimes \frac{1}{\sqrt{2}} (|0\rangle + e^{2\pi i (0.x_2 x_1)} |1\rangle)$$

Now we apply the controlled R_3 gate to this state with the control bit being $|x_0\rangle$. This acts in the same manner transforming the system into the state

$$|x_0\rangle \otimes |x_1\rangle \otimes \frac{1}{\sqrt{2}} (|0\rangle + e^{2\pi i (0.x_2 x_1 x_0)} |1\rangle)$$

Next $|x_1\rangle$ goes through the Hadamard gate and the controlled R_2 gate, transforming it as

$$|x_1\rangle \to \frac{1}{\sqrt{2}} (|0\rangle + e^{2\pi i (.x_1 x_0)} |1\rangle)$$

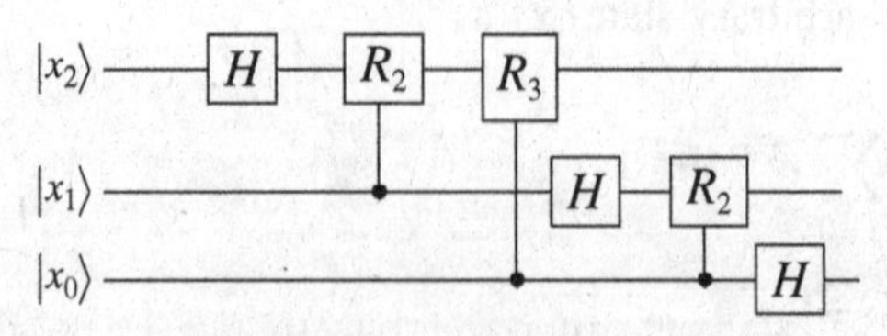

Figure 9.6 Circuit for a quantum Fourier transform applied to a three—qubit state.

The state of the system becomes

$$|x_0\rangle \otimes \frac{1}{\sqrt{2}}(|0\rangle + e^{2\pi i(0.x_1x_0)}|1\rangle) \otimes \frac{1}{\sqrt{2}}(|0\rangle + e^{2\pi i(0.x_2x_1x_0)}|1\rangle)$$

The last step is to act on $|x_0\rangle$ with the final Hadamard gate. This takes

$$|x_0\rangle \to \frac{1}{\sqrt{2}}(|0\rangle + e^{2\pi i(0.x_0)}|1\rangle)$$

The final state of the system is

$$\frac{1}{\sqrt{2}}(|0\rangle + e^{2\pi i(0.x_0)}|1\rangle) \otimes \frac{1}{\sqrt{2}}(|0\rangle + e^{2\pi i(0.x_1x_0)}|1\rangle) \otimes \frac{1}{\sqrt{2}}(|0\rangle + e^{2\pi i(0.x_2x_1x_0)}|1\rangle)$$

PHASE ESTIMATION

The quantum Fourier transform is closely related to a problem called *phase estimation*. Let U be a unitary operator on a $d = 2^n$ Hilbert space, and let $|\phi_1\rangle$, $|\phi_2\rangle$, ..., $|\phi_d\rangle$ be the eigenvectors of U, which form an orthonormal basis for the space. The eigenvalues of unitary operators are phases. Hence the eigenvalues of each $|\phi_n\rangle$ will be given by

$$U|\phi_n\rangle = e^{2\pi i\theta_n}|\phi_n\rangle \tag{9.35}$$

Notice that if we act on a given eigenvector say j times, we have

$$U^j|\phi_n\rangle = U^{j-1}(e^{2\pi i\theta_n}|\phi_n\rangle) = (e^{2\pi i\theta_n})^j|\phi_n\rangle = e^{2\pi i\theta_n j}|\phi_n\rangle \tag{9.36}$$

The problem of quantum phase estimation is the following: Given a unitary operator U and an input eigenvector $|\phi\rangle$ of U, estimate the angle θ in the associated eigenvalue that will have a form given by (9.35). We suppose that we want to know the phase angle θ to m bits of accuracy. To begin, the algorithm with the initial state given by

$$|\psi_{in}\rangle = |0\rangle^{\otimes m}|\phi\rangle \tag{9.37}$$

Here we know the eigenvector of the given unitary operator, $|\phi\rangle$ to n bits, meaning it is an n qubit state $|\phi\rangle = |\phi_{n-1}, \phi_{n-2}, \ldots, \phi_1, \phi_0\rangle$. Following a procedure you're familiar with already, the first step is to apply Hadamard gates to the $|0\rangle$ inputs to generate superposition states. Acting with m Hadamard gates on (9.37)gives

$$|\psi'\rangle = H^{\otimes m}|\psi_{in}\rangle = \left(\frac{|0\rangle + |1\rangle}{\sqrt{2}}\right)^{\otimes m}|\phi\rangle = \frac{1}{2^{m/2}}\sum_{x=0}^{2^m-1}|x\rangle|\phi\rangle \tag{9.38}$$

Now apply the unitary operator to the state in the following way:

$$|\psi''\rangle = \frac{1}{2^{m/2}}\sum_{x=0}^{2^m-1}|x\rangle(U^x|\phi\rangle) = \frac{1}{2^{m/2}}\sum_{x=0}^{2^m-1}|x\rangle(e^{2\pi i\theta x}|\phi\rangle) \tag{9.39}$$

Because $e^{2\pi i\theta x}$ is just a number, we can move it wherever we like. Notice that the state in (9.39) is a product state, with each term multiplied by $|\phi\rangle$. So we can pull it out of the sum and write

$$|\psi''\rangle = \frac{1}{2^{m/2}}\sum_{x=0}^{2^m-1}e^{2\pi i\theta x}|x\rangle|\phi\rangle = \left(\frac{1}{2^{m/2}}\sum_{x=0}^{2^m-1}e^{2\pi i\theta x}|x\rangle\right)|\phi\rangle \tag{9.40}$$

We are after an estimate of θ, so we can throw out the last n qubits (i.e., throw out $|\phi\rangle$). This leaves the state

$$|\psi_{out}\rangle = \frac{1}{2^{m/2}}\sum_{x=0}^{2^m-1}e^{2\pi i\theta x}|x\rangle \tag{9.41}$$

The algorithm we have described to this point is sometimes called a *phase kickback circuit*. This is illustrated in Figure 9.7.

To get an estimate of θ, we can use the quantum Fourier transform U_F. Specifically, we apply the inverse quantum Fourier transform, which is given by $U_F{}^{\dagger}$, to the state (9.41). The result is

$$U_F{}^{\dagger}|\psi_{out}\rangle = \frac{1}{2^m}\sum_{x=0}^{2^m-1}\sum_{y=0}^{2^m-1}e^{(-2\pi ixy/2^m)/2^m+2\pi i\theta x}|y\rangle = \frac{1}{2^m}\sum_{x,y=0}^{2^m-1}e^{2\pi ix(\theta-y/2^m)}|y\rangle \tag{9.42}$$

The probability of finding the system in the state $|y\rangle$ is

$$\Pr(y) = \left|\frac{1}{2^m}\sum_{x=0}^{2^m-1}e^{2\pi ix(\theta-y/2^m)}\right|^2 = \frac{1}{2^{2m}}\left|\sum_{x=0}^{2^m-1}e^{2\pi ix(\theta-y/2^m)}\right|^2 \tag{9.43}$$

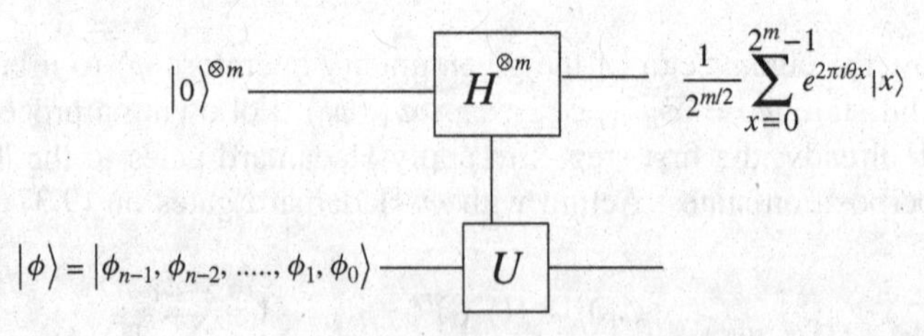

Figure 9.7 An illustration of a circuit to implement steps (9.37) through (9.41)

We can evaluate this term by considering a *geometric series*. If $|r| < 1$, then

$$\sum_{n=0}^{\infty} ar^n = \frac{a}{1-r} \tag{9.44}$$

When the sum is finite, we have

$$\sum_{n=0}^{m-1} ar^n = a\frac{r^m - 1}{r - 1} \tag{9.45}$$

Looking at (9.43), we let $a = 1$ and $r = e^{2\pi i x(\theta - y/2^m)}$. From (9.45) the probability of finding the system in the state $|y\rangle$ is

$$\Pr(y) = \frac{1}{2^{2m}} \left| \frac{r^{2m} - 1}{r - 1} \right|^2 \tag{9.46}$$

We wish to estimate a lower bound on this probability. Doing so involves looking at the terms in (9.46) in the complex plane. If you haven't had complex variables, you may wish to skip this section, which will seem obscure to many readers.

Suppose that we have an m bit approximation of θ given by $0.\theta_{m-1}\theta_{m-2}\ldots\theta_0$ that differs from θ by some error ε such that $0 < |\varepsilon| \leq 2^{m+1}$. Then we can take $r = e^{2\pi i\varepsilon}$ and estimate a lower bound on obtaining an accurate m bit precision estimate of θ. To do this, we examine the complex plane and consider that $|r^{2m} - 1|$ is the chord length from 1 to r^{2m}. We drawing a unit circle in the complex plane. As illustrated in Figure 9.8, the arc length from 1 to r^{2m} is given by $2\pi|\varepsilon|2^m$.

In Figure 9.8 the ratio of the arc length to the chord length cannot exceed $\pi/2$. Hence we obtain the bound

$$\frac{2\pi|\varepsilon|2^m}{|r^{2m} - 1|} \leq \frac{\pi}{2}, \Rightarrow |r^{2m} - 1| \geq 4|\varepsilon|2^m \tag{9.47}$$

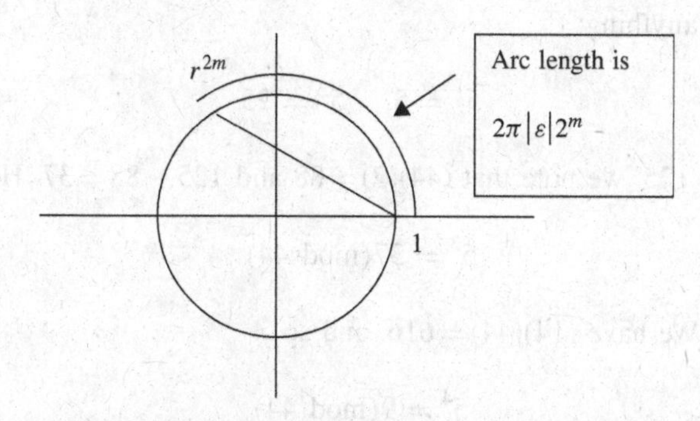

Figure 9.8 Estimating a bound on the probability (9.46) by looking at the complex plane.

Now we need a bound on the denominator in (9.46). A similar procedure for considering the chord length from 1 to r gives the bound

$$\frac{2\pi|\varepsilon|}{|r-1|} \geq 1, \quad \Rightarrow \frac{1}{|r-1|} \geq \frac{1}{2\pi|\varepsilon|} \tag{9.48}$$

Hence

$$\Pr(y) = \frac{1}{2^{2m}} \left| \frac{r^{2m}-1}{r-1} \right|^2 \geq \frac{1}{2^{2m}} \frac{|4|\varepsilon|2^m|^2}{|2\pi|\varepsilon||^2} = \frac{1}{2^{2m}} \frac{16|\varepsilon|^2 2^{2m}}{4\pi^2|\varepsilon|^2} = \frac{4}{\pi^2} \rangle 0.4 \tag{9.49}$$

The bottom line of this calculation is that the probability is greater than 0.4. that a measurement of θ to m bits of precision is obtained with every single bit correct.

SHOR'S ALGORITHM

Shor's algorithm was fundamental in demonstrating the power and importance of quantum computation. This is an algorithm that can be used to factor prime numbers—meaning that it can be used to break encryption codes if a practical quantum computer is ever built. Needless to say, this algorithm got the attention of a lot of people.

The first thing we need to know in order to do Shor's algorithm is *order finding*. Let x and N be positive integers with no common factors such that $x < N$. The *order* of x is the smallest positive integer r such that

$$x^r = 1 \bmod N \tag{9.50}$$

Let's explain what mod N means by way of an example. First of all, x and N can't have any common factors because their greatest common divisor is 1. Suppose that we let $x = 5$ and $N = 44$. To find $x^r = a \bmod N$, we compute $x^r$and subtract N until we get the last integer greater than 0. The first two cases are less than $N = 44$, so we don't do anything:

$$5^1 = 5, \quad 5^2 = 25$$

Now since $5^3 = 125$, we note that $(44)(2) = 88$ and $125 - 88 = 37$. Hence

$$5^3 = 37(\text{mod } 44)$$

Next $5^4 = 625$. We have $(14)(44) = 616$, and so

$$5^4 = 9(\text{mod } 44)$$

Finally $5^5 = 3125$. It turns out that $71 \times 44 = 3124$, which is 1 less than $5^5 = 3125$. This is where we stop. Hence

$$5^5 = 1(\text{mod } 44)$$

The order of 5 is 5 in this case.

As you can see, plugging away like this, finding the powers $x^r = 1(\text{mod} N)$ can be very time-consuming. With large numbers it will swamp the best computers available, the time required is exponential in $\log N$. This problem can be solved far more efficiently by using a quantum algorithm based on phase estimation.

To solve the problem of order finding, we consider the unitary operator

$$U_x|y\rangle = \begin{cases} |xy \text{ mod } N\rangle & 0 \leq y \leq N-1 \\ |y\rangle & N \leq y \leq 2^L - 1 \end{cases} \tag{9.51}$$

where $L = \lceil \log N \rceil$. The eigenstates of U_x are given by

$$|u_t\rangle = \frac{1}{\sqrt{r}} \sum_{k=0}^{r-1} \exp\left(-\frac{2\pi i k t}{r}\right) |x^k \text{ mod } N\rangle \tag{9.52}$$

The eigenvalues are given by

$$U_x|u_t\rangle = e^{2\pi i (t/r)}|u_t\rangle \tag{9.53}$$

Notice that we can apply the phase estimation algorithm to these states to estimate the phase t/r, and hence to find an estimate of the order. The order-finding algorithm begins by considering a superposition of these eigenstates. It turns out that the eigen states sum to $|1\rangle$, since $\sum_{k=0}^{r-1} \exp(-(2\pi i k t)/r) = r\delta_{k,0}$,

$$\frac{1}{\sqrt{r}} \sum_{t=0}^{r-1} |u_t\rangle = \frac{1}{\sqrt{r}} \sum_{t=0}^{r-1} \frac{1}{\sqrt{r}} \sum_{k=0}^{r-1} \exp\left(-\frac{2\pi i k t}{r}\right) |x^k \text{ mod } N\rangle = |1\rangle \tag{9.54}$$

So we begin order finding with the input state

$$|\psi_{in}\rangle = |1\rangle = \frac{1}{\sqrt{r}} \sum_{t=0}^{r-1} |u_t\rangle \tag{9.55}$$

We can calculate the quantity x^k mod N using quantum parallelism. The period r/t, which gives us the order r, is found by the phase estimation procedure. So if we start with $|0\rangle^{\otimes n}|1\rangle$, then we have

$$\frac{1}{\sqrt{r}} \sum_{t=0}^{r-1} \left(\frac{1}{\sqrt{2^s}} \sum_{k=0}^{2^s-1} \exp\left(-\frac{2\pi i k t}{r}\right) |k\rangle \right) |u_t\rangle = \frac{1}{\sqrt{r}} \sum_{t=0}^{r-1} |t/r\rangle |u_t\rangle \tag{9.56}$$

where we have defined

$$|t/r\rangle = \frac{1}{\sqrt{2^s}} \sum_{k=0}^{2^s-1} \exp\left(-\frac{2\pi i k t}{r}\right)|k\rangle$$

Measurement gives an estimate of a random t/r for $0 \le t \le r-1$. To obtain the exact value, continued fractions are applied.

The order-finding algorithm is the only quantum part of Shor's algorithm—the rest of it involves classical calculations. A simple way to state Shor's algorithm is that we use order finding to find the factors of some odd integer N. This can be summarized as a "Repeat... until" procedure (which I borrow from the lecture notes of John Watrous of University of Waterloo):

```
Input: odd integer  N
Repeat
  Randomly select x∈{2,...,N−1} a∈{2,...,N∨1□}

  d = gcd (x,N)
  If d≥2 Then u=d, v=N/d
  Else
   Find the order r such that x^r = 1 mod N
   y=x^(r/2) − 1 mod N
   d = gcd (y,N)
   If d≥2 Then u=d, v=N/d
Until find u,v
```

QUANTUM SEARCHING AND GROVER'S ALGORITHM

Grover's algorithm can be described as a quantum database-searching algorithm. The presentation given here is simplistic and is not something that has much real world utility. Grover's algorithm once again, demonstrates the power of a quantum computer in that the algorithm significantly reduces the number of operations necessary to solve the problem as compared to a classical computer.

Once again, we have a function $f(x)$ on bits with n inputs and $x \in \{0, 1\}^n$. The output of the function is a single bit, so we can have $f(x)=0$ or $f(x)=1$. The task at hand, which is solved by Grover's algorithm, is the following. There is a single x such that $f(x)=1$, and we want to find out what x that is. Note that it may be the case that $f(x)\equiv 0$ identically, in which case no such x exists.

To solve this problem classically, we can generate input strings x and simply test the function to find out if $f(x)=1$. It turns out that doing this will require 2^n-1 tries to solve the problem. In contrast, Grover's algorithm solves the problem with on order of $\sqrt{2^n}$ tries. Now suppose that the bit string is small, just 5 bits. The classical algorithm would require $2^5-1=31$ attempts to find the correct x, while Grover's algorithm would require $\sqrt{2^5}\approx 6$ attempts. We have a significant improvement on a small bit string. As you might imagine, as the bit strings get larger and larger, the improvement offered by Grover's algorithm becomes very significant.

Let's denote the input we seek by $|x'\rangle$. In other words, $f(x')=1$, but $f(x)=0$ for all other $|x\rangle$. So the task at hand, using Grover's algorithm, is to find $|x'\rangle$. The basic idea is we are going to create an input superposition state and rotate it into $|x'\rangle$ using the *Grover operator G*.

Following the usual procedure, we start with an initial n bit input state $|0\rangle^{\otimes n}$ and apply $H^{\otimes n}$ to this state to generate superposition states. We define the state $|\psi\rangle$ to be a superposition of all possible states $|x\rangle$, meaning

$$|\psi\rangle = \frac{1}{\sqrt{2^n}} \sum_{x\in\{0,1\}^n} |x\rangle \tag{9.57}$$

This superposition includes the state $|x'\rangle$ so that

$$\langle x'|\psi\rangle = \frac{1}{\sqrt{2^n}} \sum_{x\in\{0,1\}^n} \langle x'|x\rangle = \frac{1}{\sqrt{2^n}} \tag{9.58}$$

By excluding $|x'\rangle$ from (9.57), we can construct an othonormal basis set consisting of $|x'\rangle$ and

$$|\psi'\rangle = \frac{1}{\sqrt{2^n - 1}} \sum_{x\in\{0,1\}^n, x\neq x'} |x\rangle \tag{9.59}$$

Now we define two operators. The first is

$$U_f = \sum_{x\in\{0,1\}^n} (-1)^{f(x)} |x\rangle\langle x| = \sum_{x\in\{0,1\}^n} (-1)^{\delta_{x,x'}} |x\rangle\langle x| \tag{9.60}$$

where

$$\delta_{x,x'} = \begin{cases} 1 & x = x' \\ 0 & \text{otherwise} \end{cases} \tag{9.61}$$

is the Kronecker delta function. Next, using (9.57), we define

$$W = 2|\psi\rangle\langle\psi| - I \tag{9.62}$$

If we split $|\psi\rangle$ into two parts—the piece containing $|x'\rangle$ and the rest as defined by (9.59)—we have

$$|\psi\rangle = \sqrt{\frac{2^n - 1}{2^n}} |\psi'\rangle + \frac{1}{\sqrt{2^n}} |x'\rangle \tag{9.63}$$

Hence

$$\langle\psi|\psi'\rangle = \sqrt{\frac{2^n - 1}{2^n}} \tag{9.64}$$

Inverting (9.63) so that $|x'\rangle = \sqrt{2^n}|\psi\rangle - \sqrt{2^n - 1}|\psi'\rangle$, we see that

$$\begin{aligned}W|x'\rangle &= (2|\psi\rangle\langle\psi| - I)(\sqrt{2^n}|\psi\rangle - \sqrt{2^n - 1}|\psi'\rangle)\\ &= 2\sqrt{2^n}|\psi\rangle\langle\psi|\psi\rangle - \sqrt{2^n}|\psi\rangle - 2\sqrt{2^n - 1}|\psi\rangle\langle\psi|\psi'\rangle + \sqrt{2^n - 1}|\psi'\rangle \\ &= 2\sqrt{2^n}|\psi\rangle - \sqrt{2^n}|\psi\rangle - 2\sqrt{2^n - 1}\sqrt{\frac{2^n - 1}{2^n}}|\psi\rangle + \sqrt{2^n - 1}|\psi'\rangle\end{aligned} \tag{9.65}$$

Using (9.63) to substitute for $|\psi\rangle$ allows us to write

$$W|x'\rangle = \frac{2\sqrt{2^n - 1}}{2^n}|\psi'\rangle + \left(\frac{2}{2^n} - 1\right)|x'\rangle \tag{9.66}$$

We also find that

$$W|\psi'\rangle = -\left(\frac{2}{2^n} - 1\right)|\psi'\rangle + \frac{2\sqrt{2^n - 1}}{2^n}|x'\rangle \tag{9.67}$$

Then, if we define the angle θ as

$$\sin\theta = \frac{2\sqrt{2^n - 1}}{2^n} \tag{9.68}$$

Notice that (9.66) and (9.67) are a rotation, that is,

$$W|x'\rangle = -\cos\theta|x'\rangle + \sin\theta|\psi'\rangle \tag{9.69}$$

$$W|\psi'\rangle = \sin\theta|x'\rangle + \cos\theta|\psi'\rangle \tag{9.70}$$

If we apply the operator $G = WU_f$, which is known as the *Grover operator*, we obtain the more familiar form of a rotation, namely

$$G|x'\rangle = \cos\theta|x'\rangle - \sin\theta|\psi'\rangle \tag{9.71}$$

$$G|\psi'\rangle = \sin\theta|x'\rangle + \cos\theta|\psi'\rangle \tag{9.72}$$

The basic idea is that the Grover operator rotates the state $|\psi'\rangle$ into the state we are looking for, $|x'\rangle$. It does so only a tiny bit at a time, so we have to apply it multiple times, say, m times. In that case (9.71) and (9.72) become

$$G^m|x'\rangle = \cos m\theta|x'\rangle - \sin m\theta|\psi'\rangle \tag{9.73}$$

$$G^m|\psi'\rangle = \sin m\theta|x'\rangle + \cos m\theta|\psi'\rangle \tag{9.74}$$

If we see that $m\theta = \pi/2$, then (9.74) tells us that the application of G^m to $|\psi'\rangle$ turns it into $|x'\rangle$, since $\sin \pi/2 = 1$, $\cos \pi/2 = 0$. Hence $G^m|\psi'\rangle = |x'\rangle \ (m\theta = \pi/2)$. From our definition of θ (9.68), the condition that must be met is

$$\sin\,\theta = \frac{2\sqrt{2^n - 1}}{2^n} \tag{9.75}$$

Using the small angle approximation, (i.e., $\sin\,\theta \approx \theta$) reduces this to

$$\theta \approx \frac{2\sqrt{2^n - 1}}{2^n} \tag{9.76}$$

Using our observation that $m\theta = \pi/2$ will swap the state into the one we want, we know that the condition that must be met in order to find $|x'\rangle$ is

$$\begin{aligned} m\theta &= m\frac{2\sqrt{2^n - 1}}{2^n} = \frac{\pi}{2} \\ \Rightarrow m &= \frac{\pi}{4}\frac{2^n}{\sqrt{2^n - 1}} \approx \frac{\pi}{4}\sqrt{2^n} \end{aligned} \tag{9.77}$$

Therefore on the order of $\sqrt{2^n}$ rotations—or applications of the Grover operator G—are necessary to find the state $|x'\rangle$.

EXERCISES

9.1. *Using the matrix representation of the Hadamard gate*

$$H = \frac{1}{\sqrt{2}}\begin{pmatrix} 1 & 1 \\ 1 & -1 \end{pmatrix}$$

write down the matrix $H \otimes H$ and find $(H \otimes H)(|0\rangle \otimes |1\rangle)$. Show that this is equivalent to

$$|\phi\rangle = \left(\frac{|0\rangle + |1\rangle}{\sqrt{2}}\right)\left(\frac{|0\rangle - |1\rangle}{\sqrt{2}}\right)$$

9.2. *The Beam splitter gate has a matrix representation given by*

$$B = \frac{1}{\sqrt{2}}\begin{pmatrix} i & 1 \\ 1 & i \end{pmatrix}$$

Show that B generates superposition states out of the computational basis states $|0\rangle$ *and* $|1\rangle$. *In particular, show that*

$$B \otimes B|0\rangle|0\rangle = \left(\frac{i|0\rangle + |1\rangle}{\sqrt{2}}\right)\left(\frac{i|0\rangle + |1\rangle}{\sqrt{2}}\right)$$

Show that two applications of the beam splitter gate on the same state, namely that $B(B|\psi\rangle)$ *act analogously to the NOT gate, giving the same probabilities of finding* $|0\rangle$ *and* $|1\rangle$.

9.3. *Show that the matrix representation of* $HP(\theta)HP(\phi)$ *is given by*

$$e^{i\theta/2}\begin{pmatrix} \cos\frac{\theta}{2} & -ie^{i\phi}\sin\frac{\theta}{2} \\ -i\sin\frac{\theta}{2} & e^{i\phi}\cos\frac{\theta}{2} \end{pmatrix}$$

9.4. *Derive (9.17) through (9.19).*

9.5. *Quantum gates are universal in the sense that quantum gates can be designed that do anything a classical gate can do. Design a quantum adder, a gate that takes three inputs* $|x\rangle$, $|y\rangle$, $|z\rangle$ *and that has three output qubits* $|x\rangle$, $|x \oplus y\rangle$, $|xy\rangle$, *where* $|x \oplus y\rangle$, *is the* sum *and* $|xy\rangle$ *is the* carry.

9.6. *Consider a generalization of the qubit, a* $d = 2^n$*-dimensional system where the quantum states are called* qudits. *A qudit system is more complicated but might be considered because of the increased computing power. The computational basis for a qudit is*

$$\{|0\rangle, |1\rangle, |2\rangle, \ldots, |d-1\rangle\}$$

The qudit Hadamard gate H_d *takes the states*

$$|0_H\rangle = H_d|0\rangle = |0\rangle + |1\rangle + \cdots + |d-2\rangle + |d-1\rangle$$
$$|1_H\rangle = H_d|1\rangle = |0\rangle - |1\rangle + \cdots + |d-2\rangle - |d-1\rangle$$

(a) In $d = 4$ *dimensions, the computational basis is* $\{|0\rangle, |1\rangle, |2\rangle, |3\rangle\}$. *Show that*

$$H_d|0\rangle = |0\rangle + |1\rangle + |2\rangle + |3\rangle$$
$$H_d|1\rangle = |0\rangle - |1\rangle + |2\rangle - |3\rangle$$

and that $H_d^2|0\rangle = |0\rangle$, $H_d^2|1\rangle = |1\rangle$ *if the matrix representation of the Hadamard gate is*

$$H_d = \frac{1}{2}\begin{pmatrix} 1 & 1 & 1 & 1 \\ 1 & -1 & 1 & -1 \\ 1 & 1 & -1 & -1 \\ 1 & -1 & -1 & 1 \end{pmatrix}$$

In general,

$$H_d|x\rangle = \frac{1}{\sqrt{d}} \sum_{y=0}^{d-1} (-1)^{x \cdot y} |y\rangle$$

where

$$x \cdot y = x_o y_o \oplus x_1 y_1 \oplus \cdots \oplus x_{n-1} y_{n-1}$$

and

$$x = \sum_{i=0}^{n-1} x_i 2^i,\ y = \sum_{i=0}^{n-1} y_i 2^i, \quad x, y \in \{0, 1\}$$

A function in d dimensions is constant if

$$f(0) \oplus f(1) \oplus \cdots \oplus f(d-1) = 0$$

and it is balanced if

$$f(0) \oplus f(1) \oplus \cdots \oplus f(d-1) = \frac{d}{2}$$

where $\oplus$ *is addition mod* $d = 2^n$. *Starting with* $|\psi_{in}\rangle = |0\rangle|1\rangle$, *generalize the Deutsch-Josza algorithm to the input of two qudits with steps (b) and (c)*
(b) Apply the qudit Hadamard gate to $|\psi_{in}\rangle = |0\rangle|1\rangle$ *and show that*

$$|\psi'\rangle = H_d \otimes H_d|\psi_{in}\rangle = \sum_{x,y=0}^{d-1} (-1)^y |x\rangle|y\rangle$$

(c) Using $U_f|x, y\rangle = |x, y \oplus f(x)\rangle$, *show that*

$$U_f|\psi'\rangle = \left(\sum_{x=0}^{d-1} (-1)^x |x\rangle \right) |1_H\rangle$$

Finally, apply qudit Hadamard gates to the first set of qudits (leaving the auxiliary qudit $|1_H\rangle$ *alone). Then, if the state is all 0's, the function is constant. Otherwise, it is balanced.*

9.7. *Derive the relation (9.67).*

9.8. *Consider the eigenvectors (9.52) and show that*

$$U_x|u_t\rangle = exp\left(-\frac{2\pi i t}{r} \right) |u_t\rangle.$$

10

APPLICATIONS OF ENTANGLEMENT: TELEPORTATION AND SUPERDENSE CODING

It is often said that entanglement is a resource in quantum computation and information. This is because we can use entanglement to accomplish communications and information processing tasks that would otherwise not be possible. In this chapter we will explore two areas where entanglement can be used to do some rather unusual tasks. The first of these, known as *teleportation*, is a procedure that allows one party (Alice) to send a quantum state to her friend (Bob) without that state being transmitted in the usual sense. By using entanglement, Alice and Bob can set up a quantum communications channel that links them together in a quantum way via the EPR paradox—allowing Alice to send her state to Bob in an almost magical fashion. As we will see, however, faster than light communication isn't possible using teleportation because it is still necessary for Alice and Bob to maintain a classical communications link in order for teleportation to work.

In our second application of entanglement, we will look at *superdense coding*. This is a procedure that allows us to send two classical bits to a party using only a single qubit—demonstrating the power of quantum information processing.

TELEPORTATION

Anyone who has read science fiction or watched it on television is familiar with *teleportation*—a device or process by which people can jet around the galaxy. The basic idea is that you get scanned somehow, turned into energy, then beamed to where you want to go and rematerialized.

While such a procedure is likely to remain in the realm of science fiction, quantum mechanics does allow us to do something almost as magical. It allows us to send a quantum state from one place to another without that state traversing the space in between. Good thing Einstein wasn't around to hear about this. If entanglement itself made him feel "spooky," then teleportation is sure to be making him turn over in his grave.

While teleportation seems to work almost by magic Einstein can breathe a sigh of relief because special relativity seems to step in to prevent faster than light communication. To see how this works, let's go through the basic formalism. The task at hand is that Alice wants to transmit an unknown quantum state to Bob. Let's denote the state that Alice wants to send Bob by $|\chi\rangle$. The state is a qubit:

$$|\chi\rangle = \alpha|0\rangle + \beta|1\rangle \tag{10.1}$$

By saying the state is unknown, we are saying we don't necessarily know what α and β are. All we assume is that the state is normalized, so $|\alpha|^2 + |\beta|^2 = 1$. Teleportation takes place in a series of steps. We begin by creating an entangled EPR pair.

Teleportation Step 1: Alice and Bob Share an Entangled Pair of Particles

Alice and Bob create the entangled state

$$|\beta_{00}\rangle = \frac{|0_A\rangle|0_B\rangle + |1_A\rangle|1_B\rangle}{\sqrt{2}} = \frac{|00\rangle + |11\rangle}{\sqrt{2}} \tag{10.2}$$

Here we've decided that the first member of the pair belongs to Alice and the second member of the pair belongs to Bob. Now Alice and Bob physically separate. Alice decides that she wants to send the state (10.1) to Bob. She can do it by letting it interact with her member of the EPR pair in (10.2).

Teleportation Step 2: Alice Applies a CNOT Gate

Let's begin by writing down the state of the entire system. It's a product state of the unknown state (10.1) and the EPR pair (10.2):

$$\begin{aligned}|\psi\rangle &= |\chi\rangle \otimes |\beta_{00}\rangle = (\alpha|0\rangle + \beta|1\rangle) \otimes \left(\frac{|00\rangle + |11\rangle}{\sqrt{2}}\right) \\ &= \frac{\alpha(|000\rangle + |011\rangle) + \beta(|100\rangle + |111\rangle)}{\sqrt{2}}\end{aligned} \tag{10.3}$$

The first two qubits in this state belong to Alice, while the third rightmost qubit belongs to Bob. So $|011\rangle$ indicates that Alice has a 01 in her possession while Bob has a 1.

Alice begins interacting her member of the EPR pair, which is the second qubit in (10.3), with the unknown state—the first qubit in (10.3)—by applying a CNOT gate. She uses the unknown state $|\chi\rangle$ as the control qubit and her member of the EPR pair as the target qubit. Remember, if the control qubit is 0, nothing happens; if the control qubit is 1, the target qubit is flipped:

$$|00\rangle \mapsto |00\rangle, |01\rangle \mapsto |01\rangle, |10\rangle \mapsto |11\rangle, |11\rangle \mapsto |10\rangle \tag{10.4}$$

So, when Alice applies the CNOT gate to (10.3), the state becomes

$$\begin{aligned}
|\psi'\rangle &= U_{CNOT}|\psi\rangle \\
&= \frac{\alpha(U_{CNOT}|000\rangle + U_{CNOT}|011\rangle) + \beta(U_{CNOT}|100\rangle + U_{CNOT}|111\rangle)}{\sqrt{2}} \\
&= \frac{\alpha(|000\rangle + |011\rangle) + \beta(|110\rangle + |101\rangle)}{\sqrt{2}}
\end{aligned} \tag{10.5}$$

Teleportation Step 3: Alice Applies a Hadamard Gate

Next Alice will apply a Hadamard gate to the first qubit. Remember, a Hadamard gate acts on the computational basis states by turning them into superpositions:

$$H|0\rangle = \frac{|0\rangle + |1\rangle}{\sqrt{2}}, \quad H|1\rangle = \frac{|0\rangle - |1\rangle}{\sqrt{2}} \tag{10.6}$$

Let's rewrite the state (10.5) a bit to make things a little clearer:

$$|\psi'\rangle = \frac{\alpha|0\rangle(|00\rangle + |11\rangle)}{\sqrt{2}} + \frac{\beta|1\rangle(|10\rangle + |01\rangle)}{\sqrt{2}} \tag{10.7}$$

So Alice transforms the state into

$$\begin{aligned}
|\psi''\rangle = H|\psi'\rangle &= \frac{\alpha H|0\rangle(|00\rangle + |11\rangle)}{\sqrt{2}} + \frac{\beta H|1\rangle(|10\rangle + |01\rangle)}{\sqrt{2}} \\
&= \alpha\left(\frac{|0\rangle + |1\rangle}{\sqrt{2}}\right)\frac{(|00\rangle + |11\rangle)}{\sqrt{2}} + \beta\left(\frac{|0\rangle - |1\rangle}{\sqrt{2}}\right)\frac{(|10\rangle + |01\rangle)}{\sqrt{2}}
\end{aligned} \tag{10.8}$$

Remember, Bob is in possession of the third qubit.

Teleportation Step 4: Alice Measures Her Pair

The next step in the process is that Alice makes a measurement on both qubits in her possession. First, let's rewrite the state so that the state is written in terms of the possible measurement results on the first two qubits. These possible measurement results are $|00\rangle$, $|01\rangle$, $|10\rangle$, and $|11\rangle$. So we can write (10.8) as

$$|\psi''\rangle = \frac{1}{2}[|00\rangle(\alpha|0\rangle + \beta|1\rangle) + |01\rangle(\alpha|1\rangle + \beta|0\rangle) + |10\rangle(\alpha|0\rangle - \beta|1\rangle) + |11\rangle(\alpha|1\rangle - \beta|0\rangle)] \quad (10.9)$$

If Alice measures $|00\rangle$, then the state collapses and Bob has $|\chi\rangle = \alpha|0\rangle + \beta|1\rangle$—the unknown state that Alice wanted to send Bob in the first place—in his possession. Next, if Alice measures $|01\rangle$, then Bob has the state $\alpha|1\rangle + \beta|0\rangle$ in his possession. But he can transform this into the desired state with the application of an X gate:

$$X(\alpha|1\rangle + \beta|0\rangle) = \alpha X|1\rangle + \beta X|0\rangle = \alpha|0\rangle + \beta|1\rangle = |\chi\rangle \quad (10.10)$$

Suppose that Alice measures $|10\rangle$. The state in Bob's possession is then $\alpha|0\rangle - \beta|1\rangle$. He can turn it into the desired state using a Z gate:

$$Z(\alpha|0\rangle - \beta|1\rangle) = \alpha Z|0\rangle - \beta Z|1\rangle = \alpha|0\rangle + \beta|1\rangle = |\chi\rangle \quad (10.11)$$

Now, if Alice measures $|11\rangle$, the state Bob has is $\alpha|1\rangle - \beta|0\rangle$. This time Bob has to apply two gates, and X and a Z

$$ZX(\alpha|1\rangle - \beta|0\rangle) = \alpha ZX|1\rangle - \beta ZX|0\rangle = \alpha Z|0\rangle - \beta Z|1\rangle = \alpha|0\rangle + \beta|1\rangle = |\chi\rangle \quad (10.12)$$

So how does Bob know what to do? The answer is Alice gives him a call.

Teleportation Step 5: Alice Contacts Bob on a Classical Communications Channel and Tells Him Her Measurement Result

At this stage of the game special relativity surprisingly enters the game. Alice has to somehow tell Bob her measurement result, and she has to do it using a classical communications channel—a telephone, email message, radio wave, or something—some mechanism governed by the speed of light limit. It's this necessary step that prevents Alice and Bob from faster than light communication. But security is maintained—Alice just calls Bob and for instances, says she got 01, then Bob applies his X gate to obtain the state Alice wanted to send to Bob. Nothing about that state is communicated over the classical channel—Bob can obtain it because they shared an entangled EPR pair of particles.

The lesson here is that quantum information based communication can be characterized by two key aspects—local operations and classical communications (LOCC). That is, each party has two tasks:

- Performs local quantum mechanical (local unitaries) operations on their respective states.
- Uses classical communication to communicate measurement results.

If classical communications is not used, then the state will appear totally random to Bob.

THE PERES PARTIAL TRANSPOSITION CONDITION

Teleportation continues to be an active area of study. For us, we can use it as a vehicle to learn more tools in the toolbox of the quantum information theorist. We begin by considering the *Peres partial transposition condition*, which allows us to determine whether or not a given density operator represents an entangled state. An arbitrary density matrix can be written as

$$\rho_{AB} = \sum_{i,j,k,l} \rho_{ijkl}|i\rangle\langle j| \otimes |k\rangle\langle l| \tag{10.13}$$

The partial transposition of ρ is given by

$$\rho_{AB}^{T_B} = \sum_{i,j,k,l} \rho_{jikl}|i\rangle\langle j| \otimes |k\rangle\langle l| \tag{10.14}$$

More specifically, if we have a state

$$|\psi\rangle = \frac{1}{\sqrt{\alpha}}\sum_i |a_i b_i\rangle \quad and \quad \rho_{AB} = |\psi\rangle\langle\psi|$$

that is,

$$\rho_{AB} = \frac{1}{\alpha}\sum_{ij} |a_i b_i\rangle\langle a_j b_j| \tag{10.15}$$

Then the partial transpose is given by

$$\rho_{AB}^{T_B} = \frac{1}{\alpha}\sum_{ij} |a_i b_j\rangle\langle a_j b_i| \tag{10.16}$$

So we just swap the second qubits. For example, if we are computing the partial transpose of a density matrix,

$$|01\rangle\langle 00| \mapsto |00\rangle\langle 01|,\ |01\rangle\langle 10| \mapsto |00\rangle\langle 11|,\ |01\rangle\langle 01| \mapsto |01\rangle\langle 01|$$

Why is this useful? If ρ^T has any negative eigenvalues, then ρ is a density operator of an entangled state. If the eigenvalues are all positive, then the state is separable.

Example 10.1

We know that the Bell state

$$|\beta_{01}\rangle = \frac{|01\rangle + |10\rangle}{\sqrt{2}}$$

is entangled. Show this using the Peres partial transposition condition.

Solution

The density operator is

$$\rho = |\beta_{01}\rangle\langle\beta_{01}| = \left(\frac{|01\rangle + |10\rangle}{\sqrt{2}}\right)\left(\frac{\langle 01| + \langle 10|}{\sqrt{2}}\right)$$

$$= \frac{1}{2}(|01\rangle\langle 01| + |01\rangle\langle 10| + |10\rangle\langle 01| + |10\rangle\langle 10|)$$

In the $|00\rangle$, $|01\rangle$, $|10\rangle$, $|11\rangle$ basis the matrix representation of this density operator is

$$\rho = \begin{pmatrix} 0 & 0 & 0 & 0 \\ 0 & \frac{1}{2} & \frac{1}{2} & 0 \\ 0 & \frac{1}{2} & \frac{1}{2} & 0 \\ 0 & 0 & 0 & 0 \end{pmatrix}$$

The partial transpose is found by swapping the B qubits in each term. Hence

$$\rho^{T_B} = \frac{1}{2}(|01\rangle\langle 01| + |00\rangle\langle 11| + |11\rangle\langle 00| + |10\rangle\langle 10|)$$

The matrix representation of the partial transpose is

$$\rho^{T_B} = \begin{pmatrix} 0 & 0 & 0 & \frac{1}{2} \\ 0 & \frac{1}{2} & 0 & 0 \\ 0 & 0 & \frac{1}{2} & 0 \\ \frac{1}{2} & 0 & 0 & 0 \end{pmatrix}$$

The eigenvalues of ρ^{T_B} are $\{\frac{-1}{2}, \frac{1}{2}, \frac{1}{2}, \frac{1}{2}\}$. The presence of the negative eigenvalue $\frac{-1}{2}$ tells us that this *is* an entangled state. Note that while the specific matrix representation of the transposition ρ^{T_B} is basis dependent, the eigenvalues of the matrix are basis independent.

Example 10.2

In this example we consider a state that is obviously separable

$$|\psi\rangle = \left(\frac{|0\rangle - |1\rangle}{\sqrt{2}}\right) \otimes \left(\frac{|0\rangle - |1\rangle}{\sqrt{2}}\right) = \frac{|00\rangle - |01\rangle - |10\rangle + |11\rangle}{2}$$

Show that this is the case using the Peres partial transposition condition.

Solution

The density operator is

$$\rho = |\psi\rangle\langle\psi|$$
$$= \left(\frac{|00\rangle - |01\rangle - |10\rangle + |11\rangle}{2}\right)\left(\frac{\langle 00| - \langle 01| - \langle 10| + \langle 11|}{2}\right)$$
$$= \frac{1}{4}(|00\rangle\langle 00| - |00\rangle\langle 01| - |00\rangle\langle 10| + |00\rangle\langle 11| - |01\rangle\langle 00| + |01\rangle\langle 01| + |01\rangle\langle 10| - |01\rangle\langle 11|$$
$$-|10\rangle\langle 00| + |10\rangle\langle 01| + |10\rangle\langle 10| - |10\rangle\langle 11| + |11\rangle\langle 00| - |11\rangle\langle 01| - |11\rangle\langle 10| + |11\rangle\langle 11|)$$

The partial transpose is

$$\rho^{T_B} = \frac{1}{4}(|00\rangle\langle 00| - |01\rangle\langle 00| - |00\rangle\langle 10| + |01\rangle\langle 10| - |00\rangle\langle 01| + |01\rangle\langle 01| + |00\rangle\langle 11| - |01\rangle\langle 11|$$
$$-|10\rangle\langle 00| + |11\rangle\langle 00| + |10\rangle\langle 10| - |11\rangle\langle 10| + |10\rangle\langle 01| - |11\rangle\langle 01| - |10\rangle\langle 11| + |11\rangle\langle 11|)$$

In this case the two operators happen to be the same. So

$$\rho^{T_B} = \frac{1}{4}\begin{pmatrix} 1 & -1 & -1 & 1 \\ -1 & 1 & 1 & -1 \\ -1 & 1 & 1 & -1 \\ 1 & -1 & -1 & 1 \end{pmatrix}$$

The eigenvalues of ρ^{T_B} are {1, 0, 0, 0}. Since all $\lambda_i \geq 0$, the Peres partial transposition condition shows that this is a separable state.

Returning to teleportation, it is possible for Alice to simultaneously transmit two unknown quantum states to two parties, Bob and Charlie. Suppose that Alice wants to transmit the state

$$|\phi_1\rangle = \alpha_1|0\rangle + \beta_1|1\rangle \tag{10.17}$$

to Bob and the state

$$|\phi_2\rangle = \alpha_2|0\rangle + \beta_2|1\rangle \tag{10.18}$$

to Charlie. Alice must have two entangled pairs in her possession, sharing one EPR pair with Bob and one EPR pair with Charlie:

$$|\beta_{A_1B}\rangle = \frac{|00\rangle_{A_1B} + |11\rangle_{A_1B}}{\sqrt{2}} \tag{10.19}$$

$$|\beta_{A_2C}\rangle = \frac{|00\rangle_{A_2C} + |11\rangle_{A_2C}}{\sqrt{2}} \tag{10.20}$$

The state of the joint system of Alice, Bob, and Charlie is

$$\begin{aligned} |\psi\rangle &= (\alpha_1\alpha_2|00\rangle + \alpha_1\beta_2|01\rangle + \alpha_2\beta_1|10\rangle + \beta_1\beta_2|11\rangle) \\ &\otimes\frac{1}{2}(|0000\rangle + |0101\rangle + |1010\rangle + |1111\rangle) \end{aligned} \tag{10.21}$$

The four-qubit state in the second part of this expression is organized as $|A_1A_2BC\rangle$ so that the first and third qubits represent the entanglement between Alice and Bob and the second and fourth qubits are the entanglement between Alice and Charlie.

To simultaneously teleport the states (10.17) and (10.18), the unitary transformation

$$U_{BC} = \frac{1}{\sqrt{2}}\begin{pmatrix} 1 & 0 & 1 & 0 \\ 0 & 1 & 0 & 1 \\ 0 & 1 & 0 & -1 \\ 1 & 0 & -1 & 0 \end{pmatrix} \tag{10.22}$$

is applied to Bob and Charlie's qubits. This "locks" the quantum channel. The state (10.21) is transformed into

$$\begin{aligned} |\psi'\rangle &= (\alpha_1\alpha_2|00\rangle + \alpha_1\beta_2|01\rangle + \alpha_2\beta_1|10\rangle + \beta_1\beta_2|11\rangle) \\ &\otimes \frac{1}{2\sqrt{2}}(|0000\rangle + |0101\rangle + |0011\rangle + |0110\rangle + |1000\rangle \\ &- |1011\rangle + |1101\rangle - |1110\rangle) \end{aligned} \tag{10.23}$$

Next Alice measures her qubits using the Bell basis (7.27) through (7.30). Based on her result, which she communicates with a classical channel, Bob and Charlie perform local unitary operations using the Pauli matrices I, X, Y, and Z. However, they do not yet have the states they need. They need to "unlock" the quantum channel by applying the inverse of (10.22), namely

$$U_{BC}{}^{\dagger} = \frac{1}{\sqrt{2}}\begin{pmatrix} 1 & 0 & 0 & 1 \\ 0 & 1 & 1 & 0 \\ 1 & 0 & 0 & -1 \\ 0 & 1 & -1 & 0 \end{pmatrix} \tag{10.24}$$

Example 10.3

Show that the initial state shared between Alice and Bob (10.21) is entangled, but that the state shared between Alice and Bob after the channel has been locked (10.23) is a product state.

Solution

We begin by writing down the density matrix of the system prior to locking. We look at the state

$$\begin{aligned} |\psi\rangle &= (\alpha_1\alpha_2|00\rangle + \alpha_1\beta_2|01\rangle + \alpha_2\beta_1|10\rangle + \beta_1\beta_2|11\rangle) \\ &\otimes \frac{1}{2}(|0000\rangle + |0101\rangle + |1010\rangle + |1111\rangle) \end{aligned}$$

We only need to concern ourselves with the second term:

$$|\phi\rangle = \frac{1}{2}(|0000\rangle + |0101\rangle + |1010\rangle + |1111\rangle)$$

The density operator is

$$\rho = |\phi\rangle\langle\phi| = \frac{1}{4}(|0000\rangle\langle 0000| + |0000\rangle\langle 0101| + |0000\rangle\langle 1010| + |0000\rangle\langle 1111|$$
$$+ |0101\rangle\langle 0000| + |0101\rangle\langle 0101| + |0101\rangle\langle 1010| + |0101\rangle\langle 1111|$$
$$+ |1010\rangle\langle 0000| + |1010\rangle\langle 0101| + |1010\rangle\langle 1010| + |1010\rangle\langle 1111|$$
$$+ |1111\rangle\langle 0000| + |1111\rangle\langle 0101| + |1111\rangle\langle 1010| + |1111\rangle\langle 1111|)$$

To obtain the density operator to represent the joint state between Alice and Bob, we need to compute the partial trace over the Alice–Charlie states. These are the second and fourth qubits (A_2, C). So we compute

$$\rho_{A_1 B} = \langle 00|\rho|00\rangle_{A_2 C} + \langle 01|\rho|01\rangle_{A_2 C} + \langle 10|\rho|10\rangle_{A_2 C} + \langle 11|\rho|11\rangle_{A_2 C}$$

This gives the reduced density operator

$$\rho_{A_1 B} = \frac{1}{2}(|00\rangle\langle 00| + |00\rangle\langle 11| + |11\rangle\langle 00| + |11\rangle\langle 11|)$$
$$= \frac{1}{2}\begin{pmatrix} 1 & 0 & 0 & 1 \\ 0 & 0 & 0 & 0 \\ 0 & 0 & 0 & 0 \\ 1 & 0 & 0 & 1 \end{pmatrix}$$

Now we compute the partial transpose, which turns out to be

$$\rho^{T_B}{}_{A_1 B} = \frac{1}{2}(|00\rangle\langle 00| + |01\rangle\langle 10| + |10\rangle\langle 01| + |11\rangle\langle 11|)$$
$$= \frac{1}{2}\begin{pmatrix} 1 & 0 & 0 & 0 \\ 0 & 0 & 1 & 0 \\ 0 & 1 & 0 & 0 \\ 0 & 0 & 0 & 1 \end{pmatrix}$$

The eigenvalues of this matrix are $\{\frac{-1}{2}, \frac{1}{2}, \frac{1}{2}, \frac{1}{2}\}$. Since one of the eigenvalues is $\frac{-1}{2} < 0$, we have shown that Alice and Bob share an entangled state prior to locking.

The state after locking is given by (10.23). We are concerned with the second term in the tensor product

$$|\phi_L\rangle = \frac{1}{2\sqrt{2}}(|0000\rangle + |0101\rangle + |0011\rangle + |0110\rangle + |1000\rangle - |1011\rangle + |1101\rangle - |1110\rangle)$$

The density matrix this time is

$$\rho_L = |\phi_L\rangle\langle\phi_L|$$

Needless to say, writing this matrix down is a tedious exercise. There are 64 terms in the expansion, you will have to take my word for it. After tracing out over the Alice–Charlie terms (tracing over the $A_2 C$ qubits, which are the second and fourth qubits), we obtain the reduced density operator:

$$\rho_L' = \frac{1}{4}(|00\rangle\langle 00| + |00\rangle\langle 10| + |01\rangle\langle 01| - |01\rangle\langle 11| + |10\rangle\langle 00| + |10\rangle\langle 10| - |11\rangle\langle 01| + |11\rangle\langle 11|) \quad (10.25)$$

$$= \frac{1}{4}\begin{pmatrix} 1 & 0 & 1 & 0 \\ 0 & 1 & 0 & -1 \\ 1 & 0 & 1 & 0 \\ 0 & -1 & 0 & 1 \end{pmatrix}$$

Its easy to see that the partial transpose gives the same matrix. The eigenvalues are $\{\frac{1}{2}, \frac{1}{2}, 0, 0\}$. Since each $\lambda_i \geq 0$, we conclude that the state is separable. Therefore the locking operation has destroyed the entanglement between Alice and Bob.

ENTANGLEMENT SWAPPING

In an interesting extension of teleportation, we can cause two particles that have never interacted to become entangled. This is possible, in principle, even if the particles are light years apart.

Entanglement swapping begins with two EPR pairs. We label the qubits 1, 2, 3, and 4. Alice has qubits 1 and 4 in her possession while Bob has qubits 2 and 3 in his possession. Qubits 1 and 2 are entangled in the Bell state

$$|\beta_{00}\rangle_{12} = \frac{|00\rangle_{12} + |11\rangle_{12}}{\sqrt{2}} \quad (10.26)$$

Similarly qubits 3 and 4 are entangled:

$$|\beta_{00}\rangle_{34} = \frac{|00\rangle_{34} + |11\rangle_{34}}{\sqrt{2}} \quad (10.27)$$

The product of these two states is

$$|\beta_{00}\rangle_{12}|\beta_{00}\rangle_{34} = \left(\frac{|00\rangle_{12} + |11\rangle_{12}}{\sqrt{2}}\right)\left(\frac{|00\rangle_{34} + |11\rangle_{34}}{\sqrt{2}}\right)$$

$$= \frac{1}{2}(|00\rangle_{12}|00\rangle_{34} + |00\rangle_{12}|11\rangle_{34} + |11\rangle_{12}|00\rangle_{34} + |11\rangle_{12}|11\rangle_{34})$$

Now let's do some simple algebra. First we just rearrange the qubits in each term so that we can write qubits 1 and 4 together and qubits 2 and 3 together:

$$|\beta_{00}\rangle_{12}|\beta_{00}\rangle_{34} = \frac{1}{2}(|00\rangle_{14}|00\rangle_{23} + |01\rangle_{14}|01\rangle_{23} + |10\rangle_{14}|10\rangle_{23} + |11\rangle_{12}|11\rangle_{34})$$

Notice that

$$
\begin{aligned}
|\beta_{00}\rangle_{14}|\beta_{00}\rangle_{23} &= \left(\frac{|00\rangle_{14}+|11\rangle_{14}}{\sqrt{2}}\right)\left(\frac{|00\rangle_{23}+|11\rangle_{23}}{\sqrt{2}}\right) \\
&= \frac{1}{2}(|00\rangle_{14}|00\rangle_{23}+|00\rangle_{14}|11\rangle_{23}+|11\rangle_{14}|00\rangle_{23}+|11\rangle_{14}|11\rangle_{23})
\end{aligned}
$$

The product $|\beta_{00}\rangle_{12}|\beta_{00}\rangle_{34}$ is missing a few of these terms, but we can stick them in there if we also subtract them, giving

$$
\begin{aligned}
|\beta_{00}\rangle_{12}|\beta_{00}\rangle_{34} &= \frac{1}{2}(|00\rangle_{14}|00\rangle_{23}+|01\rangle_{14}|01\rangle_{23}+|10\rangle_{14}|10\rangle_{23}+|11\rangle_{12}|11\rangle_{34} \\
&\quad +|00\rangle_{14}|11\rangle_{23}+|11\rangle_{14}|00\rangle_{23}-|00\rangle_{14}|11\rangle_{23}-|11\rangle_{14}|00\rangle_{23}) \\
&= \frac{1}{2}(|\beta_{00}\rangle_{14}|\beta_{00}\rangle_{23}+|01\rangle_{14}|01\rangle_{23}+|10\rangle_{14}|10\rangle_{23} \\
&\quad -|00\rangle_{14}|11\rangle_{23}-|11\rangle_{14}|00\rangle_{23})
\end{aligned}
$$

Similar algebraic tricks on the other terms can be used to show that in the end

$$
\begin{aligned}
|\beta_{00}\rangle_{12}|\beta_{00}\rangle_{34} &= \frac{1}{2}(|\beta_{00}\rangle_{14}|\beta_{00}\rangle_{23}+|\beta_{10}\rangle_{14}|\beta_{10}\rangle_{23} \\
&\quad +|\beta_{01}\rangle_{14}|\beta_{01}\rangle_{23}+|\beta_{11}\rangle_{14}|\beta_{11}\rangle_{23})
\end{aligned}
$$

Remember, Alice has particles 1 and 4 in her possession. Now Alice performs a Bell state measurement on particles 1 and 4. The possible results are, of course, $|\beta_{00}\rangle_{14}$, $|\beta_{10}\rangle_{14}$, $|\beta_{01}\rangle_{14}$, and $|\beta_{11}\rangle_{14}$, each with a probability of $\frac{1}{4}$. Depending on the measurement result that Alice obtains, Bob's system collapses into one of the Bell state $|\beta_{00}\rangle_{23}$, $|\beta_{10}\rangle_{23}$, $|\beta_{01}\rangle_{23}$, or $|\beta_{11}\rangle_{23}$. Now particles 2 and 3 are entangled.

Let's summarize the procedure. Alice and Bob create two entangled states. Particles 1 and 2 are entangled together, and particles 3 and 4 are entangled together. Alice has particles 1 and 4 in her possession, and Bob has particles 2 and 3 in his possession. Bob carefully flies off to a distant land with his particles, and Alice does a Bell state measurement on her particles. The state collapses and Bob's particles 2 and 3 become entangled with each other. What is teleported here? Perhaps you can say that the entanglement has been teleported. This procedure works even if particles 2 and 3 have never interacted. To see the possible significance of this, add a twist and say that Charlie takes particle 3 and goes off somewhere else before the measurement is made. Alice does her Bell state measurement—now Bob and Charlie share an entangled pair and can use it as a resource between the two of them—so Bob can teleport a state to Charlie.

SUPERDENSE CODING

Alice and Bob are once again living in separate places and Alice wants to send Bob some information. This time Alice would like to send Bob two classical bits of information, but she only has one single qubit to work with. She can accomplish this amazing feat using a protocol known as *superdense coding*. Once again, we begin with Alice and Bob sharing an entangled pair of particles. We also agree on a code: we let the Bell states $|\beta_{xy}\rangle$ correspond to the classical bit string xy, where $xy = 00$, 01, 10, or 11. The system begins in the state

$$|\psi\rangle = \frac{|00\rangle + |11\rangle}{\sqrt{2}} \tag{10.28}$$

where Alice has the first qubit in her possession and Bob has the second qubit. Alice transmits the information to Bob, this time by actually sending Bob her qubit. Depending on what bit string Alice wants to send to Bob, she first acts on the qubit with a single-qubit quantum gate of her choosing. If she wants to send Bob the classical bit string 00, then Alice leaves her qubit alone. Then the state stays in its initial state (10.28), which we denote as the Bell state

$$|\beta_{00}\rangle = \frac{|00\rangle + |11\rangle}{\sqrt{2}} \tag{10.29}$$

What if Alice applied the X gate to her qubit before transmitting it? Then the state would become

$$(X \otimes I)|\psi\rangle = \frac{|10\rangle + |01\rangle}{\sqrt{2}} = |\beta_{01}\rangle \tag{10.30}$$

If Alice instead applies a Z gate to her qubit before sending it on over to Bob, the state (10.28) becomes

$$(Z \otimes I)|\psi\rangle = \frac{|00\rangle - |11\rangle}{\sqrt{2}} = |\beta_{10}\rangle \tag{10.31}$$

Finally, notice that if Alice applies the iY gate, the state is

$$(iY \otimes I)|\psi\rangle = \frac{|01\rangle - |10\rangle}{\sqrt{2}} = |\beta_{11}\rangle \tag{10.32}$$

After Bob gets the qubit from Alice, he does a measurement in the Bell basis and finds one of $|\beta_{00}\rangle$, $|\beta_{01}\rangle$, $|\beta_{10}\rangle$, or $|\beta_{11}\rangle$. So he obtains one of the classical two-bit strings 00, 01, 10, or 11.

Example 10.4

A W state is a three-qubit state of the form

$$|W_n\rangle = \frac{1}{\sqrt{2+2n}}(|100\rangle + \sqrt{n}e^{i\gamma}|010\rangle + \sqrt{n+1}e^{i\delta}|001\rangle) \tag{10.33}$$

Consider the particular case of $|W_1\rangle = \frac{1}{2}(|100\rangle + |010\rangle + \sqrt{2}|001\rangle)$. Show that if Alice takes qubit 1 and Bob takes qubits 2 and 3, Alice can perform a local unitary operation and send Bob her qubit, which results in the transmission of two classical bits to Bob.

Solution

If Alice does nothing to her qubit, then when she sends Bob the first qubit he has the state

$$|W_1\rangle = \frac{1}{2}(|100\rangle + |010\rangle + \sqrt{2}|001\rangle) = |\psi_{00}\rangle \tag{10.34}$$

in his possession. Now suppose that Alice applies the X gate. The state becomes

$$X \otimes I \otimes I|W_1\rangle = \frac{1}{2}(|000\rangle + |110\rangle + \sqrt{2}|101\rangle) = |\psi_{01}\rangle \tag{10.35}$$

If instead Alice applies iY, we have

$$iY \otimes I \otimes I|W_1\rangle = \frac{1}{2}(-|100\rangle + |010\rangle + \sqrt{2}|001\rangle) = |\psi_{10}\rangle \tag{10.36}$$

Finally, suppose Alice applies the Z gate, giving

$$Z \otimes I \otimes I|W_1\rangle = \frac{1}{2}(|000\rangle - |110\rangle - \sqrt{2}|101\rangle) = |\psi_{11}\rangle \tag{10.37}$$

Notice that these states are orthogonal. For example,

$$\begin{aligned}
\langle\psi_{00}|\psi_{01}\rangle &= \frac{1}{4}((\langle 100| + \langle 010| + \sqrt{2}\langle 001|)(|000\rangle + |110\rangle + \sqrt{2}|101\rangle) \\
&= \frac{1}{4}(\langle 1|0\rangle\langle 0|0\rangle\langle 0|0\rangle + \langle 0|1\rangle\langle 1|1\rangle\langle 0|0\rangle + 2\langle 0|1\rangle\langle 0|0\rangle\langle 1|1\rangle) \\
&= 0
\end{aligned}$$

This means that after Alice sends Bob the first qubit, he can distinguish among the states by making a three-qubit projective measurement. With the agreed-upon coding protocol, Bob can recover the two classical bits 00, 01, 10, and 11 when he obtains the measurement result $|\psi_{00}\rangle$, $|\psi_{01}\rangle$, $|\psi_{10}\rangle$, or $|\psi_{11}\rangle$, respectively.

EXERCISES

10.1. *Suppose that Alice and Bob use the teleportation scheme but don't use classical communication. What does Bob see? Start with the state (10.9) and form the density operator. Then trace out the first two qubits to obtain the density operator for Bob. You should see that Bob sees a set of states that are all equally likely—so Bob sees a completely random mixture.*

10.2. *Using the Peres partial transposition condition, determine whether or not the state* $|\psi\rangle = \frac{1}{2}(|00\rangle + |01\rangle + |10\rangle - |11\rangle)$ *is entangled.*

10.3. *Determine whether or not the state* $\rho = \frac{1}{5}(|01\rangle + |10\rangle) \otimes (\langle 01| + \langle 10|) + \frac{1}{5}|11\rangle\langle 11|$ *is entangled.*

10.4. *If you're brave, derive (10.25) from (10.23).*

10.5. *Consider entanglement swapping and suppose that Alice and Bob share the two EPR pairs* $|\beta_{10}\rangle_{12}|\beta_{00}\rangle_{34}$. *Alice does a Bell state measurement and measures* $|\beta_{00}\rangle_{14}$. *Show that Bob's two qubits have become entangled and that they are in the state* $|\beta_{10}\rangle_{23}$.

10.6. *The GHZ state is the three-qubit state given by*

$$|GHZ\rangle = \frac{|000\rangle + |111\rangle}{\sqrt{2}}.$$

Suppose that Alice takes the first qubit and Bob takes qubits 2 and 3. Show that this scenario can be used for superdense coding.

11

QUANTUM CRYPTOGRAPHY

Secure communication can be accomplished through *key distribution*, a method by which two parties who want to communicate privately share a key that is used to encrypt or scramble a message. Later the key can be used to recover or decrypt the message. Currently cryptographic keys are generated using mathematical algorithms that are hard but not impossible to break. In contrast, quantum cryptography is a method of key distribution that relies on the laws of physics to create a key. While not 100% safe, quantum cryptography offers huge advantages over traditional methods.

Before we get into quantum cryptography, let's look at a toy example to see how messages can be encrypted. Suppose that two parties, denoted Alice and Bob as usual, want to share a private message. A trivial way to encrypt the message is to generate a key k that is just a number used to scramble up the message. We can add k to each character in the message, for example, converting a useful message into a scrambled bunch of meaningless characters. Let's say that we are representing the capital letters of the alphabet using a binary code. Since there are 26 letters in the alphabet, and $2^5 = 32$, we need at least five bits to encode the alphabet using binary. If we start with the first four letters A, B, C, D as 0, 1, 2, 3, then we can encode

Quantum Computing Explained, by David McMahon

the letters in the following way:

$$A \to 0000$$
$$B \to 0001$$
$$C \to 0010$$
$$D \to 0011$$

If we just transmitted our message over a public communications channel (e.g., over a cell phone), then an eavesdropper (commonly denoted as *Eve*) might just tap into the line and record our conversation. So how can Alice and Bob secure the message? A very simple way to secure the message is to create a key k that we add to each character before it's transmitted. Then the scrambled message is sent over the public communications channel. When it reaches Bob, he subtracts k to decrypt or recover the original message. If Eve doesn't know the value of k, then Alice and Bob get to share their private conversation. So, if the message is m Alice encrypts it with the key k by creating the transmitted string t in the following way:

$$t = m + k$$

To be specific, suppose that $k = 3 = 0011$. We add this to each character in a given message. So the strings above become

$$A \to 0000 \to 0011$$
$$B \to 0001 \to 0100$$
$$C \to 0010 \to 0101$$
$$D \to 0011 \to 0110$$

That is, $A \to D$, $B \to E$, $C \to F$, $D \to G$ for our particular encoding scheme. If Alice wants to transmit the word BAD to Bob, she encrypts it using the key and obtains

$$E \quad D \quad G$$

This string is transmitted over the public channel. Eve taps the line and gets the meaningless string of characters EDG and has no idea what Alice is talking about. Bob, on the other hand, knows that he can decrypt the signal by subtracting the key

$$m = t - k$$

Now, if we always use the same key, then Eve can study the situation over time and eventually deduce what k is, or maybe someone will find out what it is while talking to Alice or going through Bob's hard drive. We can minimize the risk by changing the key, maybe we change it every time we send a message, say. When we regularly change the key we are using a *one-time pad* method.

A BRIEF OVERVIEW OF RSA ENCRYPTION

Of course such a simplistic system is not used in the real world. More safeguards are needed to ensure the security of our data, and one method that is very popular is an encryption scheme called *RSA*. The basic idea behind RSA is to use two keys, one that is public and one that is private. To decrypt a message you've got to have the private key, and the security of the system is based on generating very large numbers that are hard to factor. You would need to factor the numbers in order to crack the system.

We begin by choosing two *prime* numbers, which we denote p and q. A prime number is a natural number p that is divisible only by itself and by 1. For example, some small prime numbers are

$$2, 3, 5, 7, 11, 19$$

The number *9* is not a prime number because it is divisible by *9, 3, 1*. The number 13 is a prime number because its only divisible by 13 and 1. In RSA encryption, very large prime numbers $p > 10^{100}$ are used, making factoring very difficult, if not impossible. It turns out that factoring products of very large prime numbers *appears* to be so difficult, in principle (meaning that it doesn't matter how fast or powerful a computer we build), that it would take billions of years to factor them. But keep in mind that this has not been proved, so it may be possible that someday an efficient mathematical algorithm will be developed that could factor products of large prime numbers on a classical computer. Regardless, Shor's algorithm can factor prime numbers readily on a quantum computer—so other cryptographic methods will have to be used (quantum cryptography) to secure messages.

Let's see how a simple RSA scheme might work. We begin with two large prime numbers p and q and create their product, which we denote n:

$$n = pq \tag{11.1}$$

Now we compute another product that some imaginative number theorists have denoted the totient:

$$\phi(n) = (p-1)(q-1) \tag{11.2}$$

Next, let $1 < e < \phi(n)$ such that the only common factor of e and $\phi(n)$ is 1. We generate the *public key* using n and e. To create the private key, we use

$$d = \frac{1 \bmod \phi(\text{n})}{e} \tag{11.3}$$

The private key consists of d and n. Notice that if we knew d, since we know e (which is available on a public channel), we would be able to decrypt the message. A quick aside for readers who aren't familiar with the mod function. This is short for *modulus*, which is the remainder we get when we divide one number by another. For example 7 divided by 5 is 1 with a remainder of 2

$$7 \div 5 = 1 \qquad R2$$

This is

$$7 \bmod 5 = 2$$

So the mod function just gives us the remainder of a division operation.

Returning to RSA encryption, a message m is encrypted in the following way:

$$c = m^e \bmod n \tag{11.4}$$

The intended party has the private key d in their possession. They can decrypt the message, since

$$c^d = m^{de} \bmod n \tag{11.5}$$

Now it so happens that since $n = pq$,

$$m^{ed} = m \bmod p$$

$$m^{ed} = m \bmod q$$

Some results from number theory tell us that this means that

$$c^d = m \bmod n \tag{11.6}$$

Example 11.1

Let's see how the RSA system works, in principle, by working an example with some very small numbers. Let $p = 3$, $q = 11$, and show how we would encrypt and decrypt a message $m = 6$ using the RSA procedure outlined above.

Solution

First, we create the product

$$n = pq = (3)(11) = 33$$

Next, we create the totient

$$\phi = (3-1)(11-1) = (2)(10) = 20$$

Let $e > 1$ such that e and $\phi = 20$ have no common factors except 1. The smallest number that satisfies this criterion is $e = 3$. To find d, we use (11.3) and find the smallest x such that

$$de = 1 + x\phi$$

where d is a natural number. In our case we have

$$d = \frac{1 + 20x}{3}$$

The smallest value is $x = 1$, giving

$$d = \frac{1 + 20}{3} = \frac{21}{3} = 7$$

The encrypted message that is transmitted for $m = 6$ is

$$c = m^e \bmod n = 6^3 \bmod 33 = 18$$

(the mod function is best computed using your favorite program like MATLAB). Alice transmits this signal to Bob, who holds the private key $d = 7$ in his possession. He decrypts the message using

$$m = c^d \bmod n = 18^7 \bmod 33 = 6$$

The RSA encryption algorithm works for many applications in the present day. However, as we mentioned earlier, Shor's algorithm demonstrates that a workable quantum computer can easily crack an RSA encryption scheme, since $n = pq$ can be readily factored. Is there a better way to encrypt messages? It turns out that quantum mechanics shows us several ways.

BASIC QUANTUM CRYPTOGRAPHY

Quantum cryptography is a method that uses quantum mechanics rather than simple numerical algorithms to generate a secret key. This is known as *quantum key distribution*, or *QKD*. To implement QKD, two communications channels are used between Alice and Bob. This includes an ordinary public channel, which is just an ordinary classical communications link—it could be the Internet, a cell phone, or your home telephone. Encrypted messages are sent over this line. In addition a second piece of the QKD puzzle is used—a quantum communications channel over which the quantum key is distributed. In practice, this is done using individual photons in different polarization states. Quantum mechanics relies on a fundamental principle of quantum theory—that measurement disturbs a quantum state. To learn something about a key encoded as a quantum state, a measurement has to be made. So if Eve taps into the line, she has to make measurements—disturbing the system in such a way that Alice and Bob can detect her presence.

The first type of quantum cryptography we will look at is called the *BB84* protocol. It's named after its discoverers (Bennett and Brassard) and the year when the protocol was first published. There are three key principles used in BB84 QKD:

1. The no-cloning theorem–quantum states cannot be copied. As a result Eve cannot tap a quantum communications channel, copy the quantum states used to create the key for herself, and send the originals down the line to Bob.
2. Measurement leads to state collapse. A key aspect of QKD is that different bases will be used to create a bit string. When we make measurements in one of the given bases, we will cause state collapse such that measurements in the other basis are completely random. In other words, extracting some information about the state disturbs the state of the system.
3. Measurements are irreversible.

To see points 2 and 3, notice that if the system is in the state

$$|+\rangle = \frac{|0\rangle + |1\rangle}{\sqrt{2}}$$

We make a measurement in the computational basis $\{|0\rangle, |1\rangle\}$, and the original state of the system is lost. Suppose we obtain measurement result 0. Then in the $|\pm\rangle$ basis the state is now different than it was originally:

$$|0\rangle = \frac{|+\rangle + |-\rangle}{\sqrt{2}}$$

There is only a 50% chance of finding $|+\rangle$ if we measure in the $|\pm\rangle$ basis—the state has been irreversibly altered.

In the BB84 protocol we use the computational basis $\{|0\rangle, |1\rangle\}$ together with the $|\pm\rangle$ basis to create our key. To review, these two bases are related using

$$|+\rangle = \frac{|0\rangle + |1\rangle}{\sqrt{2}}, \quad |-\rangle = \frac{|0\rangle - |1\rangle}{\sqrt{2}} \tag{11.7}$$

To create the key, Alice begins by randomly creating a string of $2n$ qubits. Each qubit is created in one of the four states:

$$|0\rangle, |1\rangle, |+\rangle, |-\rangle$$

Logical 0 can be represented by $|0\rangle$, $|+\rangle$, while logical 1 is represented by $|1\rangle$, $|-\rangle$. Alice then sends the random sequence of qubits over the quantum channel to Bob. Bob measures each qubit, but he does so randomly selecting the $\{|0\rangle, |1\rangle\}$ basis or the $|\pm\rangle$ basis at each position.

We have two bases here, and Alice randomly selects what basis to use to create each bit in the string. Based on ordinary probability arguments, you can see that about n of the bits will be created in the $\{|0\rangle, |1\rangle\}$ basis and n of the bits will be created in the $|\pm\rangle$ basis. Alice and Bob compare notes, only telling each other which basis was used at each position. When Alice and Bob use a different basis, they discard that qubit. We call the key that results after discarding the bits where Alice and Bob used different bases the *sifted key*.

Example 11.2

Alice creates an 8-bit string

$$|0\rangle|1\rangle|+\rangle|0\rangle|0\rangle|-\rangle|+\rangle|-\rangle$$

If Bob randomly measures in the following order using the bases $\{|0\rangle,|1\rangle\}\{|0\rangle,|1\rangle\}\{|\pm\rangle|\pm\rangle\}|\pm\rangle$ $|0\rangle,|1\rangle|\pm\rangle|\pm\rangle$, describe the bit string that Alice and Bob keep.

Solution

We simply throw out the bits where Alice and Bob used a different basis, numbering the bits from 1 to 8. We have the following table:

Alice Bits	0	1	0	0	0	1	0	1
Alice Basis	$\{\|0\rangle, \|1\rangle\}$	$\{\|0\rangle, \|1\rangle\}$	$\|\pm\rangle$	$\{\|0\rangle, \|1\rangle\}$	$\{\|0\rangle, \|1\rangle\}$	$\|\pm\rangle$	$\|\pm\rangle$	$\|\pm\rangle$
Bob Basis	$\{\|0\rangle, \|1\rangle\}$	$\{\|0\rangle, \|1\rangle\}$	$\|\pm\rangle$	$\|\pm\rangle$	$\|\pm\rangle$	$\{\|0\rangle, \|1\rangle\}$	$\|\pm\rangle$	$\|\pm\rangle$
Match	Yes	Yes	Yes	No	No	No	Yes	Yes
Keep	Yes	Yes	Yes	No	No	No	Yes	Yes

The resulting bit string is created by keeping bits in positions 1, 2, 3, 7, 8 and discarding the bits in positions 4, 5, 6. The sifted key is then

$$s = 01001$$

Once the sifted key is created, Alice and Bob need to check it for errors. Errors can occur simply from the environment—causing bit flip and phase flip errors for example—or they can occur because an eavesdropper has tapped the channel. If the error rate is too high, this probably indicates Eve is listening in. In this case, Alice and Bob check the error rate, and if it's above an agreed-upon threshold, they discard the key and start over. In Example 11.2, let's suppose that Eve had made a measurement on bit 8. There is a 50% chance she selected the computational basis to make her measurement and a 50% chance that she selected the $|\pm\rangle$ basis. Suppose that she selected the computational basis. Then

$$|-\rangle = \frac{|0\rangle - |1\rangle}{\sqrt{2}}$$

When Eve makes her measurement, the qubit will be either in the $|0\rangle$ state or the $|1\rangle$ state. Let's say it's $|0\rangle$. In the $\{1\pm\rangle\}$ basis this can be written as:

$$|0\rangle = \frac{|+\rangle + |-\rangle}{\sqrt{2}}$$

Although Alice prepared the qubit in the $|-\rangle$ state, when Bob makes his measurement, there is only a 50% chance of seeing the correct result. If we include a large number of qubits in our string, Alice and Bob can use this kind of behavior to deduce the presence of Eve. In our example Alice knows she created the bit string

01001

But Bob has the bit string

01000

Noise on the quantum channel will also create errors. Quantum error correction can be used to fix up those qubits.

AN EXAMPLE ATTACK: THE CONTROLLED NOT ATTACK

Suppose that Alice has the state

$$|+_A\rangle = \frac{|0_A\rangle + |1_A\rangle}{\sqrt{2}}$$

Can Eve duplicate this state in any way? Let's suppose that Eve wanted to make a state that gave the same measurement result for both Alice and Eve. What Eve can do is start with the state $|0_E\rangle$ and create the product state

$$|+_A\rangle \otimes |0_E\rangle = \frac{|0_A\rangle|0_E\rangle + |1_A\rangle|0_E\rangle}{\sqrt{2}}$$

Now watch what happens if Eve applies a controlled NOT gate to the state, using Alice's qubit at the control bit and Eve's qubit as the target. The state becomes

$$\frac{|0_A\rangle|0_E\rangle + |1_A\rangle|0_E\rangle}{\sqrt{2}} \underset{CN}{\rightarrow} \frac{|0_A\rangle|0_E\rangle + |1_A\rangle|1_E\rangle}{\sqrt{2}}$$

If Alice measures 0, we've got 0 for Eve. If Alice gets 1, Eve has 1. What happens if the measurement is made in the $|\pm\rangle$ basis? Then we have

$$\frac{|0_A\rangle|0_E\rangle + |1_A\rangle|1_E\rangle}{\sqrt{2}} = \frac{1}{\sqrt{2}}\left[\left(\frac{|+_A\rangle + |-_A\rangle}{\sqrt{2}}\right)\left(\frac{|+_E\rangle + |-_E\rangle}{\sqrt{2}}\right)\right.$$
$$\left. + \left(\frac{|+_A\rangle - |-_A\rangle}{\sqrt{2}}\right)\left(\frac{|+_E\rangle - |-_E\rangle}{\sqrt{2}}\right)\right]$$

$$= \frac{1}{\sqrt{2}}\left(\frac{1}{2}\right)[|+_A\rangle|+_E\rangle + |+_A\rangle|-_E\rangle + |-_A\rangle|+_E\rangle + |-_A\rangle|-_E\rangle$$

$$+ |+_A\rangle|+_E\rangle - |+_A\rangle|-_E\rangle - |-_A\rangle|+_E\rangle + |-_A\rangle|-_E\rangle]$$

$$= \frac{1}{\sqrt{2}}(|+_A\rangle|+_E\rangle + |-_A\rangle|-_E\rangle)$$

Interestingly the correlation between Alice and Eve has been maintained! Eve's qubit assumes the same value as Alice's qubit in both bases. Suppose instead that Alice has

$$|-_A\rangle = \frac{|0_A\rangle - |1_A\rangle}{\sqrt{2}}$$

Does the correlation still hold? You can show that if we apply the same procedure, we end up with the state

$$|\psi\rangle = \frac{|+_A\rangle|-_E\rangle + |-_A\rangle|+_E\rangle}{\sqrt{2}}$$

We see that Alice and Eve get opposite measurement results. But Eve doesn't know what state Alice had ahead of time—so her measurement results are meaningless. Eve cannot, in general, form a product state with $|0_E\rangle$ and apply a controlled NOT gate to find out what Alice has.

THE B92 PROTOCOL

We now give an overview of an updated QKD protocol that is a simplification of the BB84 protocol. This time Alice and Bob use two nonorthogonal bases. As usual, we use the computational basis for one of the bases and a basis that we denote $|0'\rangle$, $|1'\rangle$. The basic procedure is the following:

1. Bob measures the qubits, randomly selecting the computational basis or the $|0'\rangle, |1'\rangle$ basis. Bob only measures using $|0\rangle\langle 0|$ or $|0'\rangle\langle 0'|$.
2. Bob creates a key from the bit positions where he obtains $|1\rangle$ or $|1'\rangle$ as his measurement result.
3. Bob uses a public channel to tell Alice what bit positions he has.

The way this works is that if Alice creates a $|0\rangle$, and Bob measures in the computational basis, he obtains $|0\rangle$. If he measures in the $|0'\rangle$, $|1'\rangle$ basis, he gets $|1'\rangle$.

If Alice creates a $|0'\rangle$, and Bob measures in the computational basis, he obtains $|0'\rangle$. If he measures in the $|0'\rangle$, $|1'\rangle$ basis, he gets $|1\rangle$. Let's say that Alice creates the following 8 bit string:

$$|0\rangle|0\rangle|0'\rangle|0\rangle|0'\rangle|0'\rangle|0\rangle|0'\rangle$$

If Bob measures using

$$|0\rangle|0'\rangle|0'\rangle|0\rangle|0\rangle|0\rangle|0'\rangle|0'\rangle$$

The result is

$$|0\rangle|1'\rangle|0'\rangle|0\rangle|1\rangle|1\rangle|1'\rangle|0'\rangle$$

Bob announces he is keeping positions 2, 5, 6, 7 for the key. Notice the differences when comparing this to the BB84 protocol. In BB84, Alice generates a key. In this case, the key is generated when Bob obtains measurements results $|1\rangle$ or $|1'\rangle$. Instead of the four states $|0\rangle$, $|1\rangle$, $|+\rangle$, $|-\rangle$ used to create the key, Alice only uses the two states $|0\rangle$, $|0'\rangle$, a simplified procedure.

THE E91 PROTOCOL (EKERT)

The final method of quantum cryptography we will examine is due to Ekert and is based on quantum entanglement. We create a Bell state, giving one member of the EPR pair to Alice and the second member of the EPR pair to Bob. Suppose that the state used for the EPR pair is

$$|\beta_{00}\rangle = \frac{|00\rangle + |10\rangle}{\sqrt{2}}$$

Then we know that Alice and Bob will have measurement results that are completely correlated. On the other hand, if the state used is

$$|\beta_{01}\rangle = \frac{|01\rangle + |10\rangle}{\sqrt{2}}$$

then Alice and Bob will have measurement results that are perfectly anticorrelated. Alice and Bob measure their respective qubits in randomly chosen bases. Then they communicate over an ordinary channel and figure out on which bit positions they used the same basis. They keep these bits to create the key.

Since the measurement results will be perfectly correlated or perfectly anticorrelated, it is easy for Alice and Bob to determine whether or not an eavesdropper is present. Ordinary errors can be corrected using quantum error correction techniques.

In all cases of QKD, a refined key is created in a process known as *privacy amplification*. Basically all this involves is throwing away qubits that Eve could have attempted to measure. If the key originally had 20 bits say, and Eve knew something about 6 of the bits, the refined key is created by simply throwing away those 6 bits and leaving a 14-bit key.

EXERCISES

11.1. *Use the RSA algorithm with $p=3$, $q=19$, to encrypt and decrypt the message $m=42$.*

11.2. *Consider BB84 QKD. Alice creates an 8-bit string*

$$|+\rangle|1\rangle|+\rangle|-\rangle|0\rangle|-\rangle|+\rangle|-\rangle$$

Use a coin to randomly determine what basis Bob uses to measure each bit position, and describe the resulting bit string that Alice and Bob keep.

11.3. *Beginning with the state*

$$|-_A\rangle = \frac{|0_A\rangle - |1_A\rangle}{\sqrt{2}}$$

let Alice form a product state with $|0_E\rangle$. Show that after applying a controlled-NOT gate, she obtains a correlation with Alice in the computational basis but not in the $|\pm\rangle$basis.

11.4. *Alice generates a bit string $|0'\rangle|0\rangle|0\rangle|0\rangle|0\rangle|0'\rangle|0\rangle|0'\rangle$. If Bob does measurements $|0'\rangle|0'\rangle|0\rangle|0\rangle|0\rangle|0\rangle|0\rangle|0'\rangle$, what positions does he keep?*

12

QUANTUM NOISE AND ERROR CORRECTION

In the treatment of quantum theory we've used so far we have been looking at *closed* systems. These are quantum systems that do not interact with the outside world. That is, an idealized model. In reality, quantum systems interact with the outside environment. The problem if that interactions with the environment can introduce noise and cause errors. To deal with this and construct, for example, real quantum computers and communications systems, we are going to need some kind of error correction.

Before we get there, we are going to have to develop a mathematical formalism to describe quantum systems that interact with the environment. We refer to systems of this type as *open systems*. Open quantum systems are important for the following reason: in an open quantum system, a pure state can evolve into a mixed state. The downside of this is that we need pure states to do quantum computation, and hence this type of evolution into mixed states is undesirable. In this chapter we will describe some of the formalism used to describe open quantum systems and then discuss error correction techniques.

SINGLE-QUBIT ERRORS

At the most basic level, it can be said that the power of quantum information processing comes from the fact that quantum states can exists in superpositions. To review, this means that while a qubit could be in the state $|0\rangle$ or the state $|1\rangle$, it can also be found in the state

$$|\psi\rangle = \alpha|0\rangle + \beta|1\rangle \tag{12.1}$$

where α, β are complex constants that satisfy $|\alpha|^2 + |\beta|^2 = 1$. We have seen throughout the book that the ability to work with superposition states (12.1) is what gives quantum computers their power—quantum algorithms often begin by using Hadamard gates to create superposition states to basically perform multiple evaluations simultaneously. Unfortunately, when a quantum system interacts with the environment (we often say the system is *coupled* to the environment), superposition can be lost. We call this process—whereby a pure state is turned into a mixed state via interactions with the environment—*decoherence* and refer to states like (12.1) where superposition is maintained as *coherent* states. The idea behind coherence is that the amplitudes in (12.1) can *interfere*. To see this, we apply a Hadamard gate to the qubit. Recall that

$$\begin{aligned} H|\psi\rangle &= \alpha H|0\rangle + \beta H|1\rangle \\ &= \alpha\left(\frac{|0\rangle + |1\rangle}{\sqrt{2}}\right) + \beta\left(\frac{|0\rangle - |1\rangle}{\sqrt{2}}\right) = \left(\frac{\alpha+\beta}{\sqrt{2}}\right)|0\rangle + \left(\frac{\alpha-\beta}{\sqrt{2}}\right)|1\rangle \end{aligned} \tag{12.2}$$

Now, if the qubit is in the pure state

$$|\psi\rangle = \frac{|0\rangle + |1\rangle}{\sqrt{2}}$$

then

$$H|\psi\rangle = |0\rangle \tag{12.3}$$

The amplitudes have interfered to take $|\psi\rangle \rightarrow |0\rangle$. Similarly you can show that if

$$|\psi\rangle = \frac{|0\rangle - |1\rangle}{\sqrt{2}}$$

then application of a Hadamard gate will cause the amplitudes to interfere and take $|\psi\rangle \rightarrow |1\rangle$. What happens in the case of mixed states? If a state is mixed (an *incoherent* mixture; see Chapter 5), the amplitudes cannot interfere. This is easy to see by considering the completely mixed state $\rho = \frac{1}{2}(|0\rangle\langle 0| + |1\rangle\langle 1|) = \frac{1}{2}I$. Since $H^2 = I$, we have $H\rho H = \frac{1}{2}HIH = \frac{1}{2}H^2 = \frac{1}{2}I$. The amplitudes did not interfere in this case.

It turns out that we can describe the interaction of a single qubit with the environment and hence single-qubit errors with the familiar operators I, X, Y, Z. The identity operator, of course, represents no error at all. We can summarize the effect of noise on a single qubit by saying that quantum noise acts on qubits via the application of one of the operators I, X, Y, Z. Which operator is applied depends on the state of the environment.

In the case of a classical bit, which can be 0 or 1, engineers are concerned with *bit flip* errors. The physical details aren't important for us, but we can imagine some stray electromagnetic field causing a bit to change from 1 to 0, for example. A similar type of error can affect qubits, where we have $|0\rangle \to |1\rangle$ and $|1\rangle \to |0\rangle$. This type of error is described by the X operator, which can be written as

$$X = |0\rangle\langle 1| + |1\rangle\langle 0| \tag{12.4}$$

We have already seen several times throughout the book that the application of (12.4) to (12.1) takes the qubit to the state

$$|\psi\rangle \to \alpha|1\rangle + \beta|0\rangle \tag{12.5}$$

In quantum systems, bit flip errors are not the only problem that we can encounter. We can also have *phase flip* errors. Let $x \in \{0, 1\}$. Then a phase flip error is one that transforms a state $|x\rangle$ as

$$|x\rangle \to (-1)^x|x\rangle \tag{12.6}$$

Looking at (12.6), we readily see that a phase flip error is described by the Z Pauli operator. We recall that Z acts on a qubit in the following way:

$$Z|\psi\rangle = \alpha|0\rangle - \beta|1\rangle \tag{12.7}$$

The Y operators is related to a phase flip followed by a bit flip. In particular, we have $-iY = -|0\rangle\langle 1| + |1\rangle\langle 0|$, which acts on a qubit (12.1) as

$$-iY|\psi\rangle = \alpha|1\rangle - \beta|0\rangle \tag{12.8}$$

As mentioned above, (12.8) can be decomposed into a sequence of two errors—a phase flip (as described by Z) followed by a bit flip (as described by X). It's easy to verify that

$$-iY = XZ \tag{12.9}$$

Later we will see how quantum error-correcting codes work on bit flip and phase flip errors. Now let's see how we can describe the evolution of a quantum system.

QUANTUM OPERATIONS AND KRAUSS OPERATORS

We now turn to a method of describing the dynamical evolution of a quantum system that is quite general and lends itself to a description of the interaction between a system of interest (the *principal* system) and the external environment. Suppose that the principal system has a density operator ρ, and let $\Phi(\rho)$ be a mapping that describes the evolution of the system. This mapping acts on density operators ρ, transforming them to new density operators ρ':

$$\rho' = \Phi(\rho) \tag{12.10}$$

We call the mapping $\Phi(\rho)$ a *quantum operation*. We are already familiar with two quantum operations—unitary (time) evolution of a closed system where

$$\Phi(\rho) = U\rho U^{\dagger} \tag{12.11}$$

and measurement

$$\Phi(\rho) = M_m \rho M_m^{\dagger} \tag{12.12}$$

Now let's imagine a more general case where we have a set of operators A_k that aren't necessarily unitary. In this case we write a general quantum operation $\Phi(\rho)$as

$$\Phi(\rho) = \sum_{k=1}^{n} A_k \rho A_k^{\dagger} \tag{12.13}$$

We say that (12.13) is the *operator-sum representation* of $\Phi(\rho)$. The A_k, which are known as *operation elements*, can satisfy a completeness relation

$$\sum_{k=1}^{n} A_k A_k^{\dagger} = I \tag{12.14}$$

When $\sum_{k=1}^{n} A_k A_k^{\dagger} = I$, we say that the operation elements are *trace-preserving*. Later we will see that we can also have *non–trace-preserving* operation elements, in which case $\sum_{k=1}^{n} A_k A_k^{\dagger} < I$. When ρ is a density matrix and the operation elements are trace-preserving, then $\Phi(\rho)$ is also a density matrix, meaning (12.10) is satisfied.

The first item of business in deriving the operator sum representation of some quantum operation is to calculate the A_k. In the following we will denote the density operator of the principal system by ρ and the density operator of the environment by σ. Quantum computation (and closed system dynamics) evolves by unitary operations—let's say some unitary operation U. What we are imagining now is the interaction of the principal system with the environment, during which the operation U is applied. After the interaction and application of U, we are interested in knowing the state of the *principal system alone*. We can do this by tracing out over the environment. Hence

$$\Phi(\rho) = Tr_E(U(\rho \otimes \sigma)U^\dagger) \tag{12.15}$$

We can calculate (12.15) in the following way: First we denote the basis states of the environment by $|e_k\rangle$ and assume that the environment is in the state $\sigma = |e_0\rangle\langle e_0|$. This assumption is justified because the environment can be "widened" until we find it in a pure state. Continuing, we find that (12.15) becomes

$$\begin{aligned}\Phi(\rho) &= Tr_E(U(\rho \otimes \sigma)U^\dagger) = \sum_k \langle e_k|(U\rho \otimes \sigma U^\dagger)|e_k\rangle \\ &= \sum_k \langle e_k|(U\rho \otimes |e_0\rangle\langle e_0|U^\dagger)|e_k\rangle = \sum_k \langle e_k|U|e_0\rangle\rho\langle e_0|U^\dagger|e_k\rangle\end{aligned} \tag{12.16}$$

Comparison with (12.13) tells us that the operation elements A_k are given by

$$A_k = \langle e_k|U|e_0\rangle \tag{12.17}$$

The operation elements A_k are also known as *Kraus operators*. Note that the Kraus operators act on the principal system.

Example 12.1

Suppose that the principal system and the environment are given by single qubits. The state of the principal system is $|\psi\rangle = \alpha|0\rangle + \beta|1\rangle$ and the environment is in the state $|0\rangle$. A unitary operation that describes measurement of principal system and discards the result is given by

$$U = P_0 \otimes I + P_1 \otimes X$$

Here $P_0 = |0\rangle\langle 0|$ and $P_1 = |1\rangle\langle 1|$ are the usual projection operators. Find the Kraus operators and write down the operator-sum representation as a matrix.

Solution

Since the environment is a single qubit, the basis for the environment is the computational basis. For clarity, we will denote the basis states for the environment by $\{|0_E\rangle, |1_E\rangle\}$. The Kraus operators are then found by calculating $A_0 = \langle 0_E|U|0_E\rangle$ and $A_1 = \langle 1_E|U|0_E\rangle$. In the first case we have

$$\begin{aligned} A_0 &= \langle 0_E|U|0_E\rangle = \langle 0_E|P_0 \otimes I + P_1 \otimes X|0_E\rangle \\ &= \langle 0_E|P_0 \otimes I|0_E\rangle + \langle 0_E|P_1 \otimes X|0_E\rangle \\ &= P_0\langle 0_E|0_E\rangle + P_1\langle 0_E|1_E\rangle = P_0 \end{aligned}$$

Notice that when doing calculations with terms like $A \otimes B$, the operator A acts on states of the principal system while B acts on states of the environment. Hence we just let A pass through and only consider the action of B when deriving the Kraus operator.

Continuing, we have

$$\begin{aligned} A_1 &= \langle 1_E|U|0_E\rangle = \langle 1_E|P_0 \otimes I + P_1 \otimes X|0_E\rangle \\ &= \langle 1_E|P_0 \otimes I|0_E\rangle + \langle 1_E|P_1 \otimes X|0_E\rangle \\ &= P_0\langle 1_E|0_E\rangle + P_1\langle 1_E|1_E\rangle = P_1 \end{aligned}$$

Since $P_j = P_j^\dagger$, the operator-sum representation of this quantum operation is

$$\Phi(\rho) = P_0\rho P_0 + P_1\rho P_1 = |0\rangle\langle 0|\rho|0\rangle\langle 0| + |1\rangle\langle 1|\rho|1\rangle\langle 1|$$

If $|\psi\rangle = \alpha|0\rangle + \beta|1\rangle$, the density operator for the system is

$$\begin{aligned} \rho &= |\psi\rangle\langle\psi| = (\alpha|0\rangle + \beta|1\rangle)(\alpha^*\langle 0| + \beta^*\langle 1|) \\ &= |\alpha|^2|0\rangle\langle 0| + \alpha\beta^*|0\rangle\langle 1| + \alpha^*\beta|1\rangle\langle 0| + |\beta|^2|1\rangle\langle 1| \end{aligned}$$

Hence

$$\begin{aligned} \rho|0\rangle &= (|\alpha|^2|0\rangle\langle 0| + \alpha\beta^*|0\rangle\langle 1| + \alpha^*\beta|1\rangle\langle 0| + |\beta|^2|1\rangle\langle 1|)|0\rangle \\ &= |\alpha|^2|0\rangle + \alpha^*\beta|1\rangle \end{aligned}$$

and

$$\begin{aligned} \rho|1\rangle &= \left(|\alpha|^2|0\rangle\langle 0| + \alpha\beta^*|0\rangle\langle 1| + \alpha^*\beta|1\rangle\langle 0| + |\beta|^2|1\rangle\langle 1|\right)|1\rangle \\ &= \alpha\beta^*|0\rangle + |\beta|^2|1\rangle \end{aligned}$$

So we have

$$\begin{aligned} P_0\rho P_0 &= |0\rangle\langle 0|\rho|0\rangle\langle 0| = |0\rangle\langle 0|\left(|\alpha|^2|0\rangle + \alpha^*\beta|1\rangle\right)\langle 0| \\ &= |\alpha|^2|0\rangle\langle 0| \\ |1\rangle\langle 1|\rho|1\rangle\langle 1| &= |1\rangle\langle 1|\left(\alpha\beta^*|0\rangle + |\beta|^2|1\rangle\right)\langle 1| \\ &= |\beta|^2|1\rangle\langle 1| \end{aligned}$$

The operator-sum representation is then

$$\Phi(\rho) = |\alpha|^2|0\rangle\langle 0| + |\beta|^2|1\rangle\langle 1|$$

Since

$$|0\rangle\langle 0| = \begin{pmatrix}1\\0\end{pmatrix}\begin{pmatrix}1 & 0\end{pmatrix} = \begin{pmatrix}1 & 0\\0 & 0\end{pmatrix}, \quad |1\rangle\langle 1| = \begin{pmatrix}0\\1\end{pmatrix}\begin{pmatrix}0 & 1\end{pmatrix} = \begin{pmatrix}0 & 0\\0 & 1\end{pmatrix}$$

the matrix representation is

$$\Phi(\rho) = \begin{pmatrix}|\alpha|^2 & 0\\0 & |\beta|^2\end{pmatrix}$$

Hence this quantum operation describes the evolution of the system to

$$\rho = \begin{pmatrix}|\alpha|^2 & \alpha\beta^*\\ \alpha^*\beta & |\beta|^2\end{pmatrix} \to \rho' = \begin{pmatrix}|\alpha|^2 & 0\\0 & |\beta|^2\end{pmatrix}$$

We can also consider more complicated interactions with the environment that are undesirable. Suppose that the environment is itself in a superposition state $|\phi_E\rangle$. In that case the operation elements are given by

$$A_k = \langle e_k|U|\phi_E\rangle \tag{12.18}$$

We illustrate this with two examples.

Example 12.2

A qubit in the state $|\psi\rangle = \alpha|0\rangle + \beta|1\rangle$ interacts with an environment which is in the state

$$|\phi_E\rangle = \frac{|0_E\rangle + |1_E\rangle}{\sqrt{2}}$$

The unitary operator that describes the coupling of the qubit to the environment is given by $U = e^{-i\theta(Z_P \otimes Z_E)/2}$, where Z_P is the Pauli Z operator acting on the principal system and Z_E is the Pauli Z operator acting on the environment. Find the density operator of the principal system after the interaction.

Solution

First we know that the density operator of the principal system starts off as

$$\rho = |\alpha|^2|0\rangle\langle 0| + \alpha\beta^*|0\rangle\langle 1| + \alpha^*\beta|1\rangle\langle 0| + |\beta|^2|1\rangle\langle 1|$$

To do this calculation, let's review how to find $e^{-iA}|a\rangle$ where $|a\rangle$ is an eigenvector of A.

If the eigenvectors of A have eigenvalues given by $A|a\rangle = a|a\rangle$, then

$$e^{-iA}|a\rangle = e^{-ia}|a\rangle$$

In the case of the Pauli Z operator, $Z|0\rangle = +|0\rangle$ and $Z|1\rangle = -|1\rangle$. So

$$e^{-i\theta Z}|0\rangle = e^{-i\theta}|0\rangle, \quad e^{-i\theta Z}|1\rangle = e^{i\theta}|1\rangle$$

The Kraus operators in this case are given by

$$A_0 = \langle 0_E|U|\phi_E\rangle, \quad A_1 = \langle 1_E|U|\phi_E\rangle$$

Now

$$U|0_E\rangle = e^{-i\theta(Z_P\otimes Z_E)/2}|0_E\rangle = e^{-i\theta Z_P/2}|0_E\rangle$$

$$U|1_E\rangle = e^{-i\theta(Z_P\otimes Z_E)/2}|1_E\rangle = e^{i\theta Z_P/2}|1_E\rangle$$

Therefore

$$A_0 = \langle 0_E|U|\phi_E\rangle = \langle 0_E|U\left(\frac{|0_E\rangle + |1_E\rangle}{\sqrt{2}}\right) = \langle 0_E|\left(\frac{e^{-i\theta(Z_P\otimes Z_E)/2}|0_E\rangle + e^{-i\theta(Z_P\otimes Z_E)/2}|1E\rangle}{\sqrt{2}}\right)$$

$$= \langle 0_E|\left(\frac{e^{-i\theta Z_P/2}|0_E\rangle + e^{i\theta Z_P/2}|1_E\rangle}{\sqrt{2}}\right) = \frac{1}{\sqrt{2}}e^{-i\theta Z_P/2}$$

and

$$A_1 = \langle 1_E|U|\phi_E\rangle = \langle 1_E|U\left(\frac{|0_E\rangle + |1_E\rangle}{\sqrt{2}}\right) = \langle 1_E|\left(\frac{e^{-i\theta(Z_P\otimes Z_E)/2}|0_E\rangle + e^{-i\theta(Z_P\otimes Z_E)/2}|1_E\rangle}{\sqrt{2}}\right)$$

$$= \langle 1_E|\left(\frac{e^{-i\theta Z_P/2}|0_E\rangle + e^{i\theta Z_P/2}|1_E\rangle}{\sqrt{2}}\right) = \frac{1}{\sqrt{2}}e^{i\theta Z_P/2}$$

The operator-sum representation is

$$\Phi(\rho) = A_0\rho A_0^\dagger + A_1\rho A_1^\dagger$$

The first term is given by

$$A_0\rho A_0^\dagger = \frac{1}{\sqrt{2}}e^{-i\theta Z_P/2}|\alpha|^2|0\rangle\langle 0| + \alpha\beta^*|0\rangle\langle 1| + \alpha^*\beta|1\rangle\langle 0| + |\beta|^2|1\rangle\langle 1|\frac{1}{\sqrt{2}}e^{i\theta Z_P/2}$$

$$= \frac{1}{2}\left(e^{-i\theta Z_P/2}|\alpha|^2|0\rangle\langle 0|e^{i\theta Z_P/2} + e^{-i\theta Z_P/2}\alpha\beta^*|0\rangle\langle 1|e^{i\theta Z_P/2}\right.$$
$$\left. + e^{-i\theta Z_P/2}\alpha^*\beta|1\rangle\langle 0|e^{i\theta Z_P/2} + e^{-i\theta Z_P/2}|\beta|^2|1\rangle\langle 1|e^{i\theta Z_P/2}\right)$$

$$= \frac{1}{2}\left(e^{-i\theta/2}|\alpha|^2|0\rangle\langle 0|e^{i\theta/2} + e^{-i\theta/2}\alpha\beta^*|0\rangle\langle 1|e^{-i\theta/2}\right.$$
$$\left. + e^{i\theta/2}\alpha^*\beta|1\rangle\langle 0|e^{i\theta/2} + e^{i\theta/2}|\beta|^2|1\rangle\langle 1|e^{-i\theta/2}\right)$$

$$= \frac{1}{2}\left(|\alpha|^2|0\rangle\langle 0| + e^{-i\theta}\alpha\beta^*|0\rangle\langle 1| + e^{i\theta}\alpha^*\beta|1\rangle\langle 0| + |\beta|^2|1\rangle\langle 1|\right)$$

And the other term is

$$A_1 \rho A_1^\dagger = \frac{1}{\sqrt{2}} e^{i\theta Z_P/2} |\alpha|^2 |0\rangle\langle 0| + \alpha\beta^* |0\rangle\langle 1| + \alpha^*\beta |1\rangle\langle 0| + |\beta|^2 |1\rangle\langle 1| \frac{1}{\sqrt{2}} e^{-i\theta Z_P/2}$$

$$= \frac{1}{2}\left(e^{i\theta Z_P/2} |\alpha|^2 |0\rangle\langle 0| e^{-i\theta Z_P/2} + e^{i\theta Z_P/2} \alpha\beta^* |0\rangle\langle 1| e^{-i\theta Z_P/2} + e^{i\theta Z_P/2} \alpha^*\beta |1\rangle\right.$$
$$\left.\langle 0| e^{-i\theta Z_P/2} + e^{i\theta Z_P/2} |\beta|^2 |1\rangle\langle 1| e^{-i\theta Z_P/2}\right)$$

$$= \frac{1}{2}\left(e^{i\theta/2} |\alpha|^2 |0\rangle\langle 0| e^{-i\theta/2} + e^{i\theta/2} \alpha\beta^* |0\rangle\langle 1| e^{i\theta/2} + e^{-i\theta/2} \alpha^*\beta |1\rangle\right.$$
$$\left.\langle 0| e^{-i\theta/2} + e^{-i\theta/2} |\beta|^2 |1\rangle\langle 1| e^{i\theta/2}\right)$$

$$= \frac{1}{2}\left(|\alpha|^2 |0\rangle\langle 0| + e^{i\theta} \alpha\beta^* |0\rangle\langle 1| + e^{-i\theta} \alpha^*\beta |1\rangle\langle 0| + |\beta|^2 |1\rangle\langle 1|\right)$$

Adding these two terms together, we see that

$$\rho' = \Phi(\rho) = \left(|\alpha|^2 |0\rangle\langle 0| + \cos\ \theta\alpha\beta^* |0\rangle\langle 1| + \cos\ \theta\alpha^*\beta |1\rangle\langle 0| + |\beta|^2 |1\rangle\langle 1|\right)$$

Example 12.3

Suppose $U = C_X$, the controlled NOT operation. Find the Krauss operators and an expression for $\Phi(\rho)$ if the environment is in the state $|\phi_E\rangle = \sqrt{1-p}|0\rangle + \sqrt{p}|1\rangle$.

Solution

Recall that the controlled NOT gate acts on two qubits. If the control qubit is $|0\rangle$, then nothing happens to the target qubit. On the other hand, if the control qubit is $|1\rangle$, then the target qubit is flipped. We can write down an expression that will implement this operation in the following way: If we write the composite state of the principal system and environment in the form $|\phi_E\rangle|\psi_P\rangle$, an operator $A \otimes B$ is one where A acts on the environment and B acts on the principal system.

In the first case, to do nothing to the target qubit, we apply the identity operator. To pick out the case when the first qubit is $|0\rangle$, we use the projection operator onto this state as applied to the first qubit. That is,

$$|0\rangle\langle 0| \otimes I = P_0 \otimes I$$

will leave the second qubit alone if the first qubit is $|0\rangle$. In the second case, we can pick out when the first qubit is $|1\rangle$ by again using a projection operator. This time to flip the bit, we apply the X Pauli operator to the second qubit. The following will do the job

$$|1\rangle\langle 1| \otimes X = P_1 \otimes X$$

The controlled NOT gate is implemented with the sum of these two operators:

$$C_X = P_0 \otimes I + P_1 \otimes X$$

Remember, we just pass through operators that act on the principal system, The action of C_X on $|\phi_E\rangle = \sqrt{1-p}|0\rangle + \sqrt{p}|1\rangle$ then is

$$\begin{aligned} C_X|\phi_E\rangle &= (P_0 \otimes I + P_1 \otimes X)(\sqrt{1-p}|0\rangle + \sqrt{p}|1\rangle) \\ &= (\sqrt{1-p}|0\rangle\langle 0|0\rangle I + \sqrt{p}|1\rangle\langle 1|1\rangle X) = \sqrt{1-p}|0\rangle I + \sqrt{p}|1\rangle X \end{aligned}$$

Hence the Kraus operators are

$$A_0 = \langle 0_E|C_X(\sqrt{1-p}|0_E\rangle + \sqrt{p}|1_E\rangle) = \langle 0_E|(\sqrt{1-p}|0\rangle I + \sqrt{p}|1\rangle X) = \sqrt{1-p}I$$

$$A_1 = \langle 1_E|C_X(\sqrt{1-p}|0_E\rangle + \sqrt{p}|1_E\rangle) = \langle 1_E|(\sqrt{1-p}|0\rangle I + \sqrt{p}|1\rangle X) = \sqrt{p}X$$

Then the operator sum representation is

$$\Phi(\rho) = A_0\rho A_0^\dagger + A_1\rho A_1^\dagger = (1-p)\rho + pX\rho X$$

If this operation represents noise in the system, it tells us the probability that nothing happens to the state of the principal system is $1-p$, while the probability of a bit flip error is p.

THE DEPOLARIZATION CHANNEL

Quantum noise is often described in terms of *channels*. The basic idea is that Alice transmits a qubit to Bob. This is done through some communications channel with noise or distortion.

Let's extend the result in the last example to describe the following disagreeable error: Imagine that there is a probability p that the principal system evolves into a completely mixed state. That is,

$$\rho \to \frac{1}{2}I = \begin{pmatrix} \frac{1}{2} & 0 \\ 0 & \frac{1}{2} \end{pmatrix}$$

with probability p. Suppose that the probability that there is no error, in other words, that the system stays the same, is $1-p$. This is known as the *depolarization channel*. What are the Kraus operators that can be used for the operator-sum representation of the depolarization channel? To leave the system alone with probability $1-p$, we just apply the identity operator:

$$A_0 = \sqrt{1-p}I$$

To obtain the other Kraus operators, first recall that the density operator for a single qubit can be written as

$$\rho = \frac{I + \vec{n} \cdot \vec{\sigma}}{2} \tag{12.19}$$

where $\vec{n} = n_x\hat{x} + n_y\hat{y} + n_z\hat{z}$ is a unit vector. Expanding out (12.19) obtains

$$\rho = \frac{1}{2}(I + n_x X + n_y Y + n_z Z)$$

Notice that

$$X\rho X = \frac{1}{2}(X^2 + n_x X^2 X + n_y XYX + n_z XZX) = \frac{1}{2}(I + n_x X - n_y Y - n_z Z)$$

We also find that

$$Y\rho Y = \frac{1}{2}\left(I - n_x X + n_y Y - n_z Z\right)$$

$$Z\rho Z = \frac{1}{2}\left(I - n_x X - n_y Y + n_z Z\right)$$

Adding all three together, results in

$$X\rho X + Y\rho Y + Z\rho Z = \frac{3}{2}I$$

We are now in a position to write down the remaining Kraus operators. If we take

$$A_1 = \sqrt{\frac{p}{3}}X, \quad A_2 = \sqrt{\frac{p}{3}}Y, \quad A_3 = \sqrt{\frac{p}{3}}Z$$

Hence

$$\rho' = \Phi(\rho) = A_0\rho A_0 + A_1\rho A_1 + A_2\rho A_2 + A_3\rho A_3$$

$$= (1-p)\rho + \frac{p}{3}X\rho X + \frac{p}{3}Y\rho Y + \frac{p}{3}Z\rho Z$$

$$= (1-p)\rho + \frac{p}{3}\left(\frac{3}{2}I\right) = (1-p)\rho + +p\frac{1}{2}I$$

This is the result we need: there is a probability p that the system will evolve into a completely mixed state and a probability $1-p$ that nothing happens to the system.

THE BIT FLIP AND PHASE FLIP CHANNELS

Now suppose that the environment is in the state $|\phi_E\rangle = |0_E\rangle$ and that it interacts with the system qubit via the Pauli X operator, where we take $|\phi_E\rangle|\psi_P\rangle$ and $U = \sqrt{p}I \otimes I + \sqrt{1-p}X \otimes X$. This operator acts on the state of the environment in the following way:

$$U|0_E\rangle = (\sqrt{p}I \otimes I)|0_E\rangle + (\sqrt{1-p}X \otimes X)|0_E\rangle = \sqrt{p}|0_E\rangle I + \sqrt{1-p}|1_E\rangle X$$

We can derive the Kraus operators

$$\begin{aligned} A_0 &= \langle 0_E|U|\phi_E\rangle = \langle 0_E|\left(\sqrt{p}|0_E\rangle I + \sqrt{1-p}|1_E\rangle X\right) \\ &= \sqrt{p}\langle 0_E|0_E\rangle I + \sqrt{1-p}\langle 0_E|1_E\rangle X = \sqrt{p}I \end{aligned} \tag{12.20}$$

$$\begin{aligned} A_1 &= \langle 1_E|U|\phi_E\rangle = \langle 1_E|\left(\sqrt{p}|0_E\rangle I + \sqrt{1-p}|1_E\rangle X\right) \\ &= \sqrt{p}\langle 1_E|0_E\rangle I + \sqrt{1-p}\langle 1_E|1_E\rangle X = \sqrt{1-p}X \end{aligned} \tag{12.21}$$

which describe the following interaction: there is a probability p that nothing happens to the qubit, while there is a probability $1-p$ that there is a bit flip error. The quantum operation is

$$\Phi(\rho) = p\rho + (1-p)X\rho X \tag{12.22}$$

This quantum operation is called the *bit flip channel*. When there is a probability p that nothing happens to the qubit, while there is a probability $1-p$ that there is a phase flip error, we have

$$A_0 = \sqrt{p}I, \quad A_1 = \sqrt{1-p}Z \tag{12.23}$$

and the quantum operation

$$\Phi(\rho) = p\rho + (1-p)Z\rho Z \tag{12.24}$$

This is called the *phase flip channel*.

AMPLITUDE DAMPING

Real physical systems lose energy. When describing a quantum system undergoing energy dissipation because of some type of interaction with the environment, we apply a quantum operation known as *amplitude damping*.

Amplitude damping describes a decay process. The example we will use is an atom decaying from an excited energy state $|1_A\rangle$ to the ground state $|0_A\rangle$. When an atom decays, it emits a photon, so the environment goes from $|0_E\rangle \rightarrow |1_E\rangle$. Suppose that there is a probability p for the process $|1_A\rangle \rightarrow |0_A\rangle$ to occur and a probability $1-p$ that the atom just stays in the state $|1_A\rangle$. This quantum process can be described by a unitary operator U that acts as follows:

$$U|0_A\rangle|0_E\rangle = |0_A\rangle|0_E\rangle \tag{12.25}$$

$$U|1_A\rangle|0_E\rangle = \sqrt{p}|0_A\rangle|1_E\rangle + \sqrt{1-p}|1_A\rangle|0_E\rangle \tag{12.26}$$

Equation (12.25) describes the situation where the atom and the environment are both in the ground state—nothing happens. On the other hand, (12.26) describes the decay of the atom and excitation of the environment with probability p together with the atom remaining in the excited state with probability $1-p$. We can use this information to derive the Kraus operators for this damping process but first let's figure out what we can use for U.

Let's introduce two new operators called the *ladder operators* $\sigma_{\pm}$. The operator σ_+ takes the ground state to the excited state and gives zero when applied to the excited state:

$$\sigma_+|0\rangle = |1\rangle, \quad \sigma_+|1\rangle = 0 \tag{12.27}$$

Sometimes this operator is called the *raising operator*. The operator σ_- takes the excited state to the ground state and eliminates or annihilates the ground state:

$$\sigma_-|0\rangle = 0, \quad \sigma_-|1\rangle = |0\rangle \tag{12.28}$$

Following previous examples, we can build U by looking at (12.25) and (12.26) and using projection operators that project onto the appropriate state of the atom. In the first case, (12.25), we want to project onto the $|0_A\rangle$ state while doing nothing to the environment. We can do this with

$$|0_A\rangle\langle 0_A| \otimes I \tag{12.29}$$

Looking at (12.26), we have two situations. In the first case, we start with the atom in the $|1_A\rangle$ state and end in the $|0_A\rangle$ state. This can be accomplished using (12.28). Note that the lowering operator can be written as $|1_A\rangle\langle 0_A|$. On the other hand, the environment goes from the ground state to the excited state, indicating the need to use (12.27). This process occurs with probability p, so this piece of the operator is

$$\sqrt{p}\,(|1_A\rangle\langle 0_A|) \otimes (|0_E\rangle\langle 1_E|) = \sqrt{p}\,\sigma_- \otimes \sigma_+ \tag{12.30}$$

Finally, with probability $1-p$, the atom stays in the excited state and the environment stays in its initial state. This is described by

$$\sqrt{1-p}|1_A\rangle\langle 1_A| \otimes I \tag{12.31}$$

Putting (12.29), (12.30), and (12.31) together gives us the desired unitary operator for this process:

$$U = |0_A\rangle\langle 0_A| \otimes I + \sqrt{p}\,\sigma_- \otimes \sigma_+ + \sqrt{1-p}|1_A\rangle\langle 1_A| \otimes I \tag{12.32}$$

It acts on the initial state of the environment as follows:

$$\begin{aligned}
U|0_E\rangle &= |0_A\rangle\langle 0_A| \otimes (I|0_E\rangle) + \sqrt{p}\sigma_- \otimes (\sigma_+|0_E\rangle) \\
&\quad + \sqrt{1-p}|1_A\rangle\langle 1_A| \otimes (I|0_E\rangle) \\
&= |0_A\rangle\langle 0_A| \otimes (|0_E\rangle) + \sqrt{p}\sigma_- \otimes (|1_E\rangle) + \sqrt{1-p}|1_A\rangle\langle 1_A| \otimes (|0_E\rangle)
\end{aligned} \tag{12.33}$$

So the first Kraus operator is

$$\begin{aligned}
A_0 = \langle 0_E|U|0_E\rangle &= |0_A\rangle\langle 0_A| \otimes (\langle 0_E|0_E\rangle) + \sqrt{p}\sigma_- \otimes (\langle 0_E|1_E\rangle) \\
&\quad + \sqrt{1-p}|1_A\rangle\langle 1_A| \otimes (\langle 0_E|0_E\rangle) \\
&= |0_A\rangle\langle 0_A| + \sqrt{1-p}|1_A\rangle\langle 1_A|
\end{aligned} \tag{12.34}$$

In matrix form, using the $\{|0_A\rangle, |1_A\rangle\}$ basis results in

$$A_0 = \begin{pmatrix} 1 & 0 \\ 0 & \sqrt{1-p} \end{pmatrix} \tag{12.35}$$

The other Kraus operator is

$$\begin{aligned}
A_1 = \langle 1_E|U|0_E\rangle &= |0_A\rangle\langle 0_A| \otimes (\langle 1_E|0_E\rangle) + \sqrt{p}\sigma_- \otimes (\langle 1_E|1_E\rangle) \\
&\quad + \sqrt{1-p}|1_A\rangle\langle 1_A| \otimes (\langle 1_E|0_E\rangle) \\
&= \sqrt{p}\sigma_- = \sqrt{p}|0_A\rangle\langle 1_A|
\end{aligned} \tag{12.36}$$

The matrix representation is

$$A_1 = \begin{pmatrix} 0 & \sqrt{p} \\ 0 & 0 \end{pmatrix} \tag{12.37}$$

In the next example we consider amplitude damping of a *harmonic oscillator*. For those not familiar with the harmonic oscillator of mass m, the total energy is

$$E = \frac{p^2}{2m} + \frac{1}{2}m\omega^2 x^2 \tag{12.38}$$

In quantum theory the harmonic oscillator is studied using the *annihilation* operator a and *creation* operator $a^\dagger$, which are defined as

$$a = \frac{1}{\sqrt{2}}\left(\sqrt{\frac{m\omega}{\hbar}}x + i\frac{1}{\sqrt{m\hbar\omega}}p\right), \quad a^\dagger = \frac{1}{\sqrt{2}}\left(\sqrt{\frac{m\omega}{\hbar}}x + i\frac{1}{\sqrt{m\hbar\omega}}p\right) \tag{12.39}$$

These operators obey the commutation relation

$$[a, a^{\dagger}] = 1 \tag{12.40}$$

Using (12.38), we can write the Hamiltonian as

$$H = a^{\dagger}a + \frac{1}{2} = N + \frac{1}{2} \tag{12.41}$$

where we have introduced the *number* operator $N = a^{\dagger}a$. The eigenstates of the number operator are

$$N|n\rangle = n|n\rangle \tag{12.42}$$

In addition

$$a^{\dagger}|n\rangle = \sqrt{n+1}|n+1\rangle, \quad a|n\rangle = \sqrt{n}|n-1\rangle \tag{12.43}$$

The states $|n\rangle = |0\rangle, |1\rangle, |2\rangle, \ldots$ give the different energy levels of the system. Note that $a|0\rangle = 0$. Therefore $|0\rangle$ is the lowest energy or ground state of the system.

The harmonic oscillator can be used to model modes of the electromagnetic field, and to construct a quantum computer. In the next problem we consider a quantum system a interacting with an environment r where both the system and the environment are harmonic oscillators (for example, the environment could be a heat reservoir). This is a more complicated example of amplitude damping.

Example 12.4

A quantum system given by a harmonic oscillator a interacts with the environment r, which is also a harmonic oscillator. If the interaction between the system and the environment is described by the Hamiltonian

$$H = a^{\dagger}r + r^{\dagger}a$$

find the Kraus operators.

Solution

First let's set up some notation. Let the number states of the system be denoted by $|n_a\rangle$ and the states of the environment by $|n_r\rangle$. We will calculate the Kraus operators $A_k = \langle k_r|U|0_r\rangle$. Note that since the states of the system range over all of the nonnegative integers, there will be an infinity of Kraus operators $A_0, A_1, A_2, \ldots$. So our goal is to derive an expression that gives the kth Kraus operator. Starting with the Hamiltonian $H = (a^{\dagger}r + r^{\dagger}a)$, the interaction will be described by the unitary operator $U = \exp(-iHt)$.

We will need several mathematical facts to derive what we need. First, note that for operators A and G,

$$e^{\lambda G} A e^{-\lambda G} = \sum_{n=0}^{\infty} \frac{\lambda^n}{n!} C_n \tag{12.44}$$

where the C_n are operators defined recursively with $C_0 = A,\ C_1 = [G, C_0],\ C_2 = [G, C_1]$, and so on. We take $G = a^\dagger r + r^\dagger a$ and $\lambda = -it$, and we have

$$\begin{aligned} C_0 &= r \\ C_1 &= [G, r] = [a^\dagger r + r^\dagger a, r] = a[r^\dagger, r] = -a \\ C_2 &= [G, C_1] = -[a^\dagger r + r^\dagger a, a] = r[a, a^\dagger] = r \end{aligned}$$

This allows us to write

$$UrU^\dagger = \sum_{n=0}^{\infty} \frac{(-it)^n}{n!} C_n = \sum_{\substack{n=0 \\ (n \text{ even})}}^{\infty} \frac{(-it)^n}{n!} r - \sum_{\substack{n=1 \\ (n \text{ odd})}}^{\infty} \frac{(-it)^n}{n!} a$$

which is nothing other than

$$UrU^\dagger = r \cos t + ia \sin t \tag{12.45}$$

A similar exercise shows that

$$UaU^\dagger = a \cos t + ir \sin t \tag{12.46}$$

Now, since $U^\dagger U = UU^\dagger = I$, we can write

$$Ur^k U^\dagger = UrIrI \ldots IrU^\dagger = UrU^\dagger UrU^\dagger \cdots UrU^\dagger UrU^\dagger = (UrU^\dagger)^k \tag{12.47}$$

Using (12.45) together with the binomial theorem, which tells us that

$$(x + y)^n = \sum_{l=0}^{n} \binom{n}{l} x^l y^{n-l}$$

we find that

$$(UrU^\dagger)^k = (r \cos t + ia \sin t)^k = \sum_{m=0}^{k} \cos^m t\, r^m (i \sin t)^{k-m} a^{k-m} \tag{12.48}$$

We are now in a position to derive the Kraus operators. The closure relation for the number states of a harmonic oscillator (the principal system a in this case) is

$$I = \sum_{m=0}^{\infty} |m_a\rangle\langle m_a| \tag{12.49}$$

So we can write

$$A_k = \left(\sum_{n_a} |n_a\rangle\langle n_a|\right) A_k = \left(\sum_{n_a} |n_a\rangle\langle n_a|\right) A_k \left(\sum_{n_a} |m_a\rangle\langle m_a|\right)$$

$$= \sum_{n_a,m_a} |n_a\rangle\langle m_a|\langle n_a|A_k|m_a\rangle$$

We use $A_k = \langle k_r|U|0_r\rangle$ to rewrite this expression as

$$A_k = \sum_{n_a,m_a} |n_a\rangle\langle m_a|\langle n_a|\langle k_r|U|0_r\rangle|m_a\rangle \tag{12.50}$$

Now consider that

$$\langle k_r| = \langle 0_r|\frac{r^k}{\sqrt{k!}}$$

Using this expression together with (12.47), (12.48), and (12.50), we arrive at

$$A_k = \sum_{n_a,m_a} |n_a\rangle\langle m_a|\langle n_a|\langle k_r|U|0_r\rangle|m_a\rangle$$

$$= \sum_{n_a,m_a} |n_a\rangle\langle m_a|\langle n_a|\langle 0_r|\frac{r^k}{\sqrt{k!}}U|0_r\rangle|m_a\rangle$$

$$= \sum_{n_a,m_a} |n_a\rangle\langle m_a|\langle n_a|\langle 0_r|UU^\dagger\frac{r^k}{\sqrt{k!}}U|0_r\rangle|m_a\rangle$$

Since $UrU^\dagger = r\cos t + ia\sin t$, it follows that $U^\dagger rU = r\cos t - ia\sin t$. Therefore

$$A_k = \sum_{n_a,m_a} |n_a\rangle\langle m_a|\langle n_a|\langle 0_r|UU^\dagger\frac{r^k}{\sqrt{k!}}U|0_r\rangle|m_a\rangle$$

$$= \sum_{n_a,m_a} |n_a\rangle\langle m_a|\langle n_a|\langle 0_r|U\frac{(r\cos t - ia\sin t)^k}{\sqrt{k!}}|0_r\rangle|m_a\rangle$$

Referring the binomial theorem, we write

$$A_k = \sum_{n_a,m_a} |n_a\rangle\langle m_a|\langle n_a|\langle 0_r|\frac{U}{\sqrt{k!}}\left(\sum_{l=0}^{k}(\cos t)^l r^l(-i\sin t)^{k-l}a^{k-l}\right)|0_r\rangle|m_a\rangle$$

Now we use the fact that $r|0_r\rangle = 0$. Hence the only term in the sum that contributes is when $l = 0$, and this becomes

$$A_k = \sum_{n_a,m_a} |n_a\rangle\langle m_a|\langle n_a|\langle 0_r|\frac{U}{\sqrt{k!}}(-i\sin t)^k a^k|0_r\rangle|m_a\rangle$$

Let's consider $a^k|m_a\rangle$. Some algebra gives

$$\begin{aligned}\frac{a^k}{\sqrt{k!}}|m_a\rangle &= \frac{\sqrt{m_a(m_a-1)\dots(m_a-(k-1))}}{\sqrt{k!}}|m_a-k\rangle \\ &= \frac{\sqrt{m_a(m_a-1)\dots(m_a-(k-1))}}{\sqrt{k!}}\left(\frac{\sqrt{(m_a-k)!}}{\sqrt{(m_a-k)!}}\right)|m_a-k\rangle \\ &= \sqrt{\frac{m_a!}{k!(m_a-k)!}}|m_a-k\rangle = \sqrt{\binom{m_a}{k}}|m_a-k\rangle\end{aligned}$$

Putting this together with our previous result, we have

$$A_k = \sum_{n_a,m_a}\sqrt{\binom{m_a}{k}}(-i\sin t)^k\langle n_a|\langle 0_r|U|0_r\rangle|m_a-k\rangle|n_a\rangle\langle m_a|$$

We can then repeat the process, this time applying $UaU^\dagger = a\cos t + ir\sin t$. First we have

$$\langle n_a| = \langle 0_a|\frac{a^{n_a}}{\sqrt{n_a!}}$$

Hence

$$\begin{aligned}\langle n_a|\langle 0_r|U|0_r\rangle|m_a-k\rangle &= \langle 0_a|\langle 0_r|\frac{a^{n_a}}{\sqrt{n_a!}}U|0_r\rangle|m_a-k\rangle \\ &= \langle 0_a|\langle 0_r|UU^\dagger\frac{a^{n_a}}{\sqrt{n_a!}}U|0_r\rangle|m_a-k\rangle \\ &= \langle 0_a|\langle 0_r|U\frac{(a\cos t - ir\sin t)^{n_a}}{\sqrt{n_a!}}|0_r\rangle|m_a-k\rangle \\ &= \langle 0_a|\langle 0_r|U\frac{1}{\sqrt{n_a!}}\sum_{l=0}^{n_a}(\cos t)^{n_a-l}a^{n_a-l}(i\sin t)^l r^l|0_r\rangle|m_a-k\rangle\end{aligned}$$

Once again, we have

$$r|0_r\rangle = 0$$

So the only term in the sum that contributes is $l = 0$, and thus

$$\begin{aligned}&\langle 0_a|\langle 0_r|U\frac{1}{\sqrt{n_a!}}\sum_{l=0}^{n_a}(\cos t)^{n_a-l}a^{n_a-l}(i\sin t)^l r^l|0_r\rangle|m_a-k\rangle \\ &= \langle 0_a|\langle 0_r|U\frac{1}{\sqrt{n_a!}}(\cos t)^{n_a}a^{n_a}|0_r\rangle|m_a-k\rangle\end{aligned}$$

Next we use

$$\frac{a^{n_a}}{\sqrt{n_a!}}|m_a-k\rangle = \sqrt{\binom{m_a-k}{n_a}}|m_a-k-n_a\rangle$$

together with $\langle 0_a|\langle 0_r|U = \langle 0_a|\langle 0_r|$ to write

$$A_k = \sum_{n_a,m_a} \sqrt{\binom{m_a}{k}}(-i \sin t)^k \langle n_a|\langle 0_r|U|0_r\rangle|m_a - k\rangle|n_a\rangle\langle m_a|$$

$$= \sum_{n_a,m_a} \sqrt{\binom{m_a}{k}}(-i \sin t)^k \langle 0_a|\langle 0_r|U\frac{1}{\sqrt{n_a!}}(\cos t)^{n_a} a^{n_a}|0_r\rangle|m_a - k\rangle|n_a\rangle\langle m_a|$$

$$= \sum_{n_a,m_a} \sqrt{\binom{m_a}{k}}\sqrt{\binom{m_a - k}{n_a}}(-i \sin t)^k (\cos t)^{n_a} \langle 0_a|\langle 0_r\|0_r\rangle|m_a - k - n_a\rangle|n_a\rangle\langle m_a|$$

Since $\langle 0_r|0_r\rangle = 1$, this is just

$$A_k = \sum_{n_a,m_a} \sqrt{\binom{m_a}{k}}\sqrt{\binom{m_a - k}{n_a}}(-i \sin t)^k (\cos t)^{n_a} \langle 0_a\|m_a - k - n_a\rangle|n_a\rangle\langle m_a|$$

Now we use the fact that

$$\langle 0_a\|m_a - k - n_a\rangle = \delta_{0,m-k-n}$$
$$\Rightarrow m - k - n = 0$$

Then we have $n = m - k$, and therefore

$$A_k = \sum_m \sqrt{\binom{m}{k}}\sqrt{\binom{m-k}{m-k}}(-i \sin t)^k (\cos t)^{m-k}|m - k\rangle\langle m|$$

$$= \sum_m \sqrt{\binom{m}{k}}(-i \sin t)^k (\cos t)^{m-k}|m - k\rangle\langle m|$$

Looking at the term $|m - k\rangle$, we see that the smallest term we can have is $m - k = 0$. The sum becomes

$$A_k = \sum_{m=k}^{\infty} \sqrt{\binom{m}{k}}(-i)^k (\sin t)^k (\cos t)^{m-k}|m - k\rangle\langle m|$$

If we let $\gamma = 1 - \cos^2 t$, then we obtain the final expression for the Kraus operators for the interaction between two harmonic oscillators

$$A_k = \sum_{m=k}^{\infty} \sqrt{\binom{m}{k}}(-i)^k \sqrt{(\sin t)^{2k}}\sqrt{(\cos t)^{2(m-k)}}|m - k\rangle\langle m|$$

$$= \sum_{m=k}^{\infty} \sqrt{\binom{m}{k}}(-i)^k \sqrt{(1 - \cos^2 t)^k}\sqrt{(\cos^2 t)^{(m-k)}}|m - k\rangle\langle m|$$

$$= \sum_{m=k}^{\infty} \sqrt{\binom{m}{k}}(-i)^k \sqrt{(1 - \gamma)^{m-k}}\sqrt{\gamma^k}|m - k\rangle\langle m|$$

where $|m - k\rangle$ and $\langle m|$ are states of the system (with annihilation operator a).

PHASE DAMPING

Phase damping is a quantum process that involves information loss, but unlike amplitude damping, it does not involve energy loss. Specifically phase damping involves the loss of information about relative phases in a quantum state. During phase damping the principal quantum system becomes entangled with the environment. Of course, this is undesirable if we are trying to use the quantum system to perform quantum computing.

To illustrate phase damping, we consider a qubit with density matrix $\rho = \begin{pmatrix} a & b \\ c & d \end{pmatrix}$ interacting with the environment, which can assume three states: $|0_E\rangle, |1_E\rangle$, and $|2_E\rangle$. We assume that the environment is initially in the state $|0_E\rangle$. A unitary operator U generates entanglement between the principal system and the environment in the following way:

$$U|0\rangle|0_E\rangle = \sqrt{1-p}|0\rangle|0_E\rangle + \sqrt{p}|0\rangle|1_E\rangle$$

$$U|1\rangle|0_E\rangle = \sqrt{1-p}|1\rangle|0_E\rangle + \sqrt{p}|1\rangle|2_E\rangle$$

Since there are three states of the environment, we need three Kraus operators. These are

$$A_0 = \sqrt{1-p}I \tag{12.51}$$

$$A_1 = \sqrt{p}|0\rangle\langle 0| \tag{12.52}$$

$$A_2 = \sqrt{p}|1\rangle\langle 1| \tag{12.53}$$

The quantum process is then

$$\rho' = \Phi(\rho) = A_0\rho A_0^\dagger + A_1\rho A_1^\dagger + A_2\rho A_2^\dagger = \begin{pmatrix} a & (1-p)b \\ (1-p)c & d \end{pmatrix} \tag{12.54}$$

where we used the matrix representations of (12.51) through (12.53) together with $\rho = \begin{pmatrix} a & b \\ c & d \end{pmatrix}$ as the original density matrix. If this operation is carried out n times, then

$$\rho \to \begin{pmatrix} a & (1-p)^n b \\ (1-p)^n c & d \end{pmatrix}$$

Now, if the probability $p = \Gamma\Delta t$ and $n = t/\Delta t$, we have

$$\lim_{\Delta t \to 0} (1-p)^n = \lim_{\Delta t \to 0} (1-\Gamma\Delta t)^{t/\Delta t} = e^{-\Gamma t}$$

So the density matrix becomes

$$\rho \to \begin{pmatrix} a & e^{-\Gamma t}b \\ e^{-\Gamma t}c & d \end{pmatrix} \underset{\rightarrow}{\text{exponentially decays}} \begin{pmatrix} a & 0 \\ 0 & d \end{pmatrix}$$

The terms b and c go to zero, leaving only the terms along the diagonal. All information about the relative phases in the original state of the principal system is lost. This is phase damping.

Example 12.5

Consider a system–environment interaction that leaves the system unchanged and applies the rotation gate

$$R_y(\theta) = \begin{pmatrix} \cos\frac{\theta}{2} & -\sin\frac{\theta}{2} \\ \sin\frac{\theta}{2} & \cos\frac{\theta}{2} \end{pmatrix}$$

to the environment if the system is in the state $|1\rangle$. If the system is in the state $|0\rangle$, the environment is unchanged. Let the initial state of the environment be $|\phi_E\rangle = |0_E\rangle$.

Solution

The interaction described in this example is a controlled-rotation gate, with the principal system playing the role of the control qubit and the environment playing the role of the target qubit. First note that

$$R_y(\theta)|0_E\rangle = \begin{pmatrix} \cos\frac{\theta}{2} & -\sin\frac{\theta}{2} \\ \sin\frac{\theta}{2} & \cos\frac{\theta}{2} \end{pmatrix}\begin{pmatrix} 1 \\ 0 \end{pmatrix} = \cos\frac{\theta}{2}|0_E\rangle + \sin\frac{\theta}{2}|1_E\rangle$$

The unitary operator that will do what we need, taking the system as the first qubit and the environment as the second qubit, is given by

$$U = |0\rangle\langle 0| \otimes I + |1\rangle\langle 1| \otimes R_y(\theta)$$

So we have

$$\begin{aligned} A_0 &= \langle 0_E|U|0_E\rangle = \langle 0_E|(|0\rangle\langle 0| \otimes I + |1\rangle\langle 1| \otimes R_y(\theta))|0_E\rangle \\ &= |0\rangle\langle 0|\langle 0_E|0_E\rangle + |1\rangle\langle 1|\langle 0_E|\left(\cos\frac{\theta}{2}|0_E\rangle + \sin\frac{\theta}{2}|1_E\rangle\right) \\ &= |0\rangle\langle 0| + \cos\frac{\theta}{2}|1\rangle\langle 1| \end{aligned}$$

Next we find that

$$\begin{aligned} A_1 &= \langle 1_E|U|0_E\rangle = \langle 1_E|(|0\rangle\langle 0|\otimes I + |1\rangle\langle 1|\otimes R_y(\theta))|0_E\rangle \\ &= |0\rangle\langle 0|\langle 1_E|0_E\rangle + |1\rangle\langle 1|\langle 1_E|\left(\cos\frac{\theta}{2}|0_E\rangle + \sin\frac{\theta}{2}|1_E\rangle\right) \\ &= \sin\frac{\theta}{2}|1\rangle\langle 1| = |0\rangle\langle 0| + \sqrt{1-\cos\frac{\theta}{2}}|1\rangle\langle 1| \end{aligned}$$

If we define the probability p by $p=\cos\theta/2$, the matrix representations of these operators are

$$A_0 = \begin{pmatrix} 1 & 0 \\ 0 & \sqrt{p} \end{pmatrix}, \quad A_1 = \begin{pmatrix} 1 & 0 \\ 0 & \sqrt{1-p} \end{pmatrix}$$

The quantum operation described here is an example of *phase damping*. Given an arbitrary density matrix,

$$\rho = \begin{pmatrix} a & b \\ c & d \end{pmatrix}$$

the off-diagonal terms c, d are damped by this operation:

$$\begin{aligned} \rho' &= A_0\rho A_0^\dagger + A_1\rho A_1^\dagger \\ &= \begin{pmatrix} 1 & 0 \\ 0 & \sqrt{p} \end{pmatrix}\begin{pmatrix} a & b \\ c & d \end{pmatrix}\begin{pmatrix} 1 & 0 \\ 0 & \sqrt{p} \end{pmatrix} + \begin{pmatrix} 1 & 0 \\ 0 & \sqrt{1-p} \end{pmatrix}\begin{pmatrix} a & b \\ c & d \end{pmatrix}\begin{pmatrix} 1 & 0 \\ 0 & \sqrt{1-p} \end{pmatrix} \\ &= \begin{pmatrix} a & \sqrt{p}b \\ \sqrt{p}c & pd \end{pmatrix} + \begin{pmatrix} a & \sqrt{1-p}b \\ \sqrt{1-p}c & (1-p)d \end{pmatrix} = \begin{pmatrix} a & (\sqrt{p}+\sqrt{1-p})b \\ (\sqrt{p}+\sqrt{1-p})c & d \end{pmatrix} \end{aligned}$$

Now we see why this is an example of phase damping. For the right value of $\cos\theta/2$, the off-diagonal terms b and c will get vanishingly small. Applying the operation many times will result in their vanishing completely from the resulting density matrix.

QUANTUM ERROR CORRECTION

So far we've seen that interaction with the environment can produce a lot of undesirable effects on a qubit. To deal with this unavoidable fact, we need to develop error correction techniques before we can construct a useful quantum computer. As we will see later, quantum mechanics does not allow you to duplicate or copy an unknown quantum state because of the *no-cloning theorem*. So we are going to have to come up with more clever error correction schemes rather than simply making multiple copies of our information—something we cannot do. Our first look into error correction will involve a look at decoherence.

Decoherence is the loss of coherence in a quantum system due to interactions with the external environment. Using qubits, we can model decoherence by the introduction of a relative phase. Specifically $|0\rangle\to|0\rangle$ and $|1\rangle\to e^{i\theta}|1\rangle$. Therefore an arbitrary qubit $|\psi\rangle = \alpha|0\rangle + \beta|1\rangle$ is transformed in the following way:

$$|\psi\rangle \to \alpha|0\rangle + e^{i\theta}\beta|1\rangle \tag{12.55}$$

For example, if

$$|\psi\rangle = \frac{|0\rangle + |1\rangle}{\sqrt{2}}$$

then the density matrix is transformed according to

$$\rho = \frac{1}{2}\begin{pmatrix} 1 & 1 \\ 1 & 1 \end{pmatrix} \to \frac{1}{2}\begin{pmatrix} 1 & e^{-i\theta} \\ e^{i\theta} & 1 \end{pmatrix}$$

This type of model used for coherence is called *collective dephasing*. Clearly, such transformations induced by the environment are not desirable. We can get around this problem, however. First, the important thing to remember is that an overall global phase (a phase that multiplies all terms in a superposition) has no effect on the predictions of measurement results. A relative phase, that is, a phase that only multiplies a single term in a superposition, does change predictions about possible measurement results. Hence the introduction of a global phase is irrelevant, but the introduction of a relative phase, what we have in (12.55), is problematic.

We can get around this problem with a little trick, a mapping to what is called a *decoherent free subspace*. This is done by introducing *logical* qubits, which are denoted by $|0_L\rangle, |1_L\rangle$. The qubits we have been using all along are *physical* qubits, so they are the states of actual physical systems such as the energy levels in an atom or the spin state of an electron. The two logical qubits are defined in the following way:

$$|0_L\rangle = \frac{|0\rangle|1\rangle - i|1\rangle|0\rangle}{\sqrt{2}}, \quad |1_L\rangle = \frac{|0\rangle|1\rangle + i|1\rangle|0\rangle}{\sqrt{2}} \tag{12.56}$$

Notice next what happens when we introduce decoherence using the collective dephasing model:

$$|0_L\rangle \to \frac{|0\rangle e^{i\theta}|1\rangle - ie^{i\theta}|1\rangle|0\rangle}{\sqrt{2}} = e^{i\theta}\frac{|0\rangle|1\rangle - i|1\rangle|0\rangle}{\sqrt{2}} = e^{i\theta}|0_L\rangle$$

$$|1_L\rangle \to \frac{|0\rangle e^{i\theta}|1\rangle + ie^{i\theta}|1\rangle|0\rangle}{\sqrt{2}} = e^{i\theta}\frac{|0\rangle|1\rangle + i|1\rangle|0\rangle}{\sqrt{2}} = e^{i\theta}|1_L\rangle$$

Each logical qubit has been altered by an overall global phase $e^{i\theta}$. Now the best part of this is that since each logical qubit is altered by the same global phase, an arbitrary logical qubit is unchanged by decoherence

$$|\psi_L\rangle = \alpha|0_L\rangle + \beta|1_L\rangle \to e^{i\theta}\alpha|0_L\rangle + e^{i\theta}\beta|1_L\rangle = e^{i\theta}|\psi_L\rangle$$

There is no change in the state whatsoever.

In this chapter we have seen that there is a large class of errors that can effect qubits—for example, bit flip and phase flip errors. While we have addressed decoherence with the introduction of the decoherent free subspace, there are a lot of problems that we have not dealt with. Error correction schemes known as *Calderbank-Shor-Steane*, or CSS codes, have been developed to deal with bit flip and phase flip errors. We will focus on one particular error correction scheme known as *Shor's nine-bit code*. This is based on a simple idea, to deal with errors that are inevitable in the real world and make the system redundant. Like we did when creating the decoherent free subspace, we will define logical qubits, but this time the logical qubits consist of three physical qubits. The logical qubits are

$$|0_L\rangle = |000\rangle, \quad |1_L\rangle = |111\rangle \tag{12.57}$$

Let's consider an arbitrary logical qubit $|\psi\rangle = \alpha|0_L\rangle + \beta|1_L\rangle = \alpha|000\rangle + \beta|111\rangle$. It is possible that nothing happens to the state—that is, there are no errors at all. This operation is, of course, described by the application of the identity operator to each physical qubit:

$$(I \otimes I \otimes I)|\psi\rangle = \alpha|000\rangle + \beta|111\rangle$$

We can carry this procedure further and figure out how to represent bit flips and phase flips.

Considering bit flips first, we suppose that we can flip the first qubit, the second qubit, or the third qubit. Bit flips, as you recall, are represented by the application of the Pauli operator X. Hence to flip the first qubit, we have

$$(X \otimes I \otimes I)|\psi\rangle = \alpha(X \otimes I \otimes I)|000\rangle + \beta(X \otimes I \otimes I)|111\rangle = \alpha|100\rangle + \beta|011\rangle$$

Similarly flipping the second and third qubits, we have

$$(I \otimes X \otimes I)|\psi\rangle = \alpha(I \otimes X \otimes I)|000\rangle + \beta(I \otimes X \otimes I)|111\rangle = \alpha|010\rangle + \beta|101\rangle$$
$$(I \otimes I \otimes X)|\psi\rangle = \alpha(I \otimes I \otimes X)|000\rangle + \beta(I \otimes I \otimes X)|111\rangle = \alpha|001\rangle + \beta|110\rangle$$

How do we correct for such errors? This can be done using a quantum circuit that basically works in two steps:

- Figure out which error occurred.
- Use a unitary transformation to fix it.

By using the correct unitary operations, it turns out that we can do the first step without in any way damaging the superposition of the logical qubit state. For example, let's suppose that the second qubit was flipped giving

$$\alpha|010\rangle + \beta|101\rangle$$

To detect the error, we first form the product state

$$(\alpha|010\rangle + \beta|101\rangle)|00\rangle = \alpha|010\rangle|00\rangle + \beta|101\rangle|00\rangle \tag{12.58}$$

We refer to the extra qubits as *ancillary* qubits. We pass the entire state through a series of five controlled NOT gates, and then measure the ancillary qubits at the end. The ancillary qubits will encode the position of the error for the first, second, or third qubit as binary 01, 10, or 11. The result of the two ancillary bits is known as the *syndrome measurement*.

Step One

Passing (12.58) through the first controlled NOT, we take the first qubit as the control bit and the second ancillary qubit as the targe bit. The state becomes

$$\alpha|010\rangle|00\rangle + \beta|101\rangle|00\rangle \rightarrow \alpha|010\rangle|00\rangle + \beta|101\rangle|01\rangle \tag{12.59}$$

Step Two

Now we pass (12.59) through a controlled NOT, using the second qubit as the control bit and the second ancillary bit as the target. The state becomes

$$\alpha|010\rangle|00\rangle + \beta|101\rangle|01\rangle \rightarrow \alpha|010\rangle|01\rangle + \beta|101\rangle|01\rangle \tag{12.60}$$

Step Three

This time the control qubit is the second qubit, while the target is the first ancillary:

$$\alpha|010\rangle|01\rangle + \beta|101\rangle|01\rangle \rightarrow \alpha|010\rangle|11\rangle + \beta|101\rangle|01\rangle \tag{12.61}$$

Step Four

Now we transform (12.61) by passing the state through yet another controlled-NOT gate. This time the control bit is the third qubit, and the target is the first ancillary, giving

$$\alpha|010\rangle|11\rangle + \beta|101\rangle|01\rangle \rightarrow \alpha|010\rangle|11\rangle + \beta|101\rangle|11\rangle \tag{12.62}$$

Step Five

Finally, we apply one more controlled NOT, using the first ancillary bit as the control bit and the second ancillary bit as the target, giving

$$\alpha|010\rangle|11\rangle + \beta|101\rangle|11\rangle \rightarrow \alpha|010\rangle|10\rangle + \beta|101\rangle|10\rangle \tag{12.63}$$

The ancillary qubits are measured, giving 10. This tells us that the error is in the second qubit. The superposition of the state has been maintained throughout the procedure, since the qubits of our state were only used as control bits in the controlled NOT gate.

We can fix the error with an X gate, by applying it to the qubit which has been flipped, giving

$$(I \otimes X \otimes I)(\alpha|010\rangle + \beta|101\rangle) = \alpha|000\rangle + \beta|111\rangle$$

The original state has been recovered.

We can also detect single bit flips using the following projection operators:

$$\begin{aligned} P_0 &= |000\rangle\langle 000| + |111\rangle\langle 111| \\ P_1 &= |100\rangle\langle 100| + |011\rangle\langle 011| \\ P_2 &= |010\rangle\langle 010| + |101\rangle\langle 101| \\ P_3 &= |001\rangle\langle 001| + |110\rangle\langle 110| \end{aligned} \tag{12.64}$$

Measurement of the four projection operators reveals where the error is and does not alter the state. Suppose that a bit flip has occurred on the second qubit so that $|\psi\rangle = \alpha|010\rangle + \beta|101\rangle$. Its easy to see that

$$\begin{aligned} \langle\psi|P_0|\psi\rangle &= 0 \\ \langle\psi|P_1|\psi\rangle &= 0 \\ \langle\psi|P_2|\psi\rangle &= 1 \\ \langle\psi|P_3|\psi\rangle &= 0 \end{aligned} \tag{12.65}$$

To correct phase flip errors, we can again use a three-qubit code to encode logical states. This is done using the $|\pm\rangle$ basis instead of the computational basis. We generate a logical qubit that takes $\alpha|0\rangle + \beta|111\rangle$, and then apply Hadamard gates to each qubit. The result is the state $\alpha|+++\rangle + \beta|---\rangle$. If the state has been altered by a phase flip, say, on the second qubit, it will now be $\alpha|+-+\rangle + \beta|-+-\rangle$. We can invert the process by applying Hadamard gates again, which converts this to $\alpha|010\rangle + \beta|101\rangle$. Now we can use the procedure developed in the last section to recover the original qubit, $\alpha|000\rangle + \beta|111\rangle$.

The price we pay is that we can't correct bit flip errors in this case. To be able to correct both types of errors, we need a modified approach, which is Shor's nine-bit code. We begin by defining two more states, that we will call the *up* and *down* states:

$$|\uparrow\rangle = \frac{|000\rangle + |111\rangle}{\sqrt{2}}, \quad |\downarrow\rangle = \frac{|000\rangle - |111\rangle}{\sqrt{2}} \tag{12.66}$$

What the up and down states do is combine the bit flip error correction scheme we described in detail with the phase flip correction scheme that uses the $|\pm\rangle$ basis. A physical qubit $\alpha|0\rangle + \beta|1\rangle$ is encoded as

$$\alpha|\uparrow\uparrow\uparrow\rangle + \beta|\downarrow\downarrow\downarrow\rangle \tag{12.67}$$

where we have defined the logical qubits

$$|0_L\rangle = |\uparrow\uparrow\uparrow\rangle, \quad |1_L\rangle = |\downarrow\downarrow\downarrow\rangle \tag{12.68}$$

All together, there are 27 different errors that can occur with a bit flip or phase flip on each of the nine physical qubits used to create the state (12.67). Notice the effect of any single phase flip on the basis states (12.66) is the same. For example, we can phase flip the second qubit, giving

$$I \otimes Z \otimes I|\uparrow\rangle = \frac{|000\rangle - |111\rangle}{\sqrt{2}} = |\downarrow\rangle, \quad I \otimes Z \otimes I|\downarrow\rangle = \frac{|000\rangle + |111\rangle}{\sqrt{2}} = |\uparrow\rangle$$

The correction procedure is then just a generalization of the error correction methods we've described so far. We can detect errors on single qubits by projecting onto the appropriate subspace and then applying the corresponding unitary transformation to get back to the original state.

EXERCISES

12.1. *Show that the matrix representation of $C_X = P_0 \otimes I + P_1 \otimes X$ is given by*

$$C_X = \begin{pmatrix} 1 & 0 & 0 & 0 \\ 0 & 1 & 0 & 0 \\ 0 & 0 & 0 & 1 \\ 0 & 0 & 1 & 0 \end{pmatrix}$$

12.2. *The environment is in the state*

$$|\phi_E\rangle = \sqrt{\frac{2}{3}}|0\rangle + \frac{1}{\sqrt{3}}|1\rangle$$

What is the probability that there is a phase flip error on a principal system with density matrix ρ?

12.3. *Let $\vec{n} = n_x\hat{x} + n_y\hat{y} + n_z\hat{z}$. Noting that we can write any density operator on qubits as*

$$\rho = \frac{I + \vec{n} \cdot \vec{\sigma}}{2}$$

show that

$$X\rho X + Y\rho Y + Z\rho Z = \frac{3}{2}I$$

12.4. *Consider a system–environment interaction that leaves the system unchanged and applies the rotation gate*

$$R_z(\theta) = \begin{pmatrix} e^{-i\theta/2} & 0 \\ 0 & e^{i\theta/2} \end{pmatrix}$$

to the environment if the system is in the state $|1\rangle$. If the system is in the state $|0\rangle$, the environment is unchanged. Let the initial state of the environment be $|\phi_E\rangle = |0_E\rangle$.

12.5. *Show that if the state of the environment is $|0_E\rangle$ and $U = \sqrt{p}I \otimes I + \sqrt{1-p}X \otimes Z$, this is the phase flip channel (12.24).*

12.6. *Consider phase damping and derive (12.51) through (12.53).*

12.7. *Verify that if the third qubit in a logical state has been flipped so that the state is $\alpha|001\rangle + \beta|110\rangle$, the algorithm described in (12.59) through (12.63) will reveal that the third qubit is flipped by setting the ancillary bits to 11.*

12.8. *What result is returned for the ancillary bits in the bit flip correcting algorithm?*

13

TOOLS OF QUANTUM INFORMATION THEORY

In this chapter we consider several important aspects of quantum information theory that are important for a thorough understanding of quantum computers. First we discuss the *no-cloning theorem*, which shows that you cannot make copies of an unknown quantum state. After recognizing this fact, we see how to measure the closeness of two states to each other. We will do this by looking at trace distance and fidelity. We can characterize the amount of entanglement in a state by looking at the concurrence, and we can determine the resources needed to create a given entangled state by calculating the entanglement of formation.

We conclude the chapter by considering how to characterize the information content in a state. This is done by calculating *entropy*.

THE NO-CLONING THEOREM

A routine task performed in information processing is making copies of data. We take it for granted that we can make as many copies as we like of something—whether it's a word processing file or a bit of music. As we have seen, the remarkable power of a quantum computer comes from the fact a qubit can exist in a superposition $|\psi\rangle = \alpha|0\rangle + \beta|1\rangle$. Given this fact, can we make an exact copy of an arbitrary qubit?

It turns out the answer is no. This result, which we state below, is known as the no-cloning theorem, and it was derived by Wooters and Zurek in 1982.

Consider two pure states $|\psi\rangle$ and $|\phi\rangle$, and suppose that there exists a unitary operator U such that

$$U(|\psi\rangle \otimes |\chi\rangle) = |\psi\rangle \otimes |\psi\rangle \tag{13.1}$$

$$U(|\phi\rangle \otimes |\chi\rangle) = |\phi\rangle \otimes |\phi\rangle \tag{13.2}$$

for some target state $|\chi\rangle$. We take the inner product of the left-hand side of (13.1) with the left-hand side of (13.2) and use the fact that $U^\dagger U = I$ to get

$$(\langle\psi| \otimes \langle\chi|U^\dagger)(U|\phi\rangle \otimes |\chi\rangle) = \langle\psi|\phi\rangle\langle\chi|\chi\rangle = \langle\psi|\phi\rangle \tag{13.3}$$

However, taking the inner product of the right-hand sides of (13.1) and (13.2) gives

$$(\langle\psi|\phi\rangle)^2 \tag{13.4}$$

Equating these two results gives us the equation

$$\langle\psi|\phi\rangle = (\langle\psi|\phi\rangle)^2 \tag{13.5}$$

This equation can only be true in two cases—if $\langle\psi|\phi\rangle = 0$, in which case the states are orthogonal, or if $|\phi\rangle = |\psi\rangle$. What this result means is that there is no unitary operator U that can be used to clone arbitrary quantum states.

Here is a second proof, this time a simple proof by contradiction. In quantum mechanics we use linear operators. If U is linear, then

$$U(\alpha|\psi\rangle \otimes |\chi\rangle) = \alpha(U(|\psi\rangle \otimes |\chi\rangle)) = \alpha|\psi\rangle \otimes |\psi\rangle \tag{13.6}$$

However, we can let $|\omega\rangle = \alpha|\psi\rangle$ and then apply (13.1), giving

$$U(|\omega\rangle \otimes |\chi\rangle) = |\omega\rangle \otimes |\omega\rangle = \alpha|\psi\rangle \otimes \alpha|\psi\rangle = \alpha^2|\psi\rangle \otimes |\psi\rangle \tag{13.7}$$

Comparison of (13.6) and (13.7) gives a contradiction. Hence general cloning is not possible. So a question we may ask, given that we can't make a perfect copy of a quantum state in general, is how close is one quantum state to another? Is it possible to make imperfect copies?

TRACE DISTANCE

Because we cannot, in principle, make an exact copy of an unknown quantum state, the next question we might ask is, can we make an *approximate* copy? Before we look into the answer, let's see what tools are at our disposal that can be used to determine how similar two states are.

The first measure we will consider is the *trace distance*. Let ρ and σ be two density matrices. The trace distance $\delta(\rho, \sigma)$ is defined to be

$$\delta(\rho, \sigma) = \frac{1}{2} Tr|\rho - \sigma| \tag{13.8}$$

Note that $|\rho| = \sqrt{\rho^\dagger \rho}$. Suppose that the states ρ and σ are equally likely and that we want to do a measurement to distinguish between the two states. The average probability of success is

$$\Pr = \frac{1}{2} + \frac{1}{2}\delta(\rho, \sigma) \tag{13.9}$$

The trace distance acts like a metric on the Hilbert space. For example, the trace distance is nonnegative

$$0 \le \delta(\rho, \sigma) \tag{13.10}$$

with equality if and only if $\rho = \sigma$. The trace distance is symmetric,

$$\delta(\rho, \sigma) = \delta(\sigma, \rho) \tag{13.11}$$

and it satisfies the triangle inequality

$$\delta(\rho, \sigma) \le \delta(\rho, \vartheta) + \delta(\vartheta, \sigma) \tag{13.12}$$

If $\rho = |\psi\rangle\langle\psi|$ is a pure state, then $\delta(\rho, \sigma)$ is given by

$$\delta(\rho, \sigma) = \sqrt{1 - \langle\psi|\sigma|\psi\rangle} \tag{13.13}$$

If ρ and σ commute, meaning $[\rho, \sigma] = 0$, and they are both diagonal with respect to some basis $\{|u_i\rangle\}$ such that the eigenvalues of ρ are r_i and the eigenvalues of σ are s_i, then

$$\delta(\rho, \sigma) = \frac{1}{2} Tr\left|\sum_i (r_i - s_i)|u_i\rangle\langle u_i|\right| \tag{13.14}$$

Example 13.1

Compute the trace distance between

$$\rho = \frac{3}{4}|0\rangle\langle 0| + \frac{1}{4}|1\rangle\langle 1|$$

and each of

$$\sigma = \frac{2}{3}|0\rangle\langle 0| + \frac{1}{3}|1\rangle\langle 1|, \quad \pi = \frac{1}{8}|0\rangle\langle 0| + \frac{7}{8}|1\rangle\langle 1|$$

Solution

Looking at each of these states. Intuitively you would expect that ρ and σ are closer together than ρ and π, since π is more weighted toward $|1\rangle$. First we have

$$\rho - \sigma = \frac{3}{4}|0\rangle\langle 0| + \frac{1}{4}|1\rangle\langle 1| - \left(\frac{2}{3}|0\rangle\langle 0| + \frac{1}{3}|1\rangle\langle 1|\right)$$
$$= \frac{1}{12}|0\rangle\langle 0| - \frac{1}{12}|1\rangle\langle 1|$$

Let's recall a couple of facts about the trace. It's linear, meaning $Tr(\alpha A + \beta B) = \alpha Tr(A) + \beta Tr(B)$. Second, the trace turns outer products into inner products, meaning $Tr(|\psi\rangle\langle\psi|) = \langle\psi|\psi\rangle$. So we can write

$$\delta(\rho, \sigma) = \frac{1}{2} Tr|\rho - \sigma|$$
$$= \frac{1}{2} Tr\left|\frac{1}{12}|0\rangle\langle 0| - \frac{1}{12}|1\rangle\langle 1|\right|$$
$$= \frac{1}{2}\left(\frac{1}{12}\right)(Tr(|0\rangle\langle 0|) + Tr(|1\rangle\langle 1|)) = \frac{1}{2}\left(\frac{1}{12}\right)(\langle 0|0\rangle + \langle 1|1\rangle)$$
$$= \frac{1}{2}\left(\frac{1}{12}\right)(2) = \frac{1}{12}$$

Now let's see what the trace distance $\delta(\rho, \pi)$ is. We have

$$\rho - \pi = \frac{3}{4}|0\rangle\langle 0| + \frac{1}{4}|1\rangle\langle 1| - \left(\frac{1}{8}|0\rangle\langle 0| + \frac{7}{8}|1\rangle\langle 1|\right)$$
$$= \frac{5}{8}(|0\rangle\langle 0| - |1\rangle\langle 1|)$$

We find that

$$\delta(\rho,\pi) = \frac{1}{2}Tr|\rho - \pi|$$
$$= \frac{1}{2}Tr\left|\frac{5}{8}(|0\rangle\langle 0| - |1\rangle\langle 1|)\right|$$
$$= \frac{1}{2}\left(\frac{5}{8}\right)(Tr(|0\rangle\langle 0|) + Tr(|1\rangle\langle 1|)) = \frac{1}{2}\left(\frac{5}{8}\right)(\langle 0|0\rangle + \langle 1|1\rangle)$$
$$= \frac{1}{2}\left(\frac{5}{8}\right)(2) = \frac{5}{8}$$

As expected, since ρ and σ are more heavily weighted toward $|0\rangle$, we have $\delta(\rho,\pi)\rangle\delta(\rho,\sigma)$. This tells us the states ρ and σ are more similar than the states ρ and π are.

You Try It

Write down the matrix representations of $\rho = \frac{3}{4}|0\rangle\langle 0| + \frac{1}{4}|1\rangle\langle 1|, \sigma = \frac{2}{3}|0\rangle\langle 0| + \frac{1}{3}|1\rangle\langle 1|$, and $\pi = \frac{1}{8}|0\rangle\langle 0| + \frac{7}{8}|1\rangle\langle 1|$, and calculate (13.8). Verify the result obtained in Example 13.1.

A simple way to calculate the trace distance is to use the eigenvalues of the matrix $\rho - \sigma$. If we denote the eigenvalues by λ_i, then the trace distance is

$$\delta(\rho,\sigma) = \frac{1}{2}\sum_i |\lambda_i| = \frac{1}{2}\sum_i \sqrt{\lambda_i^* \lambda_i} \tag{13.15}$$

If we know the Bloch vectors of each density matrix, then we can calculate the trace distance easily. Suppose that $\vec{r}$ is the Bloch vector of ρ and $\vec{s}$ is the Bloch vector of σ. Then the trace distance $\delta(\rho,\sigma)$ can be calculated as

$$\delta(\rho,\sigma) = \frac{1}{2}|\vec{r} - \vec{s}| \tag{13.16}$$

Example 13.2

Find the trace distance between the states

$$\rho = \begin{pmatrix} \frac{5}{8} & \frac{i}{4} \\ \frac{-i}{4} & \frac{3}{8} \end{pmatrix}, \quad \sigma = \begin{pmatrix} \frac{2}{5} & \frac{-i}{8} \\ \frac{i}{8} & \frac{3}{5} \end{pmatrix}$$

Solution

Let's do it using (13.8) first. We have

$$\rho - \sigma = \begin{pmatrix} \frac{5}{8} & \frac{i}{4} \\ \frac{-i}{4} & \frac{3}{8} \end{pmatrix} - \begin{pmatrix} \frac{2}{5} & \frac{-i}{8} \\ \frac{i}{8} & \frac{3}{5} \end{pmatrix} = \begin{pmatrix} \frac{9}{40} & \frac{i3}{8} \\ \frac{-i3}{8} & \frac{-9}{40} \end{pmatrix}$$

Now $(\rho - \sigma)^{\dagger} = \rho - \sigma$, so

$$(\rho - \sigma)^{\dagger} \rho - \sigma = \begin{pmatrix} \frac{9}{40} & \frac{i3}{8} \\ \frac{-i3}{8} & \frac{-9}{40} \end{pmatrix} \begin{pmatrix} \frac{9}{40} & \frac{i3}{8} \\ \frac{-i3}{8} & \frac{-9}{40} \end{pmatrix} = \begin{pmatrix} \frac{153}{800} & 0 \\ 0 & \frac{153}{800} \end{pmatrix}$$

Next we find

$$|\rho - \sigma| = \sqrt{(\rho - \sigma)^{\dagger}(\rho - \sigma)} = \sqrt{\begin{pmatrix} \frac{153}{800} & 0 \\ 0 & \frac{153}{800} \end{pmatrix}} = \frac{1}{20} \begin{pmatrix} 3\sqrt{\frac{17}{2}} & 0 \\ 0 & 3\sqrt{\frac{17}{2}} \end{pmatrix}$$

Hence

$$\delta(\rho, \sigma) = \frac{1}{2} \left(\frac{1}{20}\right) (2) \left(3\sqrt{\frac{17}{2}}\right) \approx 0.437$$

The Bloch vector for ρ was found in Example 5.12:

$$S_x = Tr(X\rho) = Tr \left[\begin{pmatrix} 0 & 1 \\ 1 & 0 \end{pmatrix} \begin{pmatrix} \frac{5}{8} & \frac{i}{4} \\ \frac{-i}{4} & \frac{3}{8} \end{pmatrix} \right] = Tr \begin{pmatrix} \frac{-i}{4} & \frac{3}{8} \\ \frac{5}{8} & \frac{i}{4} \end{pmatrix} = 0$$

$$S_y = Tr(Y\rho) = Tr \left[\begin{pmatrix} 0 & -i \\ i & 0 \end{pmatrix} \begin{pmatrix} \frac{5}{8} & \frac{i}{4} \\ \frac{-i}{4} & \frac{3}{8} \end{pmatrix} \right] = Tr \begin{pmatrix} \frac{-1}{4} & \frac{-i3}{8} \\ \frac{i5}{8} & \frac{-1}{4} \end{pmatrix} = \frac{-1}{2}$$

$$S_z = Tr(Z\rho) = Tr \left[\begin{pmatrix} 1 & 0 \\ 0 & -1 \end{pmatrix} \begin{pmatrix} \frac{5}{8} & \frac{i}{4} \\ \frac{-i}{4} & \frac{3}{8} \end{pmatrix} \right] = Tr \begin{pmatrix} \frac{5}{8} & \frac{i}{4} \\ \frac{i}{4} & \frac{-3}{8} \end{pmatrix} = \frac{1}{4}$$

$$\vec{r} = -\frac{1}{2}\hat{y} + \frac{1}{4}\hat{z}$$

The Bloch vector for σ, which was calculated as Exercise 5.10, is

$$\vec{s} = \frac{1}{4}\hat{y} - \frac{1}{5}\hat{z}$$

Therefore

$$\vec{r} - \vec{s} = -\frac{3}{4}\hat{y} + \frac{9}{20}\hat{z}$$

The magnitude of this vector is

$$|\vec{r} - \vec{s}| = \sqrt{\left(-\frac{3}{4}\right)^2 + \left(\frac{9}{20}\right)^2} = \frac{\sqrt{306}}{20}$$

Hence

$$\delta(\rho, \sigma) = \frac{1}{2}|\vec{r} - \vec{s}| = \frac{1}{2}\frac{\sqrt{306}}{20} \approx 0.437$$

Example 13.3

A system is in the pure state

$$\rho = \frac{3}{4}|+\rangle\langle+| + \frac{1}{4}|-\rangle\langle-|$$

Find the trace distance between ρ and $\sigma = |\psi\rangle\langle\psi|$, where

$$|\psi\rangle = \frac{1}{\sqrt{5}}|0\rangle + \frac{2}{\sqrt{5}}|1\rangle$$

Solution

Both density matrices need to be written with respect to the same basis. Let's start by rewriting ρ in terms of the computational basis. In Example 5.5 we found that

$$\begin{aligned}\rho &= \frac{3}{4}|+\rangle\langle+| + \frac{1}{4}|-\rangle\langle-| \\ &= \left(\frac{3}{4}\right)\left(\frac{1}{2}\right)(|0\rangle\langle0| + |0\rangle\langle1| + |1\rangle\langle0| + |1\rangle\langle1|) + \left(\frac{1}{4}\right)\left(\frac{1}{2}\right)((|0\rangle\langle0| - |0\rangle\langle1| - |1\rangle\langle0| + |1\rangle\langle1|)) \\ &= \frac{1}{2}|0\rangle\langle0| + \frac{1}{4}|0\rangle\langle1| + \frac{1}{4}|1\rangle\langle0| + \frac{1}{2}|1\rangle\langle1|\end{aligned}$$

The matrix representation of this density operator is

$$\rho = \frac{1}{4}\begin{pmatrix} 2 & 1 \\ 1 & 2 \end{pmatrix}$$

Now for

$$|\psi\rangle = \frac{1}{\sqrt{5}}|0\rangle + \frac{2}{\sqrt{5}}|1\rangle$$

we found in Example 5.4 that

$$\begin{aligned}\sigma = |\psi\rangle\langle\psi| &= \left(\frac{1}{\sqrt{5}}|0\rangle + \frac{2}{\sqrt{5}}|1\rangle\right)\left(\frac{1}{\sqrt{5}}\langle 0| + \frac{2}{\sqrt{5}}\langle 1|\right) \\ &= \frac{1}{5}|0\rangle\langle 0| + \frac{2}{5}|0\rangle\langle 1| + \frac{2}{5}|1\rangle\langle 0| + \frac{4}{5}|1\rangle\langle 1|\end{aligned}$$

The matrix representation is

$$\sigma = \begin{pmatrix} \frac{1}{5} & \frac{2}{5} \\ \frac{2}{5} & \frac{4}{5} \end{pmatrix}$$

The matrix $\rho - \sigma$ is given by

$$\rho - \sigma = \frac{1}{20}\begin{pmatrix} 6 & -3 \\ -3 & -6 \end{pmatrix}$$

This matrix has two eigenvalues, namely

$$\lambda_1 = -\frac{3}{4\sqrt{5}}, \quad \lambda_2 = \frac{3}{4\sqrt{5}}$$

Using (13.15), we find the trace distance to be

$$\delta(\rho, \sigma) = \frac{1}{2}\sum_i |\lambda_i| = \frac{1}{2}\left(\left|-\frac{3}{4\sqrt{5}}\right| + \left|\frac{3}{4\sqrt{5}}\right|\right) = \frac{3}{4\sqrt{5}}$$

FIDELITY

Another measure that can be used to determine how close one state is to another is based on the notion of the amount of statistical overlap between two distributions, called the *fidelity*. Once again, let ρ and σ be two density operators. Then the fidelity is given by

$$F(\rho, \sigma) = Tr\left(\sqrt{\sqrt{\rho}\,\sigma\,\sqrt{\rho}}\right) \tag{13.17}$$

In short, fidelity is a concept that comes from the inner product of two quantum states. Let $|\psi\rangle$and $|\phi\rangle$ be two states. The inner product $|\langle\phi|\psi\rangle|^2$ gives the probability of finding the system in the state $|\phi\rangle$ if it is known to be in the state $|\psi\rangle$, and vice versa. Hence this is a kind of measure of how similar the two states are or how much overlap there is between them. Suppose that they are pure states and with density operators $\rho = |\psi\rangle\langle\psi|$ and $\sigma = |\phi\rangle\langle\phi|$. Since these are pure states, $\rho^2 = \rho$, $\sigma^2 = \sigma$, and hence $\rho = \sqrt{\rho}, \sigma = \sqrt{\sigma}$. Then

$$F(\rho,\sigma) = Tr\left(\sqrt{\sqrt{\rho}\sigma\sqrt{\rho}}\right) = Tr\sqrt{(|\psi\rangle\langle\psi|)(|\phi\rangle\langle\phi|)(|\psi\rangle\langle\psi|)}$$

$$= Tr\sqrt{(|\langle\phi|\psi\rangle|^2)(|\psi\rangle\langle\psi|)} = |\langle\phi|\psi\rangle|\sqrt{\langle\psi|\psi\rangle} = |\langle\phi|\psi\rangle| \tag{13.18}$$

From (13.18), a few general properties of fidelity can be seen. The first is that fidelity is a number that ranges between 0 and 1,

$$0 \le F(\rho,\sigma) \le 1 \tag{13.19}$$

with unity if the states ρ and σ are the same state and 0 if there is no overlap whatsoever. We can also see from (13.18) that the fidelity of two pure states is symmetric. In fact this is true in general, that is,

$$F(\rho,\sigma) = F(\sigma,\rho) \tag{13.20}$$

The fidelity is futher invariant under unitary operations, that is,

$$F(U\rho U^\dagger, U\sigma U^\dagger) = F(\rho,\sigma) \tag{13.21}$$

If ρ and σ commute and are hence diagonal in the same basis, which we denote by $|u_i\rangle$, then we can write the fidelity in terms of the eigenvalues of ρ and σ. Suppose that $\rho = \sum_i r_i |u_i\rangle\langle u_i|$ and $\sigma = \sum_i s_i |u_i\rangle\langle u_i|$, then

$$F(\rho,\sigma) = \sum_i \sqrt{r_i s_i} \tag{13.22}$$

Example 13.4

Compute the fidelity between

$$\rho = \frac{3}{4}|0\rangle\langle 0| + \frac{1}{4}|1\rangle\langle 1|$$

and each of

$$\sigma = \frac{2}{3}|0\rangle\langle 0| + \frac{1}{3}|1\rangle\langle 1|, \quad \pi = \frac{1}{8}|0\rangle\langle 0| + \frac{7}{8}|1\rangle\langle 1|$$

Can the fidelity be calculated using (13.18) or (13.22)? Compare with Example 13.1.

Solution

First we compute

$$\rho^2 = \begin{pmatrix} \frac{3}{4} & 0 \\ 0 & \frac{1}{4} \end{pmatrix}\begin{pmatrix} \frac{3}{4} & 0 \\ 0 & \frac{1}{4} \end{pmatrix} = \begin{pmatrix} \frac{9}{16} & 0 \\ 0 & \frac{1}{16} \end{pmatrix}$$

Since $Tr(\rho^2) = 10/16 < 1, \rho$ is not a pure state. Similar calculations show that $Tr(\sigma^2) = 5/9 < 1$ and $Tr(\pi^2) = 50/64 < 1$, so σ and π are also mixed states and (13.18) does not apply. However, since all three density operators are diagonal in the computational basis, we can use (13.22). Notice that

$$\rho\sigma = \begin{pmatrix} \frac{3}{4} & 0 \\ 0 & \frac{1}{4} \end{pmatrix}\begin{pmatrix} \frac{2}{3} & 0 \\ 0 & \frac{1}{3} \end{pmatrix} = \begin{pmatrix} \frac{1}{2} & 0 \\ 0 & \frac{1}{12} \end{pmatrix}$$

$$\sigma\rho = \begin{pmatrix} \frac{2}{3} & 0 \\ 0 & \frac{1}{3} \end{pmatrix}\begin{pmatrix} \frac{3}{4} & 0 \\ 0 & \frac{1}{4} \end{pmatrix} = \begin{pmatrix} \frac{1}{2} & 0 \\ 0 & \frac{1}{12} \end{pmatrix}$$

Therefore $[\rho, \sigma] = 0$. Using (13.22), we find that the fidelity is

$$F(\rho, \sigma) = \sum_i \sqrt{r_i s_i} = \sqrt{\left(\frac{3}{4}\right)\left(\frac{2}{3}\right)} + \sqrt{\left(\frac{1}{4}\right)\left(\frac{1}{3}\right)} = \frac{1}{\sqrt{2}} + \frac{1}{\sqrt{12}} = \frac{1+\sqrt{6}}{\sqrt{12}} = 0.996$$

Since the fidelity is close to 1, this tells us that the two states are very similar—they have a lot of overlap. In Example 13.1 we found that the trace distance between the states was 1/12, a small number indicating that there is not much "distance" between the two states. Hence states that are similar have a high fidelity and a small trace distance.

For the other state we find that

$$F(\rho, \pi) = \sum_i \sqrt{r_i s_i} = \sqrt{\left(\frac{3}{4}\right)\left(\frac{1}{8}\right)} + \sqrt{\left(\frac{1}{4}\right)\left(\frac{7}{8}\right)} = \sqrt{\frac{3}{32}} + \sqrt{\frac{7}{32}} = \frac{\sqrt{3}+\sqrt{7}}{\sqrt{32}} = 0.774$$

The smaller fidelity indicates there is not nearly as much overlap between these two states as there is in the first case. In Example 13.1 we found that the trace distance was 5/8. So a state that is not as similar results in a smaller fidelity and a larger trace distance.

Fidelity can be looked at as a transition probability. In other words, the probability that ρ evolves into σ is

$$\Pr(\rho \to \sigma) = (F(\rho, \sigma))^2 \tag{13.23}$$

Example 13.5

What is the probability that ρ evolves into each of the states σ and π in the previous example?

Solution

The probability that ρ evolves into σ is

$$\Pr(\rho \to \sigma) = (F(\rho, \sigma))^2 = (0.996)^2 = 0.992$$

The probability that ρ evolves into π is

$$\Pr(\rho \to \pi) = (F(\rho, \pi))^2 = (0.774)^2 = 0.599$$

The *Bures distance function* is a distance measure between quantum states that makes use of the fidelity. It is

$$d_B^2(\rho, \sigma) = 2(1 - F(\rho, \sigma)) \tag{13.24}$$

The *Bures distance function* or *modified Bures metric* is a distance measure between quantum states that makes use of the fidelity. It is

$$d_B^2(\rho, \sigma) = 2(1 - F(\rho, \sigma)) \tag{13.25}$$

The *Bures metric* is given by

$$d_B(\rho, \sigma) = 2 - 2\sqrt{F(\rho, \sigma)} \tag{13.26}$$

Example 13.6

Consider the states in Example 13.4 and show that the Bures distance between ρ and π is much larger than the Bures distance between ρ and σ.

Solution

In the first case we find that

$$d_B^2(\rho, \sigma) = 2(1 - F(\rho, \sigma)) = 2(1 - 0.996) = 0.008$$

For the other two states we have

$$d_B^2(\rho,\pi) = 2(1 - F(\rho,\pi)) = 2(1 - 0.774) = 0.452$$

$d_B^2(\rho,\pi) \gg d_B^2(\rho,\sigma)$ once again indicating that ρ and σ are far more similar than ρ and π.

In practical cases of interest we often want to find the *minimum* fidelity for a given channel. This is because we do not know the quantum state $|\psi\rangle$, so finding the minimum fidelity gives us a worst-case analysis of a given quantum channel.

Example 13.7

On a certain quantum channel there is a probability $p = 1/9$ that there is a bit flip error. What is the minimum fidelity of the bit flip channel in this case? Assume that system starts in some pure state $\rho = |\psi\rangle\langle\psi|$.

Solution

The bit flip channel was described in (12.22). Recall that there is a probability p that nothing happens to the qubit, while there is a probability $1 - p$ that there is a bit flip error. The quantum operation is

$$\rho' = \Phi(\rho) = p\rho + (1 - p)X\rho X$$

The fidelity between this state and $\rho = |\psi\rangle\langle\psi|$ is given by

$$F(\rho,\rho') = F(\rho',\rho) = Tr(\sqrt{\sqrt{\rho'}\rho\sqrt{\rho'}}) = tr\sqrt{\sqrt{\rho'}(|\psi\rangle\langle\psi|)\sqrt{\rho'}}$$
$$= \sqrt{\langle\psi|\sqrt{\rho'}\sqrt{\rho'}|\psi\rangle} = \sqrt{\langle\psi|\rho'|\psi\rangle}$$

Using (12.22) simplifies the fidelity to

$$F(\rho',\rho) = \sqrt{\langle\psi|(p\rho + (1 - p)X\rho X)|\psi\rangle} = \sqrt{\langle\psi|(p|\psi\rangle\langle\psi| + (1 - p)X|\psi\rangle\langle\psi|X)|\psi\rangle}$$
$$= \sqrt{p + (1 - p)\langle\psi|X|\psi\rangle\langle\psi|X|\psi\rangle}$$
$$= \sqrt{p + (1 - p)\langle\psi|X|\psi\rangle^2}$$

Now we want to find the state where F is a minimum, which gives the worst-case scenario. Since it is always the case that $0 \le p \le 1$, it follows that$1 - p \ge 0$ and F will assume the smallest value when $(1 - p)\langle\psi|X|\psi\rangle^2 = 0$. So we need to find the state for which $\langle\psi|X|\psi\rangle = 0$. Notice that if

$$|\psi\rangle = \frac{|0\rangle + i|1\rangle}{\sqrt{2}}$$

Then

$$\langle\psi|X|\psi\rangle = \left(\frac{\langle 0| - i\langle 1|}{\sqrt{2}}\right) X \left(\frac{|0\rangle + i|1\rangle}{\sqrt{2}}\right) = \left(\frac{\langle 0| - i\langle 1|}{\sqrt{2}}\right)\left(\frac{|1\rangle + i|0\rangle}{\sqrt{2}}\right)$$
$$= \frac{i\langle 0 \mid 0\rangle - i\langle 1 \mid 1\rangle}{2} = 0$$

Let's verify that $|\psi\rangle$ is a pure state. We find that

$$\rho = |\psi\rangle\langle\psi| = \frac{1}{2}\begin{pmatrix} 1 & -i \\ i & 1 \end{pmatrix}$$
$$\Rightarrow \rho^2 = \frac{1}{4}\begin{pmatrix} 2 & -2i \\ 2i & 2 \end{pmatrix}$$

So we have $Tr(\rho^2) = \frac{1}{4}(2+2) = 1$, and this is a pure state. So the minimum fidelity occurs when $\langle\psi|X|\psi\rangle = 0$, in which case

$$F(\rho, \rho') = \sqrt{p}$$

For the case where $p = 1/9$ the minimum fidelity is $F_{\min} = \sqrt{1/9} \approx 0.33$.

ENTANGLEMENT OF FORMATION AND CONCURRENCE

Here we return to the examination of entangled states of two qubits. Two questions we can ask are how much entanglement does a state have, and second, what is the cost of creating a given entangled state? One way to characterize entanglement is by calculating the *concurrence*. To characterize the resources required to create a given entangled state, we can calculate the *entanglement of formation*.

First let's consider the concurrence. Basically this is just the amount of overlap between a state $|\psi\rangle$ and a state $|\tilde{\psi}\rangle$:

$$C(\psi) = |\langle\psi|\tilde{\psi}\rangle| \tag{13.27}$$

where $|\tilde{\psi}\rangle = Y \otimes Y|\psi^*\rangle$ and ψ^* is the complex conjugate of the state. The concurrence can also be calculated using the density operator by considering the quantity given by $\rho\,(Y \otimes Y)\rho^\dagger(Y \otimes Y)$.

Example 13.8

Concurrence can be a measure of entanglement. Consider the product state

$$|\psi\rangle = |0\rangle \otimes |1\rangle$$

and show that the concurrence is zero.

Solution

We have

$$|\tilde{\psi}\rangle = Y \otimes Y|\psi\rangle = Y|0\rangle \otimes Y|1\rangle = -i|1\rangle \otimes i|0\rangle = |1\rangle \otimes |0\rangle$$

So we have

$$\langle\tilde{\psi}|\psi\rangle = ((\langle 1| \otimes \langle 0|)(|0\rangle \otimes |1\rangle) = \langle 1|0\rangle\langle 0|1\rangle = 0$$

So the concurrence is zero. We can also see the concurrence vanishes by using the matrix representations of the operators. First, we have

$$Y \otimes Y = \begin{pmatrix} 0 & -iY \\ iY & 0 \end{pmatrix} = \begin{pmatrix} 0 & 0 & 0 & -1 \\ 0 & 0 & 1 & 0 \\ 0 & 1 & 0 & 0 \\ -1 & 0 & 0 & 0 \end{pmatrix}$$

The density operator for this state is

$$\rho = |01\rangle\langle 01| = \begin{pmatrix} 0 & 0 & 0 & 0 \\ 0 & 1 & 0 & 0 \\ 0 & 0 & 0 & 0 \\ 0 & 0 & 0 & 0 \end{pmatrix}$$

Hence

$$\rho(Y \otimes Y)\rho^{\dagger}(Y \otimes Y)$$

$$= \begin{pmatrix} 0 & 0 & 0 & 0 \\ 0 & 1 & 0 & 0 \\ 0 & 0 & 0 & 0 \\ 0 & 0 & 0 & 0 \end{pmatrix}\begin{pmatrix} 0 & 0 & 0 & -1 \\ 0 & 0 & 1 & 0 \\ 0 & 1 & 0 & 0 \\ -1 & 0 & 0 & 0 \end{pmatrix}\begin{pmatrix} 0 & 0 & 0 & 0 \\ 0 & 1 & 0 & 0 \\ 0 & 0 & 0 & 0 \\ 0 & 0 & 0 & 0 \end{pmatrix}\begin{pmatrix} 0 & 0 & 0 & -1 \\ 0 & 0 & 1 & 0 \\ 0 & 1 & 0 & 0 \\ -1 & 0 & 0 & 0 \end{pmatrix}$$

$$= \begin{pmatrix} 0 & 0 & 0 & 0 \\ 0 & 1 & 0 & 0 \\ 0 & 0 & 0 & 0 \\ 0 & 0 & 0 & 0 \end{pmatrix}\begin{pmatrix} 0 & 0 & 0 & -1 \\ 0 & 0 & 1 & 0 \\ 0 & 1 & 0 & 0 \\ -1 & 0 & 0 & 0 \end{pmatrix}\begin{pmatrix} 0 & 0 & 0 & 0 \\ 0 & 0 & 1 & 0 \\ 0 & 0 & 0 & 0 \\ 0 & 0 & 0 & 0 \end{pmatrix}$$

$$= \begin{pmatrix} 0 & 0 & 0 & 0 \\ 0 & 0 & 0 & 0 \\ 0 & 0 & 0 & 0 \\ 0 & 0 & 0 & 0 \end{pmatrix}$$

We can find the concurrence by looking at the eigenvalues of the resulting matrix. The eigenvalues of this matrix all vanish, hence the concurrence is zero.

A second way to define concurrence is to look at the eigenvalues of the matrix

$$R = \sqrt{\sqrt{\rho}\tilde{\rho}\sqrt{\rho}} \tag{13.28}$$

which are denoted by λ_1, λ_2, λ_3, λ_4. The concurrence is

$$C(\rho) = \max\{0, \lambda_1 - \lambda_2 - \lambda_3 - \lambda_4\} \tag{13.29}$$

where $\lambda_1 \geq \lambda_2 \geq \lambda_3 \geq \lambda_4$. In the next two examples we consider the concurrence of entangled states.

Example 13.9

Find the concurrence of

$$|S\rangle = \frac{|01\rangle - |10\rangle}{\sqrt{2}}$$

Solution

Again, note that

$$Y \otimes Y = \begin{pmatrix} 0 & -iY \\ iY & 0 \end{pmatrix} = \begin{pmatrix} 0 & 0 & 0 & -1 \\ 0 & 0 & 1 & 0 \\ 0 & 1 & 0 & 0 \\ -1 & 0 & 0 & 0 \end{pmatrix}$$

The density operator is

$$\rho = |S\rangle\langle S| = \left(\frac{|01\rangle - |10\rangle}{\sqrt{2}}\right)\left(\frac{\langle 01| - \langle 10|}{\sqrt{2}}\right)$$

$$= \frac{1}{2}(|01\rangle\langle 01| - |01\rangle\langle 10| - |10\rangle\langle 01| + |10\rangle\langle 10|)$$

The matrix representation is

$$\rho = \frac{1}{2}\begin{pmatrix} 0 & 0 & 0 & 0 \\ 0 & 1 & -1 & 0 \\ 0 & -1 & 1 & 0 \\ 0 & 0 & 0 & 0 \end{pmatrix}$$

So we have

$\rho(Y \otimes Y)\rho^{\dagger}(Y \otimes Y)$

$$= \frac{1}{4}\begin{pmatrix} 0 & 0 & 0 & 0 \\ 0 & 1 & -1 & 0 \\ 0 & -1 & 1 & 0 \\ 0 & 0 & 0 & 0 \end{pmatrix}\begin{pmatrix} 0 & 0 & 0 & -1 \\ 0 & 0 & 1 & 0 \\ 0 & 1 & 0 & 0 \\ -1 & 0 & 0 & 0 \end{pmatrix}\begin{pmatrix} 0 & 0 & 0 & 0 \\ 0 & 1 & -1 & 0 \\ 0 & -1 & 1 & 0 \\ 0 & 0 & 0 & 0 \end{pmatrix}\begin{pmatrix} 0 & 0 & 0 & -1 \\ 0 & 0 & 1 & 0 \\ 0 & 1 & 0 & 0 \\ -1 & 0 & 0 & 0 \end{pmatrix}$$

$$= \frac{1}{4}\begin{pmatrix} 0 & 0 & 0 & 0 \\ 0 & 1 & -1 & 0 \\ 0 & -1 & 1 & 0 \\ 0 & 0 & 0 & 0 \end{pmatrix}\begin{pmatrix} 0 & 0 & 0 & -1 \\ 0 & 0 & 1 & 0 \\ 0 & 1 & 0 & 0 \\ -1 & 0 & 0 & 0 \end{pmatrix}\begin{pmatrix} 0 & 0 & 0 & 0 \\ 0 & -1 & 1 & 0 \\ 0 & 1 & -1 & 0 \\ 0 & 0 & 0 & 0 \end{pmatrix}$$

$$= \frac{1}{2}\begin{pmatrix} 0 & 0 & 0 & 0 \\ 0 & 1 & -1 & 0 \\ 0 & -1 & 1 & 0 \\ 0 & 0 & 0 & 0 \end{pmatrix}$$

The eigenvalues of this matrix are

$$\lambda_1 = 1, \lambda_2 = \lambda_3 = \lambda_4 = 0$$

Using (13.29), we find the concurrence to be

$$C(\rho) = \max\{0, \lambda_1 - \lambda_2 - \lambda_3 - \lambda_4\} = \max\{0, 1\} = 1$$

Example 13.10

Find the concurrence of

$$|\psi\rangle = \frac{|00\rangle + |11\rangle}{\sqrt{2}}$$

Solution

The density operator in this case is

$$\rho = |\psi\rangle\langle\psi| = \frac{1}{2}(|00\rangle\langle 00| + |00\rangle\langle 11| + |11\rangle\langle 00| + |11\rangle\langle 11|)$$
$$= \frac{1}{2}\begin{pmatrix} 1 & 0 & 0 & 1 \\ 0 & 0 & 0 & 0 \\ 0 & 0 & 0 & 0 \\ 1 & 0 & 0 & 1 \end{pmatrix}$$

Therefore

$$\rho(Y \otimes Y)\rho^\dagger(Y \otimes Y)$$
$$= \frac{1}{4}\begin{pmatrix} 1 & 0 & 0 & 1 \\ 0 & 0 & 0 & 0 \\ 0 & 0 & 0 & 0 \\ 1 & 0 & 0 & 1 \end{pmatrix}\begin{pmatrix} 0 & 0 & 0 & -1 \\ 0 & 0 & 1 & 0 \\ 0 & 1 & 0 & 0 \\ -1 & 0 & 0 & 0 \end{pmatrix}\begin{pmatrix} 1 & 0 & 0 & 1 \\ 0 & 0 & 0 & 0 \\ 0 & 0 & 0 & 0 \\ 1 & 0 & 0 & 1 \end{pmatrix}\begin{pmatrix} 0 & 0 & 0 & -1 \\ 0 & 0 & 1 & 0 \\ 0 & 1 & 0 & 0 \\ -1 & 0 & 0 & 0 \end{pmatrix}$$
$$= \frac{1}{4}\begin{pmatrix} 1 & 0 & 0 & 1 \\ 0 & 0 & 0 & 0 \\ 0 & 0 & 0 & 0 \\ 1 & 0 & 0 & 1 \end{pmatrix}\begin{pmatrix} 0 & 0 & 0 & -1 \\ 0 & 0 & 1 & 0 \\ 0 & 1 & 0 & 0 \\ -1 & 0 & 0 & 0 \end{pmatrix}\begin{pmatrix} -1 & 0 & 0 & -1 \\ 0 & 0 & 0 & 0 \\ 0 & 0 & 0 & 0 \\ -1 & 0 & 0 & -1 \end{pmatrix}$$
$$= \frac{1}{2}\begin{pmatrix} 1 & 0 & 0 & 1 \\ 0 & 0 & 0 & 0 \\ 0 & 0 & 0 & 0 \\ 1 & 0 & 0 & 1 \end{pmatrix}$$

The eigenvalues are {1, 0, 0, 0}, and hence the concurrence is 1.

In the next section we will see that we can write the *Shannon entropy* as

$$h(p) = -p \log_2 p - (1 - p) \log_2(1 - p) \tag{13.30}$$

The *entanglement of formation* is defined in terms of the concurrence as

$$E(\rho) = h\left(\frac{1+\sqrt{1-C(\rho)^2}}{2}\right) \tag{13.31}$$

This is a mathematical characterization of the resources required to create an entangled state.

Example 13.11

Find the entanglement of formation for the Werner state

$$\rho = \frac{5}{6}|\phi^+\rangle\langle\phi^+| + \frac{1}{24}I_4 = \begin{pmatrix} \frac{11}{24} & 0 & 0 & \frac{5}{12} \\ 0 & \frac{1}{24} & 0 & 0 \\ 0 & 0 & \frac{1}{24} & 0 \\ \frac{5}{12} & 0 & 0 & \frac{11}{24} \end{pmatrix}$$

For $\rho\ (Y\otimes Y)\rho^{\dagger}(Y\otimes Y)$.

Solution

First we have

$$\rho(Y\otimes Y)\rho^{\dagger}(Y\otimes Y) \Rightarrow \rho^2 = \begin{pmatrix} \frac{221}{576} & 0 & 0 & \frac{55}{144} \\ 0 & \frac{1}{576} & 0 & 0 \\ 0 & 0 & \frac{1}{576} & 0 \\ \frac{55}{144} & 0 & 0 & \frac{221}{576} \end{pmatrix}$$

The eigenvalues of this matrix are

$$\lambda_i = \left\{\frac{49}{64}, \frac{1}{576}, \frac{1}{576}, \frac{1}{576}\right\}$$

From (13.29) the concurrence is

$$C(\rho) = 0.76$$

From (13.31) the entanglement of formation is

$$E(\rho) = -\frac{1+\sqrt{1-C(\rho)^2}}{2}\log_2\frac{1+\sqrt{1-C(\rho)^2}}{2} - \frac{1-\sqrt{1-C(\rho)^2}}{2}\log_2\frac{1-\sqrt{1-C(\rho)^2}}{2}$$
$$= 0.67$$

You Try It

Show that the entanglement of formation for the case considered in Example 13.9 is 1 and the entanglement of formation for the case in Example 13.10 is also 1.

INFORMATION CONTENT AND ENTROPY

Entropy is a way to quantify the information content in a signal. Specifically, suppose that there exists a random variable X. Entropy tells us the amount of ignorance we have about the random variable X prior to measurement. Or put another way, entropy provides the answer on how much information we will gain when we measure X.

We define entropy by using the probabilities that each possible outcome occurs. Suppose that p_j is the probability of the j th outcome where there are n total possible outcomes. The Shannon entropy H is given by

$$H = -\sum_{j=1}^{n} p_j \log_2 p_j \tag{13.32}$$

An example of an entropy function is shown in Figure 13.1, where we have a plot of the *binary entropy function* $H_2(x) = -x\log x - (1-x)\log(1-x)$. In the plot it is clear that entropy is a *concave* function. This means that for $0 \le \lambda \le 1$,

$$\lambda H(p) + (1-\lambda)H(q) \le H(\lambda p + (1-\lambda)q) \tag{13.33}$$

The maximum entropy occurs for the case where we have the least amount of knowledge. In the case of discrete probabilities p_j, we have the least amount of knowledge about the outcome of a measurement when each of the possible outcomes is equally likely. That is, with n possible outcomes, each of the probabilities is given by

$$p_j = \frac{1}{n}$$

A simple example is the binary entropy function. There are two possible outcomes. We first suppose that both are equally likely so that $x = 1/2$. Then we have

$$-x\log x - (1-x)\log(1-x) = 1$$

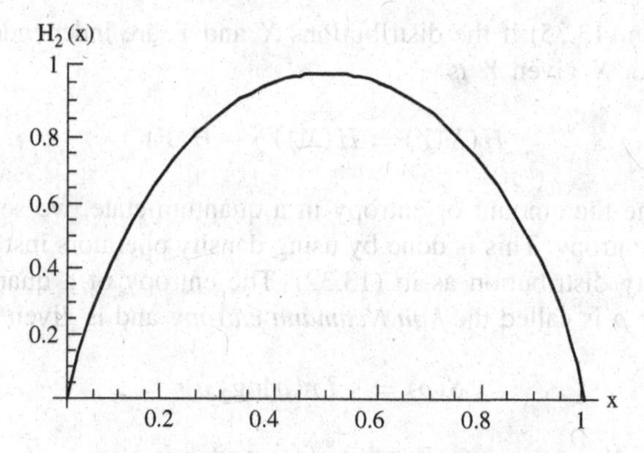

Figure 13.1 The binary entropy function

(note we are using base 2 logarithms). Now suppose that one outcome is substantially more likely. For example, if $x = 0.2$, the probability of finding the alternative is $1 - x = 0.8$. This is a situation where we have more knowledge about the state prior to measurement because one alternative is far more likely than the other. In this case

$$-x \log x - (1 - x) \log(1 - x) = 0.72$$

If $x = 0.05$, meaning that there is a 95% chance of seeing the alternative, then

$$-x \log x - (1 - x) \log(1 - x) = 0.29$$

So, if we are completely uncertain as to what outcome will occur—meaning all possible outcomes are equally likely—then with two possible outcomes

$$H(x) = 1$$

This case represents maximum entropy. For all other cases

$$H(x) < 1$$

The general rule is that the larger the entropy, the more ignorance you have about the outcome.

Now suppose that there are two random variables, X and Y. If the probability that we find result $X = x_i$ and $Y = y_i$ is $p(x, y)$, then the joint Shannon entropy is

$$H(X, Y) = -\sum_{x,y} p(x, y) \log(p(x, y)) \tag{13.34}$$

The following inequality called subadditivity is obeyed in general:

$$H(X, Y) \leq H(X) + H(Y) \tag{13.35}$$

Equality holds in (13.35) if the distributions X and Y are independent. The *conditional entropy* for X given Y is

$$H(X|Y) = H(X, Y) - H(Y) \tag{13.36}$$

To determine the amount of entropy in a quantum state, we seek an analogue to the Shannon entropy. This is done by using density operators instead of elements of the probability distribution as in (13.32). The entropy of a quantum state with density operator ρ is called the *Von Neumann entropy* and is given by

$$S(\rho) = -Tr(\rho \log_2 \rho) \tag{13.37}$$

The *relative Von Neumann entropy* of states ρ and σ is

$$S(\rho\|\sigma) = Tr(\rho \log \rho) - Tr(\rho \log \sigma) \tag{13.38}$$

Note that $S(\rho\|\sigma) \geq 0$ with equality if and only if $\rho = \sigma$.

Suppose that the eigenvalues of the density operator ρ are given by λ_i. We can write the Von Neumann entropy in terms of the eigenvalues as

$$S(\rho) = -\sum_i \lambda_i \log_2 \lambda_i \tag{13.39}$$

Example 13.12

The maximally mixed state for a qubit is

$$\rho = \begin{pmatrix} \frac{1}{2} & 0 \\ 0 & \frac{1}{2} \end{pmatrix}$$

What is the Von Neumann entropy for this state?

Solution

The eigenvalues are

$$\{\lambda_1, \lambda_2\} = \left\{\frac{1}{2}, \frac{1}{2}\right\}$$

Using (13.39), we find the entropy to be

$$S(\rho) = -\sum_i \lambda_i \log_2 \lambda_i = -\frac{1}{2}\log_2\left(\frac{1}{2}\right) - \frac{1}{2}\log_2\left(\frac{1}{2}\right) = -\log_2\left(\frac{1}{2}\right) = \log_2 2 = 1$$

In general, if a quantum system is in a Hilbert space of dimension n, the completely mixed state has entropy

$$\log_2 n \tag{13.40}$$

Example 13.13

Find the entropy of the two states

$$\rho = \begin{pmatrix} \frac{3}{4} & 0 \\ 0 & \frac{1}{4} \end{pmatrix}, \quad \sigma = \begin{pmatrix} \frac{9}{10} & 0 \\ 0 & \frac{1}{10} \end{pmatrix}$$

Solution

Let's look at the two states. We might guess that ρ has higher entropy because we have a little less information about the possible outcomes. Using the eigenvalues of ρ together with (13.39), we see that the entropy is

$$S(\rho) = -\frac{3}{4}\log_2\left(\frac{3}{4}\right) - \frac{1}{4}\log_2\left(\frac{1}{4}\right) = 0.81$$

For σ we find that

$$S(\sigma) = -\frac{9}{10}\log_2\left(\frac{9}{10}\right) - \frac{1}{10}\log_2\left(\frac{1}{10}\right) = 0.47$$

Our intuition is confirmed, we have more knowledge about the state σ before a measurement is made because its far more certain that the outcome is $|0\rangle$.

Example 13.14

Find the entropy of the state

$$\rho = \begin{pmatrix} \frac{1}{2} & \frac{1}{4} \\ \frac{1}{4} & \frac{1}{2} \end{pmatrix}$$

Solution

The eigenvalues of this matrix are

$$\lambda_{1,2} = \left\{\frac{3}{4}, \frac{1}{4}\right\}$$

Is this a pure state or a mixed state? We met this state in Example 5.5 where we found that

$$\rho^2 = \begin{pmatrix} \frac{1}{2} & \frac{1}{4} \\ \frac{1}{4} & \frac{1}{2} \end{pmatrix} \begin{pmatrix} \frac{1}{2} & \frac{1}{4} \\ \frac{1}{4} & \frac{1}{2} \end{pmatrix} = \begin{pmatrix} \frac{5}{16} & \frac{1}{4} \\ \frac{1}{4} & \frac{5}{16} \end{pmatrix}$$

$$\Rightarrow Tr(\rho^2) = \frac{5}{16} + \frac{5}{16} = \frac{10}{16} = \frac{5}{8}$$

Hence this is a mixed state. The entropy is the same as the matrix in the previous example, even though it looks very different:

$$S(\rho) = -\frac{3}{4}\log_2\left(\frac{3}{4}\right) - \frac{1}{4}\log_2\left(\frac{1}{4}\right) = 0.81$$

While a completely mixed state has the highest entropy, a pure state has the smallest possible entropy.

Example 13.15

Find the entropy of the state

$$|\psi\rangle = \frac{|0\rangle + |1\rangle}{\sqrt{2}}$$

Solution

First let's write down the density operator. We find that

$$\rho = |\psi\rangle\langle\psi| = \left(\frac{|0\rangle + |1\rangle}{\sqrt{2}}\right)\left(\frac{\langle 0| + \langle 1|}{\sqrt{2}}\right)$$

$$= \frac{1}{2}(|0\rangle\langle 0| + |0\rangle\langle 1| + |1\rangle\langle 0| + |1\rangle\langle 1|)$$

The matrix representation of this density operator is thus

$$\rho = \frac{1}{2}\begin{pmatrix} 1 & 1 \\ 1 & 1 \end{pmatrix}$$

The eigenvalues of this matrix are

$$\lambda_{1,2} = \{1, 0\}$$

To calculate the entropy, we use the fact that

$$\lim_{x \to 0} x \log x = 0$$

So we only need to consider $\lambda = 1$, and we find that

$$S(\rho) = -\log_2(1) = 0$$

This state is a pure state and it has zero entropy. We know with certainty what the state is prior to measurement—or put another way—there is no ignorance as to what the state of the system is.

In n dimensions, the entropy of a quantum state obeys the following inequality:

$$\log_2 n \geq S(\rho) \geq 0 \tag{13.41}$$

We have seen examples of both extrema.The completely mixed state—which is characterized by equally probable outcomes $p_i = 1/n$, has entropy given by $\log_2 n$— The pure state with zero entropy is the lower bound in (13.41). The Von Neumann entropy is invariant under a change of basis as we show in the next example.

Example 13.16

Let $\rho = \frac{3}{4}|+\rangle\langle+| + \frac{1}{4}|-\rangle\langle-|$. Show that the entropy of the state is invariant under a change of basis.

Solution

The matrix representation of this state is

$$\rho = \begin{pmatrix} \frac{3}{4} & 0 \\ 0 & \frac{1}{4} \end{pmatrix}$$

Note that this matrix is written in the $\{|\pm\rangle\}$ basis. The eigenvalues are $\lambda_{1,2} = \left\{\frac{3}{4}, \frac{1}{4}\right\}$ and we have already seen that the entropy in this case is

$$S(\rho) = -\frac{3}{4}\log_2\left(\frac{3}{4}\right) - \frac{1}{4}\log_2\left(\frac{1}{4}\right) = 0.81$$

What happens if we write the state in the computational basis? In that case the matrix representation is given by

$$\rho = \begin{pmatrix} \frac{1}{2} & \frac{1}{4} \\ \frac{1}{4} & \frac{1}{2} \end{pmatrix}$$

We met this matrix earlier—in Example 13.14. It has the same eigenvalues, and we find that once again,

$$S(\rho) = -\frac{3}{4}\log_2\left(\frac{3}{4}\right) - \frac{1}{4}\log_2\left(\frac{1}{4}\right) = 0.81$$

Now let's consider states formed by tensor products of qubits. If a composite state is separable, that is, a product state of the form $\rho \otimes \sigma$, entropy is additive in the sense that

$$S(\rho \otimes \sigma) = S(\rho) + S(\sigma) \tag{13.42}$$

In general, entropy is *subadditive*. Let ρ_A and ρ_B be the reduced density matrices of a composite system ρ. The subadditivity inequality states that

$$S(\rho) \leq S(\rho_A) + S(\rho_B) \tag{13.43}$$

The result (13.43) indicates that in order to have the most information about an entangled system, you need to consider the system as a whole— that is, $S(\rho)$ is smaller than the entropies of the reduced density matrices, because ignorance decreases when considering the system as a whole. Alice and Bob, in possession of the reduced density matrices ρ_A and ρ_B, have more ignorance about the state when considering their parts of the system alone.

Example 13.17

Alice and Bob each share one member of an EPR pair that is in the Bell state:

$$|\beta_{10}\rangle = \frac{|0_A\rangle|0_B\rangle - |1_A\rangle|1_B\rangle}{\sqrt{2}}$$

Determine the entropy of the entire system, and the entropy as seen by Alice and Bob alone.

Solution

We showed in (5.15) that the density operator for this state is

$$\begin{aligned}
\rho &= |\beta_{10}\rangle\langle\beta_{10}| \\
&= \left(\frac{|0_A\rangle|0_B\rangle - |1_A\rangle|1_B\rangle}{\sqrt{2}}\right)\left(\frac{\langle 0_A|\langle 0_B| - \langle 1_A|\langle 1_B|}{\sqrt{2}}\right) \\
&= \frac{|0_A\rangle|0_B\rangle\langle 0_A|\langle 0_B| - |0_A\rangle|0_B\rangle\langle 1_A|\langle 1_B| - |1_A\rangle|1_B\rangle\langle 0_A|\langle 0_B| + |1_A\rangle|1_B\rangle\langle 1_A|\langle 1_B|}{2}
\end{aligned}$$

In matrix form we have

$$\rho = \frac{1}{2}\begin{pmatrix} 1 & 0 & 0 & -1 \\ 0 & 0 & 0 & 0 \\ 0 & 0 & 0 & 0 \\ -1 & 0 & 0 & 1 \end{pmatrix}$$

The eigenvalues of this matrix are

$$\lambda_{1,2,3,4} = \{1, 0, 0, 0\}$$

So we quickly see that the entropy is $S(\rho) = -\log 1 = 0$. To get the states as seen by Alice and Bob individually, we compute the partial trace of this density operator. For example, in Chapter 5 we found that the reduced density matrix for Bob is

$$\rho_B = \frac{1}{2}\begin{pmatrix} 1 & 0 \\ 0 & 1 \end{pmatrix}$$

This is the completely mixed state with entropy given by

$$S_B(\rho_B) = -\log_2\left(\frac{1}{2}\right) = 1$$

We find a similar result for Alice, and clearly (13.43) is satisfied.

EXERCISES

13.1. *Verify that $\delta(\rho, \sigma) = 3/4\sqrt{5}$ for the density matrices in Example 13.3 by calculating $\frac{1}{2}Tr\sqrt{(\rho-\sigma)^\dagger(\rho-\sigma)}$ with the matrices written in the $\{|\pm\rangle\}$ basis.*

13.2. *Let $\rho = \sum_i r_i |u_i\rangle\langle u_i|$ and $\sigma = \sum_i s_i |u_i\rangle\langle u_i|$. Prove (13.22).*

13.3. *Compute the trace distance and fidelity between*

$$\rho = \frac{4}{7}|0\rangle\langle 0| + \frac{3}{7}|1\rangle\langle 1|$$

and each of

$$\sigma = \frac{2}{3}|0\rangle\langle 0| + \frac{1}{3}|1\rangle\langle 1|, \quad \pi = \frac{3}{5}|0\rangle\langle 0| + \frac{2}{5}|1\rangle\langle 1|$$

13.4. *On a certain quantum channel there is a probability p = 1/11 that there is a phase flip error. What pure state leads to the minimum fidelity in this case? What is the minimum fidelity of the bit flip channel in this case? Assume that the system starts in some pure state $\rho = |\psi\rangle\langle\psi|$.*

13.5. *Determine the concurrence for the Werner state where*

$$\rho = \begin{pmatrix} \frac{7}{16} & 0 & 0 & \frac{3}{8} \\ 0 & \frac{1}{16} & 0 & 0 \\ 0 & 0 & \frac{1}{16} & 0 \\ \frac{3}{8} & 0 & 0 & \frac{7}{16} \end{pmatrix}$$

13.6. *Find the entanglement of formation for the state in Exercise 13.5.*

13.7. *Find the entropy of the state*

$$\rho = \begin{pmatrix} \frac{5}{6} & 0 \\ 0 & \frac{1}{6} \end{pmatrix}$$

13.8. *Find the entropy of the state*

$$|\psi\rangle = \frac{2}{3}|0\rangle + \frac{\sqrt{5}}{3}|1\rangle$$

13.9. *Consider the product state $|A\rangle|B\rangle$ used in Example 5.11 where*

$$|A\rangle = \frac{|0\rangle - i|1\rangle}{\sqrt{2}}, \quad |B\rangle = \sqrt{\frac{2}{3}}|0\rangle + \frac{1}{\sqrt{3}}|1\rangle$$

The density matrix is

$$\rho = \begin{pmatrix} \frac{1}{3} & \frac{1}{\sqrt{18}} & \frac{i}{3} & \frac{i}{\sqrt{18}} \\ \frac{1}{\sqrt{18}} & \frac{1}{6} & \frac{i}{\sqrt{18}} & \frac{i}{6} \\ \frac{-i}{3} & \frac{-i}{\sqrt{18}} & \frac{1}{3} & \frac{1}{\sqrt{18}} \\ \frac{-i}{\sqrt{18}} & \frac{-i}{6} & \frac{1}{\sqrt{18}} & \frac{1}{6} \end{pmatrix}$$

(A) Show that the entropy of the density operator for $|A\rangle|B\rangle$ is zero.
(B) Find the entropy for the density operator seen only by Alice, ρ_A.

13.10. *Consider the density operators for the state*

$$|\psi\rangle = \sqrt{\frac{3}{7}}|0\rangle + \frac{2}{\sqrt{7}}|1\rangle$$

and the state

$$|\phi\rangle = \sqrt{\frac{2}{3}}|0\rangle + \frac{1}{\sqrt{3}}|1\rangle$$

Which state has a higher entropy?

14

ADIABATIC QUANTUM COMPUTATION

Adiabatic quantum computation is an alternative approach to quantum computation based on the time evolution of a quantum system. Before describing adiabatic processes, let's quickly review the dynamics of a quantum system. The time evolution of a quantum state $|\psi(t)\rangle$ is described by the Schrödinger equation

$$i\hbar\frac{\partial|\psi(t)\rangle}{\partial t} = H|\psi(t)\rangle \tag{14.1}$$

where H is the Hamiltonian operator. This operator is the total energy of the system and can be expressed in terms of kinetic and potential energies

$$H = \frac{p^2}{2m} + V = -\frac{\hbar^2}{2m}\nabla^2 + V \tag{14.2}$$

where ∇^2 is the *Laplacian* operator and V is the potential energy function. In one dimension (14.2) becomes

$$H = \frac{p^2}{2m} + V = -\frac{\hbar^2}{2m}\frac{\partial^2}{\partial x^2} + V \tag{14.3}$$

Quantum Computing Explained, by David McMahon

In many cases of interest, the potential V is time-independent. Where this is the case, then solutions of (14.1) are products of a function that depends only on the spatial coordinates and a function that depends only on time. In one dimension the solution can be written in the form

$$\Psi(x,t) = \psi(x)f(t) \tag{14.4}$$

By inserting (14.4) into (14.1) and using (14.3), its not too hard to show that the solution can be written as

$$\Psi(x,t) = \psi(x)e^{-iEt/\hbar} \tag{14.5}$$

Solutions of Schrödingers equation are known as *wave functions*. Using (14.5) in (14.1), we also arrive at a *time-independent* Schrodinger's equation that gives a solution for the spatial part of the wave function:

$$-\frac{\hbar^2}{2m}\frac{d^2\psi}{dx^2} + V(x)\psi = E\psi \tag{14.6}$$

where E is the energy of the system. The modulus squared of the wave function is a probability density, where $|\Psi(x,t)|^2 dx$ is the probability of finding the particle in a volume dx. The total probability over all space must be equal to one, so the normalization condition for a wave function is

$$\int_{-\infty}^{\infty} |\Psi(x,t)|^2\, dx = 1 \tag{14.7}$$

Solutions of the type (14.5) are called *stationary*. This is because when we calculate the modulus squared of (14.5), we see that the probability density does not change with time:

$$|\Psi(x,t)|^2 = |\psi(x)e^{-iEt/\hbar}|^2 = (\psi^*(x)e^{iEt/\hbar})(\psi(x)e^{-iEt/\hbar}) = |\psi(x)|^2 \tag{14.8}$$

The eigenstates of the Hamiltonian $\Phi_n(x,t) = \phi_n(x)f_n(t)$ have eigenvalues that are the energies E_n that the system can assume. Bound states have discrete energy levels; that is, the system can only take on specific energy values like the rungs on a ladder. Unbound systems have continuous energy spectra (they can assume any energy value). The general solution to the time-dependent Schrödinger equation is a superposition of energy eigenstates

$$\Psi(x,t) = \sum_n c_n\phi_n(x)\exp\left(-i\frac{E_n t}{\hbar}\right) \tag{14.9}$$

The lowest energy that the system can assume is called the *ground state*.

Example 14.1

Find the energy eignstates and most general solution to the Schrödinger equation for an infinite unsymmetric square well:

$$V(x) = \begin{cases} +\infty & x<0 \\ 0 & 0 \le x \le a \\ +\infty & x \rangle a \end{cases}$$

Solution

A particle confined to an infinite square well is a bound system. Hence it will have a discrete energy spectrum. At the boundaries of the well, the potential is infinite. The wave function must vanish outside this region as well as at the boundaries for continuity. Inside the well, the time-independent Schrödinger equation (14.6) can be written as

$$\frac{d^2\psi}{dx^2} + k^2\psi = 0 \tag{14.10}$$

where we define k^2 as

$$k^2 = \frac{2mE}{\hbar^2} \tag{14.11}$$

The solutions of (14.10) are

$$\psi(x) = A\sin kx + B\cos kx \tag{14.12}$$

The constants A and B are determined by the boundary conditions. The first of these is the vanishing of the wave function at the left boundary—that is, $\psi(0) = 0$. This tells us that $B = 0$. The other boundary condition is $\psi(a) = 0$, from which we conclude that

$$k_n a = n\pi \tag{14.13}$$

where $n = 1, 2, 3, \ldots$. Using this together with (14.11) allows us to determine the energies that the system can assume, which are

$$E_n = \frac{\hbar^2}{2m}k_n^2 = \frac{\hbar^2\pi^2}{2ma^2}n^2, \quad n = 1, 2, 3, \ldots \tag{14.14}$$

Therefore the wave functions are of the form

$$\psi_n(x) = \sqrt{\frac{2}{a}}\sin\left(\frac{n\pi}{a}x\right), \quad n = 1, 2, 3, \ldots \tag{14.15}$$

The constant $\sqrt{2/a}$ is called the normalization constant. It is determined from

$$1 = \int_{-\infty}^{\infty} \psi_n^*(x)\,\psi_n(x)\,dx \tag{14.16}$$

The most general solution to (14.10) is a superposition of all possible solutions:

$$\Psi(x,t) = \sum_{n=1}^{\infty} \psi_n(x) e^{-iE_n t/\hbar} = \sqrt{\frac{2}{a}} \sum_{n=1}^{\infty} \sin\left(\frac{n\pi x}{a}\right) \exp\left(-i\frac{n^2\pi^2\hbar}{2ma^2}t\right) \tag{14.17}$$

ADIABATIC PROCESSES

An *adiabatic process* is one for which the rate of change of the Hamiltonian is slow with respect to the characteristic time of the system (i.e., the time scale over which changes to the system occur). If we denote the characteristic time scale for changes in the system by T_c and the time change over which the Hamiltonian changes as T_H, then we denote the condition for an adiabatic process as

$$T_H \gg T_c \tag{14.18}$$

In quantum mechanics, an adiabatic process is one for which the initial Hamiltonian H_{init} slowly changes to some different, final Hamiltonian H_{final}:

$$H_{init} \underset{slowly}{\rightarrow} H_{final} \tag{14.19}$$

with the time scale over which this occurs being T_H. The *adiabatic theorem* tells us that if a system is in the nth energy eigenstate of H_{init}, it will also be in the nth energy eigenstate of H_{final} if the change is adiabatic. For a change to be adiabatic, the time scale of the change must be proportional to the energy gap between the ground state and the lowest excited state:

$$T \propto \frac{1}{\Delta E} \tag{14.20}$$

Example 14.2

A particle is in an infinite unsymmetric square well of width a; the potential is

$$V(x) = \begin{cases} +\infty & x<0 \\ 0 & 0 \le x \le a \\ +\infty & x \rangle a \end{cases}$$

It is known that the particle is in the ground state. What is the state of the system if the width of the well is very slowly widened to $3a$?

Solution

In this case H_{init} is just the Hamiltonian for Example 14.1. The system starts off in the ground state for which we set $n = 1$ in (14.15), giving

$$\psi_{init}(x) = \sqrt{\frac{2}{a}} \sin\left(\frac{\pi}{a}x\right) \tag{14.21}$$

The energy of the particle is found using (14.14), which is

$$E_{init} = \frac{\hbar^2\pi^2}{2ma^2}$$

For H_{final}, the energy eigenstates are found by letting $a \to 3a$ in (14.15). These are

$$\psi_n(x) = \sqrt{\frac{2}{3a}} \sin\left(\frac{n\pi}{3a}x\right), \quad n = 1, 2, 3, \ldots \tag{14.22}$$

with energies

$$E_n = \frac{\hbar^2\pi^2}{18\,ma^2}n^2, \quad n = 1, 2, 3, \ldots \tag{14.23}$$

We see that widening the well lowers the energies. The adiabatic theorem tells us that if the well is widened slowly, that a system in the nth energy eigenstate of H_{init} will go to the nth energy eigenstate of H_{final}. Since we started in the ground state, we end up in the ground state of the new Hamiltonian, which is found from (14.22). We obtain

$$\psi_{final}(x) = \sqrt{\frac{2}{3a}} \sin\left(\frac{\pi}{3a}x\right)$$

The final energy of the particle is

$$E_{final} = \frac{\hbar^2\pi^2}{18ma^2}$$

What sort of time scale should be used to implement the change? We can give the time required by looking at (14.5) where we have an expression of the form $e^{-iEt/\hbar}$. The characteristic time can be estimated from

$$t_c = \frac{\hbar}{\Delta E} = \frac{\hbar}{(4\hbar^2\pi^2/2ma^2) - (\hbar^2\pi^2/2ma^2)} = \frac{2ma^2}{3\hbar\pi^2} \tag{14.24}$$

where we took the energy gap to be between the ground state and the first excited state, which is found by setting $n = 2$ in (14.14) giving $E_2 = 4\hbar^2\pi^2/2\,ma^2$. So, if the time over which the walls of the potential well are expanded satisfies

$$T \gg \frac{2\,ma^2}{3\hbar\pi^2}$$

the process will be adiabatic.

ADIABATIC QUANTUM COMPUTING

Adiabatic quantum computation uses adiabatic processes to do a computation using the following steps:

- Create an initial state of qubits that will be used to calculate the result of a computation.
- Start with an initial Hamiltonian, and very it slowly (adiabatically) so that it transforms into a final Hamiltonian whose eigenstates encode the solution.

Let's look at the process more formally. The adiabatic process starts with a known easy to prepare Hamiltonian. This Hamiltonian should have a ground state that is easy to create. The quintessential example is a Hamiltonian consisting of Pauli operators

$$H_{init} = -\sum_j X^{(j)} \tag{14.25}$$

Next let's select a *problem Hamiltonian*. The ground state of the problem Hamiltonian encodes the solution to the problem we are trying to solve:

$$H_{final} = \sum_x c_x |x\rangle\langle x| \tag{14.26}$$

An interpolation Hamiltonian is used to connect H_{init} and H_{final}. If T is the total time over which the computation is completed, and we let $s = t/T$ and $0 \leq s \leq 1$, then

$$\tilde{H} = (1-s)H_{init} + sH_{final} \tag{14.27}$$

Example 14.3

Describe how the Hadamard gate can be implemented using adiabatic quantum computation.

Solution

First we recall the action of the Hadamard gate. This gate creates superposition states out of the computational basis in the following way:

$$H|0\rangle = \frac{|0\rangle + |1\rangle}{\sqrt{2}}, \quad H|1\rangle = \frac{|0\rangle - |1\rangle}{\sqrt{2}}$$

In the computational basis the Hadamard gate has the representation

$$H = \frac{1}{\sqrt{2}}\begin{pmatrix} 1 & 1 \\ 1 & -1 \end{pmatrix}$$

In order to implement the Hadamard gate using adiabatic quantum computation, we need to find a Hamiltonian that will create superposition states out of the computational basis, and we need to find an initial Hamiltonian with a ground state that is easy to prepare. One such Hamiltonian is

$$H_{init} = -|0\rangle\langle 0| + |1\rangle\langle 1|$$

The matrix representation of this Hamiltonian is

$$H_{init} \doteq \begin{pmatrix} \langle 0|H_{init}|0\rangle & \langle 0|H_{init}|1\rangle \\ \langle 1|H_{init}|0\rangle & \langle 1|H_{init}|1\rangle \end{pmatrix} = \begin{pmatrix} -1 & 0 \\ 0 & 1 \end{pmatrix}$$

The characteristic equation is

$$0 = \det|H_{init} - I| = \det\left|\begin{pmatrix} -1 & 0 \\ 0 & 1 \end{pmatrix} - \begin{pmatrix} \lambda & 0 \\ 0 & \lambda \end{pmatrix}\right| = \det\begin{vmatrix} -1-\lambda & 0 \\ 0 & 1-\lambda \end{vmatrix} = -1 + \lambda^2$$

Therefore the eigenvalues of the matrix (the Hamiltonian H_{init}) are

$$\lambda_{0,1} = \pm 1$$

The eigenvalues of the Hamiltonian are the energy levels that the system can assume. The ground state has the lowest energy, so the ground state in this case corresponds to $\lambda_0 = -1$. The eigenvector corresponding to this eigenvalue is found from

$$\begin{pmatrix} -1 & 0 \\ 0 & 1 \end{pmatrix}\begin{pmatrix} a \\ b \end{pmatrix} = -\begin{pmatrix} a \\ b \end{pmatrix}$$

We obtain the equations

$$-a = -a, \quad \Rightarrow a = 1$$

$$b = -b, \quad \Rightarrow b = 0$$

We chose $a = 1$so that the eigenvector is normalized (has unit norm). The eigenvector corresponding to $\lambda_0 = -1$ is the ground state of H_{init}:

$$|u_0\rangle = \begin{pmatrix} 1 \\ 0 \end{pmatrix} = |0\rangle$$

A similar procedure taking $\lambda_1 = 1$ gives the second eigenvector of H_{init}:

$$|u_1\rangle = \begin{pmatrix} 0 \\ 1 \end{pmatrix} = |1\rangle$$

Now let's look at the final Hamiltonian. We can create a superposition state with

$$H_{final} = -|0\rangle\langle 1| - |1\rangle\langle 0| = \begin{pmatrix} 0 & -1 \\ -1 & 0 \end{pmatrix} = -X$$

Let's explicitly calculate the eigenvectors even though some readers may know what they are. First we have the characteristic equation

$$0 = \det\left|H_{final} - I\right| = \det\left|\begin{pmatrix} 0 & -1 \\ -1 & 0 \end{pmatrix} - \begin{pmatrix} \lambda & 0 \\ 0 & \lambda \end{pmatrix}\right| = \det\begin{vmatrix} -\lambda & -1 \\ -1 & -\lambda \end{vmatrix} = \lambda^2 - 1$$

$$\Rightarrow \lambda_{0,1} = \pm 1$$

The ground state is found by taking the lowest energy, which is the smallest eigenvalue. So choosing $\lambda_0 = -1$, we have

$$H_{final}|v_0\rangle = \begin{pmatrix} 0 & -1 \\ -1 & 0 \end{pmatrix}\begin{pmatrix} a \\ b \end{pmatrix} = -\begin{pmatrix} a \\ b \end{pmatrix}$$

This yields the equations

$$-b = -a$$

$$-a = -b$$

$$\Rightarrow a = b$$

We can compute the value of the coefficients by "normalizing" the vector

$$1 = \begin{pmatrix} a^* & a^* \end{pmatrix}\begin{pmatrix} a \\ a \end{pmatrix} = 2\,|a|^2$$

$$\Rightarrow a = \frac{1}{\sqrt{2}}$$

Hence the eigenvector is

$$|v_0\rangle = \frac{1}{\sqrt{2}}\begin{pmatrix} 1 \\ 1 \end{pmatrix} = \frac{1}{\sqrt{2}}\begin{pmatrix} 1 \\ 0 \end{pmatrix} + \frac{1}{\sqrt{2}}\begin{pmatrix} 0 \\ 1 \end{pmatrix} = \frac{|0\rangle + |1\rangle}{\sqrt{2}}$$

A similar procedure applied to $\lambda_1 = 1$ give the eigenvector corresponding to the excited state of H_{final}

$$|v_1\rangle = \frac{1}{\sqrt{2}}\begin{pmatrix} 1 \\ -1 \end{pmatrix} = \frac{1}{\sqrt{2}}\begin{pmatrix} 1 \\ 0 \end{pmatrix} - \frac{1}{\sqrt{2}}\begin{pmatrix} 0 \\ 1 \end{pmatrix} = \frac{|0\rangle - |1\rangle}{\sqrt{2}}$$

So we have our initial and final states that correspond to the output of the Hadamard gate. To implement a Hadamard gate $H|0\rangle = (1/\sqrt{2})(|0\rangle + |1\rangle)$ using adiabatic quantum computation, we proceed as follows:

- We prepare the system in the ground state of H_{init}, which is $|\psi_0\rangle = |0\rangle$.
- We evolve the Hamiltonian using $H(s) = (1-s)H_{init} + sH_{final}$ by which the process evolves slowly.

- By the adiabatic theorem, when $t = T(s = 1)$, the system is in the ground state of the final Hamiltonian:

$$|v_0\rangle = \frac{|0\rangle + |1\rangle}{\sqrt{2}}$$

So we have implemented a Hadamard gate.

EXERCISES

14.1. *Derive (14.5) by solving the differential equation that results from (14.1). Assume that (14.5) is a separable solution of the form (14.4), and the constant of integration is the energy E.*

14.2. *Verify that the unit of time in (14.24) is seconds by taking a to be in meters and doing dimensional analysis. The units of $\hbar$ are joule-seconds.*

14.3. *A square well is gradually decreased from width a to $\frac{1}{2}a$. If the change is adiabatic and the system starts off with the particle in the first excited state, what is the energy of the particle when the width is $\frac{1}{2}a$?*

14.4. *Consider the implementation of a CNOT gate using adiabatic computation. Take*

$$H_0 = 3|00\rangle\langle 00| + 2|01\rangle\langle 01| + |10\rangle\langle 10|$$
$$H_1 = 3|00\rangle\langle 00| + 2|01\rangle\langle 01| + |11\rangle\langle 11|$$
$$H_{01} = |10\rangle\langle 11| + |11\rangle\langle 10|$$

and let

$$H(s) = (1 - s)H_0 + sH_1 + s(1 - s)H_{01}$$

Follow the procedure used in Example 14.3 and show this system can be used to implement a CNOT gate.

15

CLUSTER STATE QUANTUM COMPUTING

Cluster state quantum computing is a model that does not rely on quantum gates to do its processing. This measurement based model can instead simulate the unitary dynamics of quantum mechanics. Since quantum gates are implemented using unitary operators, quantum computation can be simulated or implemented using the cluster state model.

A cluster state is a multiple-qubit state that is processed by a series of measurements. Each time a measurement is made, the result of the measurement is used to select the basis used for the next measurement—so this type of quantum computation involves a feedback loop. Cluster state quantum computing can be summarized in two steps:

- Initialize a set of qubits in some state. For example, start with $|+\rangle$ and then apply controlled phase gates to the states.
- Measure the qubits in some basis. As the measurement process is repeated, the choice of basis each time is determined by the previous measurement results, creating a feedback process.

Quantum Computing Explained, by David McMahon

A controlled phase operation is the controlled-Z gate with matrix representation in the computational basis

$$CZ = \begin{pmatrix} 1 & 0 & 0 & 0 \\ 0 & 1 & 0 & 0 \\ 0 & 0 & 1 & 0 \\ 0 & 0 & 0 & -1 \end{pmatrix} \tag{15.1}$$

The effect of applying the controlled phase gate is to entangle the states.

CLUSTER STATES

Cluster states are represented by graphs, which are a set of nodes or vertices and edges. Each vertex is a qubit, while the edges represent controlled phase gates. An example is shown in Figure 15.1.

As stated above, we begin cluster state quantum computation by initializing each of the qubits in the $|+\rangle$ state. Then we apply controlled phase gates to qubits that are neighbors in the graph.

Cluster State Preparation

The first step required in cluster state preparation is to prepare a product state of the form

$$|+\rangle_C = |+\rangle^{\otimes n} \tag{15.2}$$

For example, we can prepare the cluster state

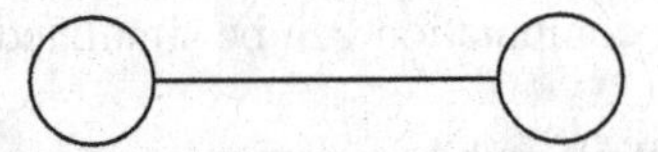

which is two qubits, represented by the two vertices in the graph and a single controlled phase operation applied to them, represented by the line connecting the qubits. The initial product state is then

$$|+\rangle_C = |+\rangle \otimes |+\rangle \tag{15.3}$$

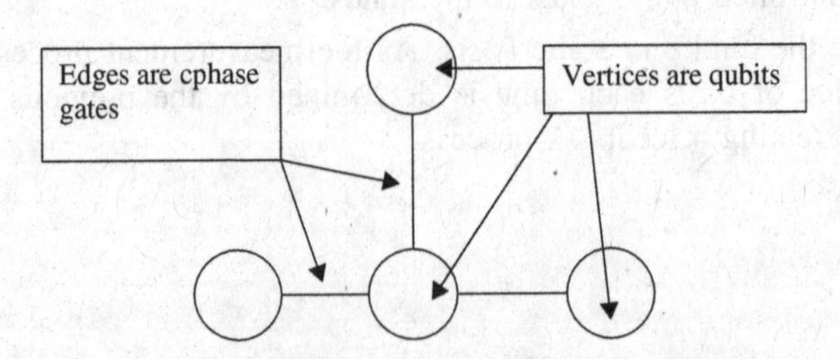

Figure 15.1 Cluster state quantum computation is represented by a graph

Step two is to apply the controlled phase gate. For example, we can apply a controlled-Z gate given by

$$S = \frac{1}{2}(I \otimes I + Z \otimes I + I \otimes Z - Z \otimes Z) \tag{15.4}$$

to obtain

$$\begin{aligned}
S|+\rangle \otimes |+\rangle &= \frac{1}{2}(I \otimes I + Z \otimes I + I \otimes Z - Z \otimes Z)|+\rangle \otimes |+\rangle \\
&= \frac{1}{2}\left[\left(\frac{|0\rangle + |1\rangle}{\sqrt{2}}\right)\left(\frac{|0\rangle + |1\rangle}{\sqrt{2}}\right) + (Z \otimes I)\left(\frac{|0\rangle + |1\rangle}{\sqrt{2}}\right)\left(\frac{|0\rangle + |1\rangle}{\sqrt{2}}\right)\right. \\
&\quad + I \otimes Z\left(\frac{|0\rangle + |1\rangle}{\sqrt{2}}\right)\left(\frac{|0\rangle + |1\rangle}{\sqrt{2}}\right) \\
&\quad \left. - Z \otimes Z\left(\frac{|0\rangle + |1\rangle}{\sqrt{2}}\right)\left(\frac{|0\rangle + |1\rangle}{\sqrt{2}}\right)\right] \\
&= \frac{1}{2}[|00\rangle + |01\rangle + |10\rangle - |11\rangle]
\end{aligned}$$

This entangled state is represented by

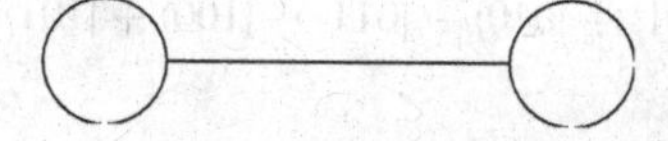

Example 15.1

Consider three qubits arranged in a triangle. What states result in creating the graphs?

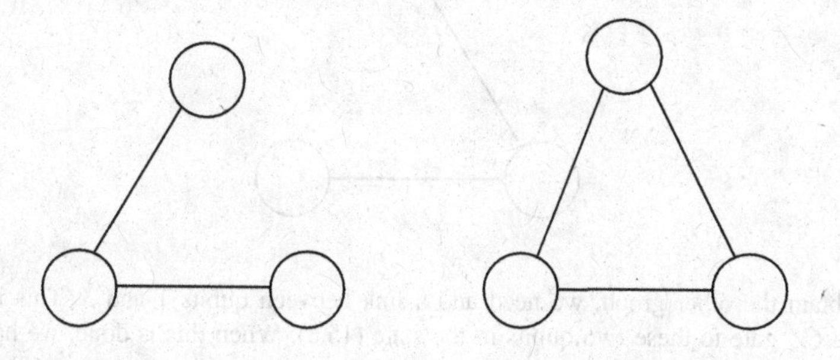

Solution

We denote the upper qubit as qubit 1, the left lower qubit as qubit 2, and the right lower qubit as qubit 3. The initial state is the product state

$$|+\rangle \otimes |+\rangle \otimes |+\rangle = \left(\frac{|0\rangle + |1\rangle}{\sqrt{2}}\right)\left(\frac{|0\rangle + |1\rangle}{\sqrt{2}}\right)\left(\frac{|0\rangle + |1\rangle}{\sqrt{2}}\right)$$
$$= \frac{1}{\sqrt{8}}(|000\rangle + |001\rangle + |010\rangle + |011\rangle + |100\rangle + |101\rangle + |110\rangle + |111\rangle) \tag{15.5}$$

Now let's review the operation of the controlled-Z gate. The first qubit acts as the control qubit and the second qubit acts as the target qubit. If the first qubit is 0, we do nothing to the second qubit. If the first qubit is 1, we apply a Z gate to the second qubit. Remember,

$$Z|0\rangle = |0\rangle, \quad Z|1\rangle = -|1\rangle \tag{15.6}$$

To obtain the first graph, we need to apply the controlled-Z gate to qubits 1 and 2 and then to 2 and 3. Applying the CZ gate to qubits 1 and 2 to (15.5), we obtain

$$|\psi'\rangle = \frac{1}{\sqrt{8}}(|000\rangle + |001\rangle + |010\rangle + |011\rangle + |100\rangle + |101\rangle - |110\rangle - |111\rangle) \tag{15.7}$$

The first qubit in each term acts as the control. In the first four terms the control is 0, so we do nothing to the target. It's a 1 in the last four terms, but the target is a 1 giving a sign change only for the last two terms.

Now we apply the CZ gate to qubits 2 and 3. So the second qubit is the control and the third qubit is the target. We obtain

$$|\psi_\Lambda\rangle = \frac{1}{\sqrt{8}}(|000\rangle + |001\rangle + |010\rangle - |011\rangle + |100\rangle + |101\rangle - |110\rangle + |111\rangle) \tag{15.8}$$

We have labeled this the *lambda* state, since the graph takes the form of a (tilted) lambda:

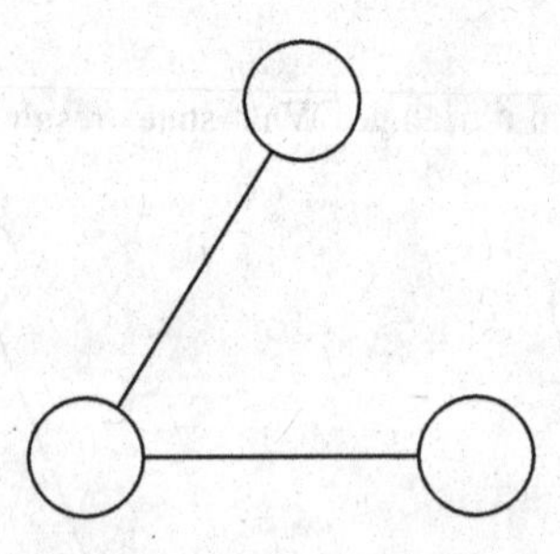

To obtain the other graph, we need add a link between qubits 1 and 3. This is done by applying a CZ gate to these two qubits in the state (15.8). When this is done, we have

$$|\psi_\Delta\rangle = \frac{1}{\sqrt{8}}(|000\rangle + |001\rangle + |010\rangle - |011\rangle + |100\rangle - |101\rangle - |110\rangle - |111\rangle) \tag{15.9}$$

We obtained this state by applying a Z gate to the third qubit whenever the first qubit was a 1. This state corresponds to the graph

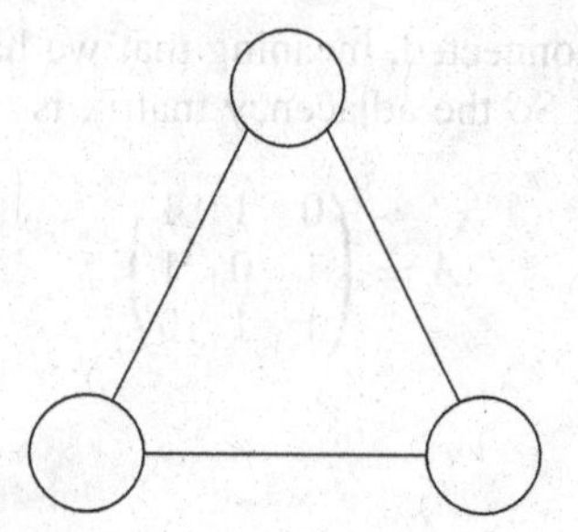

ADJACENCY MATRICES

An *adjacency matrix* is a way to represent a graph with a matrix. The rows and columns of the matrix are labeled by the vertices of the graph. If there is an edge joining the vertices represented by a given row, column entry, then that matrix element is a 1; otherwise, it is a zero. For example, consider

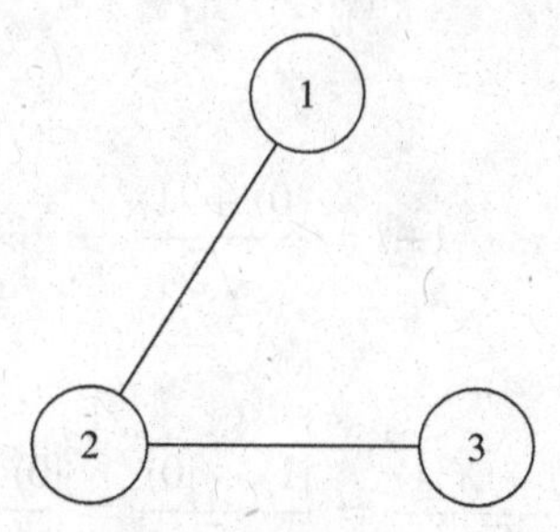

where we have added labels to each vertex for convenience. With three vertices, the adjacency matrix will be 3×3. There are edges between 1 and 2 and 2 and 3. This tells us that there will be 1's in matrix elements (1, 2), (2, 1), (2, 3), and (3, 2). All other matrix elements will be zero, giving

$$A = \begin{pmatrix} 0 & 1 & 0 \\ 1 & 0 & 1 \\ 0 & 1 & 0 \end{pmatrix}$$

In the case examined in the previous example, we had

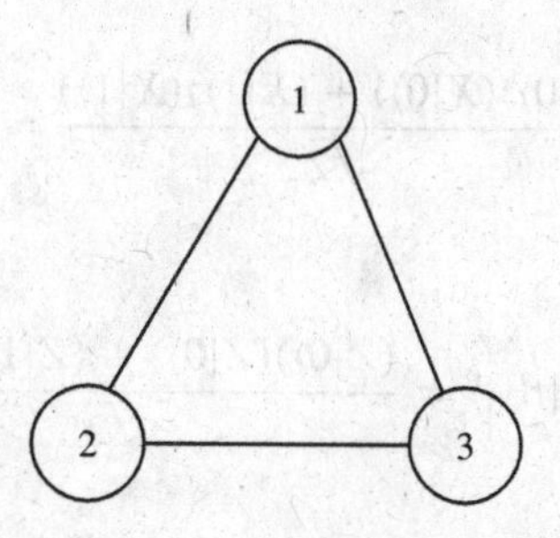

Now vertices 1 and 3 are connected, meaning that we have to add 1's to the (1, 3) and (3, 1) matrix elements. So the adjacency matrix is

$$A = \begin{pmatrix} 0 & 1 & 1 \\ 1 & 0 & 1 \\ 1 & 1 & 0 \end{pmatrix}$$

STABILIZER STATES

An operator A is a *stabilizer* of a state $|\psi\rangle$ if

$$A|\psi\rangle = |\psi\rangle \tag{15.10}$$

So a stabilizer is just an operator with an eigenvector that has eigenvalue +1. This suggests immediately that the identity operator is a stabilizer, since

$$I|\psi\rangle = |\psi\rangle$$

Now consider the state

$$|+\rangle = \frac{|0\rangle + |1\rangle}{\sqrt{2}}$$

Notice that

$$X|+\rangle = \frac{X|0\rangle + X|1\rangle}{\sqrt{2}} = \frac{|1\rangle + |0\rangle}{\sqrt{2}} = \frac{|0\rangle + |1\rangle}{\sqrt{2}} = |+\rangle$$

Hence stablizer of the $|+\rangle$ state are the operators X, I. This is commonly denoted by writing $\{X, I\}$. This single-qubit state has two operators in the stabilizer. Generally, if the state contains n qubits, then the stabilizer will be made up of 2^n commuting operators.

Now consider the Bell state

$$|\beta_{00}\rangle = \frac{|00\rangle + |11\rangle}{\sqrt{2}}$$

Notice that

$$(X \otimes X)|\beta_{00}\rangle = \frac{(X|0\rangle)(X|0\rangle) + (X|1\rangle)(X|1\rangle)}{\sqrt{2}} = \frac{|11\rangle + |00\rangle}{\sqrt{2}} = |\beta_{00}\rangle$$

It's also true that

$$(Z \otimes Z)|\beta_{00}\rangle = \frac{(Z|0\rangle)(Z|0\rangle) + (Z|1\rangle)(Z|1\rangle)}{\sqrt{2}}$$

$$= \frac{(|0\rangle)(|0\rangle) + (-|1\rangle)(-|1\rangle)}{\sqrt{2}}$$

$$= \frac{|00\rangle + |11\rangle}{\sqrt{2}} = |\beta_{00}\rangle$$

Therefore two stablizers of the Bell state $|\beta_{00}\rangle$ are

$$\pm X \otimes X, \quad \pm Z \otimes Z$$

Consider an array of qubits, which we denote by A. A cluster state on and element a of this array is defined by a stabilizer of the form

$$-1^k X^a \otimes Z^i \tag{15.11}$$

where i identifies a neighbor of a. Stabilizers for an n-qubit cluster state can be written as

$$\begin{aligned} S_1 &= X^{(1)} Z^{(2)} \\ &\vdots \\ S_j &= Z^{(j-1)} X^j Z^{j+1} \qquad j = 2, 3, \ldots, n-1 \\ S_n &= Z^{(n-1)} X^{(n)} \end{aligned} \tag{15.12}$$

where we use the notation that $A^{(j)}$ means the operator A is applied to the jth qubit. For example, the stabilizers for a 2-qubit cluster state are

$$S_1 = X^{(1)} Z^{(2)}, \quad S_2 = Z^{(1)} X^{(2)} \tag{15.13}$$

For an example let's find the stabilizer for the state

$$|\psi_-\rangle = \frac{1}{2}[|00\rangle + |01\rangle + |10\rangle - |11\rangle] \tag{15.14}$$

Now

$$\begin{aligned} |\psi_-\rangle = CZ|{+}{+}\rangle &= \frac{1}{2} CZ(|0\rangle + |1\rangle)(|0\rangle + |1\rangle) \\ &= \frac{1}{2} CZ(|00\rangle + |01\rangle + |10\rangle + |11\rangle) \\ &= \frac{1}{2}(|00\rangle + |01\rangle + |10\rangle - |11\rangle) \end{aligned}$$

To find the stabilizer, we look for operators that satisfy

$$A^{(1)}B^{(2)}CZ|{+}{+}\rangle = A^{(1)}B^{(2)}|\psi_-\rangle = |\psi_-\rangle$$

The Pauli operators square to the identity, so $CZ = CZI = CZXX$. It can be shown that $CZXX = YYCZ$, This means that

$$|\psi_-\rangle = CZ|{+}{+}\rangle = CZXX|{+}{+}\rangle = (YY)CZ|{+}{+}\rangle = YY|\psi_-\rangle$$

So YY is a stabilizer. We can also show that

$$CZ = CZXI = XZCZ$$

$$CZ = CZXI = CZIX = ZXCZ$$

This tells us that XZ, ZX also satisfy $XZ|\psi_-\rangle = |\psi_-\rangle$, $ZX|\psi_-\rangle = |\psi_-\rangle$. Of course,

$$CZ = CZII = IICZ$$

So the stabilizer for the $|\psi_-\rangle$ state is $\{YY, XZ, ZX, II\}$.

ASIDE: ENTANGLEMENT WITNESS

Since cluster states involve entangled states of n qubits where n can be greater than 2, we might ask how can we determine whether or not a multiple qubit state is entangled? We can attempt to answer this question using an observable known as an *entanglement witness*. The way that we can use an entanglement witness W is to examine its expectation value for a given state $|\psi\rangle$. If the state is separable, then

$$\langle W\rangle = \langle\psi|W|\psi\rangle > 0 \tag{15.15}$$

Some of the time the entanglement witness will tell us if the state is entangled. This is when the expectation value is negative

$$\langle W\rangle = \langle\psi|W|\psi\rangle < 0 \tag{15.16}$$

Let's examine this in detail for two states. The GHZ state is an entangled state of three qubits

$$|GHZ\rangle = \frac{|000\rangle + |111\rangle}{\sqrt{2}} \tag{15.17}$$

An entanglement witness is

$$W = \frac{1}{2}I - |GHZ\rangle\langle GHZ| \tag{15.18}$$

$$= \frac{1}{2}I - \frac{1}{2}(|000\rangle\langle 000| + |000\rangle\langle 111| + |111\rangle\langle 000| + |111\rangle\langle 111|)$$

Next consider the separable state

$$|\psi\rangle = \left(\frac{|001\rangle + |111\rangle}{\sqrt{2}}\right) = \left(\frac{|00\rangle + |11\rangle}{\sqrt{2}}\right)|1\rangle$$

Hence qubits 1 and 2 are entangled, but qubit 3 is not. We are interested in what is the expectation value of W as given by (15.18) for this state We have

$$\langle W\rangle = \langle\psi|W|\psi\rangle$$

$$= \left(\frac{\langle 001| + \langle 111|}{\sqrt{2}}\right) W \left(\frac{|001\rangle + |111\rangle}{\sqrt{2}}\right)$$

Now

$$W\left(\frac{|001\rangle + |111\rangle}{\sqrt{2}}\right) = \left[\frac{1}{2}I - \frac{1}{2}(|000\rangle\langle 000| + |000\rangle\langle 111| + |111\rangle\langle 000|\right.$$

$$\left. + |111\rangle\langle 111|)\right]\left(\frac{|001\rangle + |111\rangle}{\sqrt{2}}\right)$$

$$= \left(\frac{|001\rangle + |111\rangle}{\sqrt{2}}\right) - \frac{1}{2}\left(\frac{|000\rangle + |111\rangle}{\sqrt{2}}\right)$$

So the expectation value is

$$\langle W\rangle = \left(\frac{\langle 001| + \langle 111|}{\sqrt{2}}\right)\left[\left(\frac{|001\rangle + |111\rangle}{\sqrt{2}}\right) - \frac{1}{2}\left(\frac{|000\rangle + |111\rangle}{\sqrt{2}}\right)\right]$$

$$= \frac{\langle 001|001\rangle}{\sqrt{2}} + \frac{\langle 111|111\rangle}{\sqrt{2}} - \frac{\langle 111|111\rangle}{\sqrt{2}} = \frac{\langle 001|001\rangle}{\sqrt{2}} = \frac{1}{\sqrt{2}}$$

Since

$$\langle W\rangle = \frac{1}{\sqrt{2}} > 0$$

we know that the state is separable. What about the GHZ state? In that case

$$\langle W \rangle = \langle GHZ| \left(\frac{1}{2} I - |GHZ\rangle\langle GHZ| \right) |GHZ\rangle = \langle GHZ| \left(\frac{1}{2}|GHZ\rangle - |GHZ\rangle \right)$$

$$= \langle GHZ| \left(-\frac{1}{2}|GHZ\rangle \right) = -\frac{1}{2}\langle GHZ \mid GHZ\rangle = -\frac{1}{2} < 0$$

This result confirms that the GHZ state is entangled. So W seems to be a good entanglement witness.

CLUSTER STATE PROCESSING

To get an idea of how cluster state computation works, we consider measurements in the Z eigenbasis, the X eigenbasis, and the Y eigenbasis. We will do so considering a linear cluster, that is,

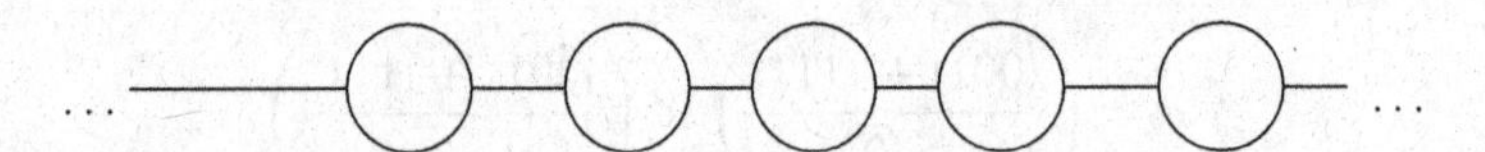

In each of the following examples we will apply a measurement to the middle qubit, which is represented by the third circle from the left. First we do a measurement in the Z eigenbasis. We represent this by writing the operator in the graph vertex

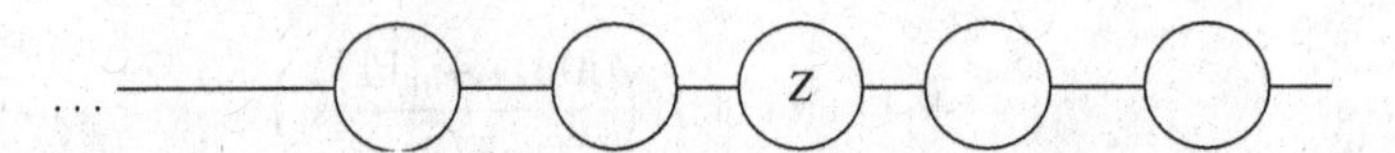

A measurement in the Z eigenbasis has two effects on the cluster:

- Break the connections (edges) between that qubit and the rest of the cluster.
- Remove the qubit.

After step one, we have

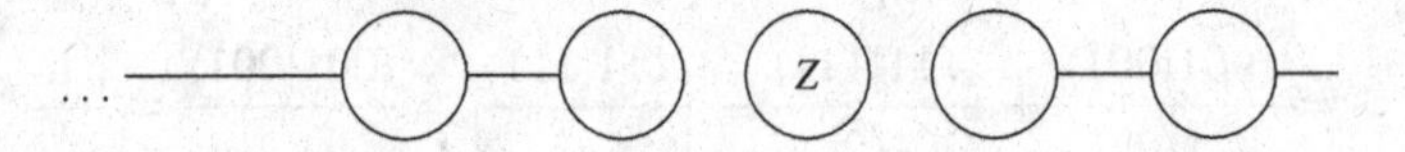

After step two, the new cluster is

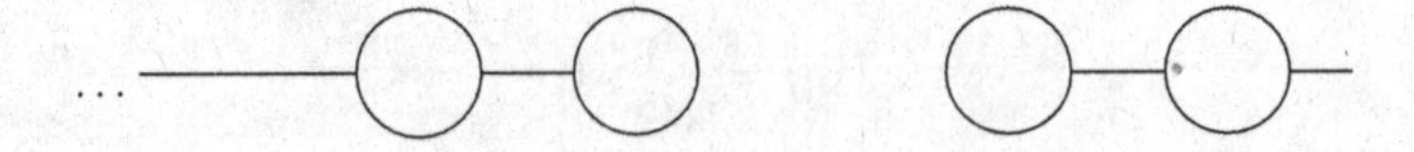

Now consider a measurement in the X eigenbasis

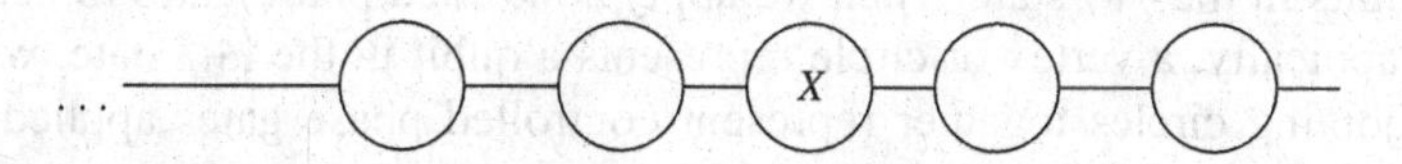

The two steps in this case are as follows:

- Remove the qubit.
- Join the neighboring qubits together into a *single logical* qubit.

Step one, remove the qubit.

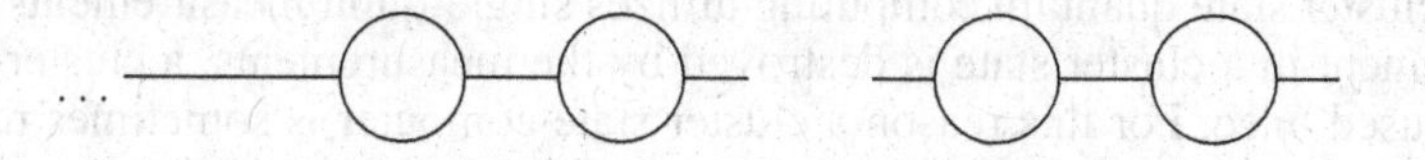

Step two, join the neighboring qubits into a single logical qubit.

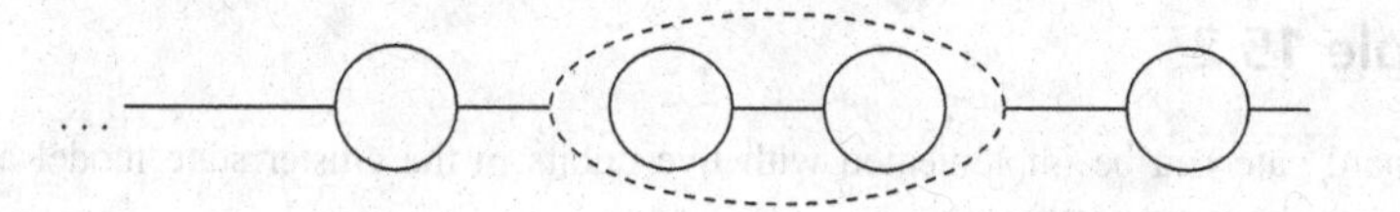

The single logical qubit represents 0 with $|++\rangle$ and 1 with $|--\rangle$.

Finally, we consider processing by measurement in the Y eigenbasis.

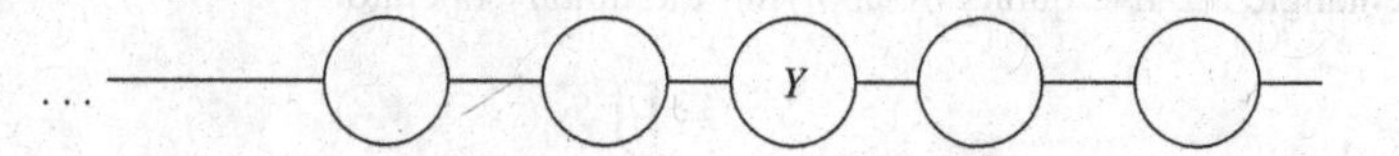

The two steps this time are

- Remove the qubit from the cluster.
- Link together the neighboring qubits left behind.

Implementing the first step, we have

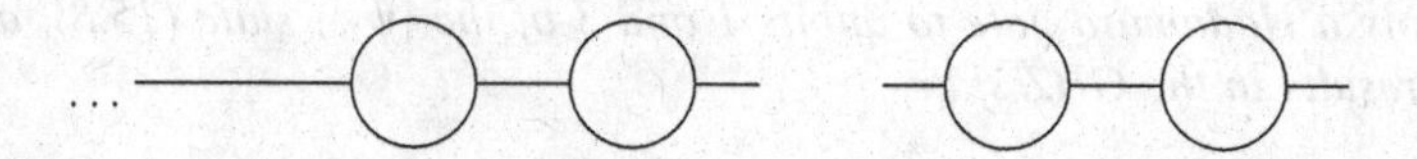

Now we link the two neighbors:

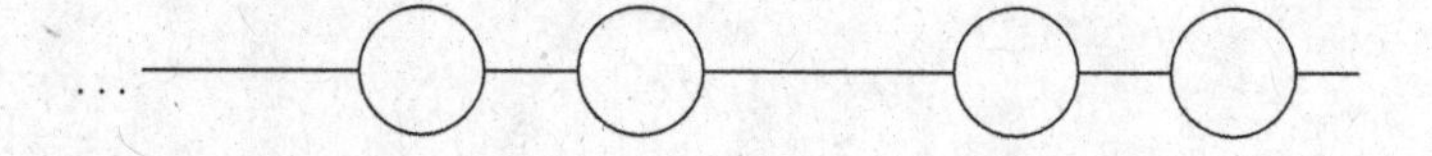

To summarize, we prepare a cluster state by creating an initial product state of several qubits in the $|+\rangle$ state. Then we apply controlled phase gates to neighboring states. Graphically, a vertex or circle represents a qubit in the $|+\rangle$ state, while lines or edges joining circles together represent controlled phase gates applied between the qubits that are joined by that line.

Processing can be done with measurements in the Z, X, and Y bases. In each case, when a measurement is made, we remove that qubit from the cluster. If the measurement was made in the Z basis, then we break the connections that existed between that qubit and the rest of the cluster. If the measurement was in the X basis, we join the neighboring qubits together to form a new logical qubit in the $|++\rangle$, $|--\rangle$ basis. Finally, if the measurement is made in the Y basis, we simply join together the left behind neighbors.

So cluster state quantum computing utilizes single-qubit measurements. Because entanglement in a cluster state is destroyed by the measurements, a cluster state can only be used once. For this reason a cluster state computer is sometimes referred to as a *one-way* quantum computer.

Example 15.2

A Hadamard gate can be implemented with five qubits in the cluster state model as follows: We start with the input state

$$|\psi_{in}\rangle = |+\rangle^{\otimes 5} = |+++++\rangle$$

Then we entangle the five qubits by applying the unitary operator

$$S^{(5)} = \prod S$$

where S is given by (15.4). Last we measure the first four qubits using

$$X \otimes Y \otimes Y \otimes Y$$

EXERCISES

15.1. *Apply a Hadamard gate to qubits 1 and 3 of the $|\psi_\Lambda\rangle$ state (15.8), and show that this results in the GHZ state*

$$|\psi\rangle = \frac{1}{\sqrt{2}}(|000\rangle + |111\rangle)$$

15.2. *What is the adjacency matrix for*

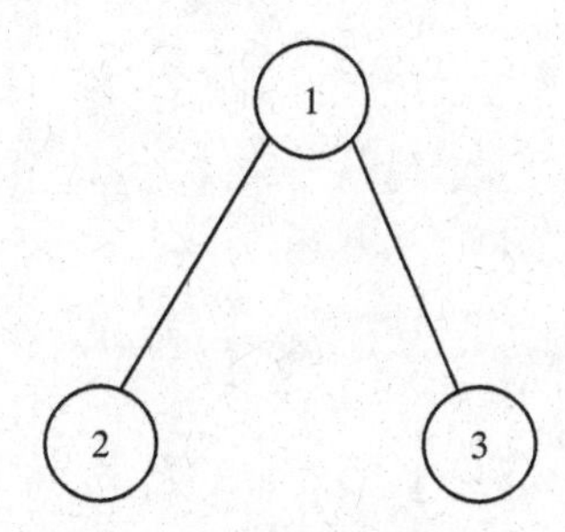

15.3. *If , show that* $|0\rangle \otimes |0\rangle$, $|0\rangle \otimes |1\rangle$, $|1\rangle \otimes |0\rangle$ *do not acquire a phase under the action of S but that* $|1\rangle \otimes |1\rangle$ *acquires a phase of* π.

15.4. *Following Example 15.2, describe how to implement a* $\pi/2$ *phase gate.*

15.5. *Are* $\pm X \otimes X$, $\pm Z \otimes Z$ *stabilizers of all the Bell states?*

15.6. *Write down the stabilizers for the GHZ state:*

$$|GHZ\rangle = \frac{|000\rangle + |111\rangle}{\sqrt{2}}.$$

REFERENCES

BOOKS

Bouwmeester, D., A. Ekert, A. Zeilinger (eds.) *The Physics of Quantum Information*, Springer, Berlin (2001).

Brown, J., *The Quest for the Quantum Computer*, Simon and Schuster, New York (2000).

Cohen-Tannoudji, C., B. Dui, F. Laloë, *Quantum Mechanics*, Vol I, Wiley, New York, (1977).

McMahon, D., *Quantum Mechanics Demystified*, McGraw Hill, New York (2005).

Nielsen, M., and I. L. Chuang, *Quantum Computation and Quantum Information*, Cambridge University Press, Cambridge (2000)

Steeb, W. H. and Y. Hardy, *Problems and solutions in Quantum Computing and Quantum Information*, World Scientific, New Jersey (2004).

Zettili, N., *Quantum Mechanics: Concepts and Applications*, Wiley New York, (2001).

ONLINE RESOURCES

Quantum Information Course notes by David Bacon
http://www.cs.washington.edu/education/courses/cse599d/06wi/

Quantum Information Lecture notes by John Watrous
http://www.cs.uwaterloo.ca/~watrous/lecture-notes.html

Wikipedia contributors. POVM. Wikipedia, The Free Encyclopedia. November 16, 2006, 19:30 UTC. Available at:
http://en.wikipedia.org/w/index.php?title=POVM&oldid=88260272. Accessed December 29, 2006.

Wikipedia contributors. Quantum entanglement. Wikipedia, The Free Encyclopedia. December 10, 2006, 08:58 UTC. Available at:
http://en.wikipedia.org/w/index.php?title=Quantum_entanglement&oldid=93333870. Accessed December 31, 2006.

Z. Meglicki, *Introduction to Quantum Computing*
http://beige.ovpit.indiana.edu/M743/

Quantum Computing Explained, by David McMahon

PAPERS AND JOURNAL ARTICLES

Agrawal, P., and Arun Pati, e-print quant-ph/0610001

Aharonov, Y., J. Anandan, G. J. Maclay, J. Suzuki, Phys. Rev. A 70: 052114 (2004).

Aharonov, D., W. van Dam, J. Kempe, Z. Landau, S. Lloyd, O. Regev, Adiabatic quantum computation is equivalent to standard quantum computation, e-print quant-ph/0405098.

Andrecut, M., and M. K. Ali, Adiabatic quantum gates and boolean functions, *J. Phys. A: Math. Gen*. 37 (2004) L267–L273.

Cereceda, J. L., Generalization of the Deutsch algorithm using two qudits. quant-ph/0407253, (2004).

Fujii, K., Coherent states and some topics in quantum information theory: Review, 2002, e-print quant-ph/0207178.

Wang, M.-Y., and Feng-Li Yan, e-print quant-ph/0612019.

Nielsen, M. A., Cluster-state quantum computation, e-print quant-ph/0504097.

Raussendorf, R., H. J. Briegel, A one-way quantum computer, *Phys Rev Lett* 86 (22): 5188–5191 (2001).

Sarandy, M. S., and D. A. Lidar, Adiabatic quantum computation in open systems, *Phys. Rev. Lett*. 95, 250503 (2005).

Shannon, C., A mathematical theory of communication, *Bell. Syst. Tech. J*. (1948)

Tessier, T. E., Complimentary relations for multi-qubit systems, *Found. Phys. Lett*. 18(2): 107 (2005).

Uhlmann, A., The “transition probability“ in the state space of a C*-algebra. *Rep. Mathem. Phys*., 9: 273–279, 1976.

Wootters, W. K., and W. H. Zurek, *A Single Quantum Cannot be Cloned*, Nature 299 (1982), pp. 802–803.

Wootters, W. K., Entanglement of formation and concurrence, *Quan. Info. Comput*. 1: 27 (2001).

Wang, J., Q. Zhang, and C.-j. Tang, quant-ph/0603236.

INDEX

Quantum Computing Explained, by David McMahon